# The Place of Geography

# The Place of Geography

Tim Unwin

LONGMAN

Addison Wesley Longman Limited
Edinburgh Gate, Harlow
Essex CM20 2JE, England
*and Associated Companies throughout the world.*



*First published 1992*
*Reprinted 1994 (twice), 1996, 1997*

**British Library Cataloguing in Publication Data**
A catalogue record for this book is available from the British Library

ISBN 0 582 05107 X

**Library of Congress Cataloging-in-Publication Data**

A catalog record for this book is
available from the Library of Congress

Set in 10/11 pt Palatino by 9
Produced through Longman Malaysia, VVP

*For*
*Pam*
*Jenny, Juliette and Jake*
*with thanks for their understanding*

# Contents

Preface xi
Acknowledgements xiv

*Chapter 1 Geography: the social construction of a discipline* 1
1.1 Geography in the public domain 2
1.2 The construction of disciplines 5
1.3 Geographical education 7
1.3.1 Geography in the secondary curriculum in North America 8
1.3.2 Geography in the educational system of England and Wales 11
1.3.3 Geography from primary to higher education 15
1.4 Geographical self-reflection 17

*Chapter 2 The place of theory* 20
2.1 Science and society 20
2.1.1 Definitions of science 21
2.1.2 Kuhn: paradigms and scientific revolutions 24
2.1.3 Foucault: truth instead of knowledge; power instead of ideology 26
2.1.4 Habermas: power, knowledge and truth 28
2.2 Science, knowledge and interest 30
2.2.1 Empirical–analytic science 31
2.2.2 Historical–hermeneutic science 35
2.2.3 Towards a critical science 39
2.3 Theory, practice and geographical interest 42

*Chapter 3 Geography and society: classical context and a world of discovery* 45
3.1 Greek and Roman geography 46
3.1.1 The origins of classical geography 46

3.1.2 The advent of classical formal geography: Strabo and Ptolemy 50
3.1.3 The Greek and Roman concept of geography 54
3.2 Chinese and Islamic geography 55
3.2.1 Chinese geography: a separate tradition 55
3.2.2 The Islamic contribution to geographical understanding 56
3.3 The resurgence of European geography 59
3.3.1 A world of 'discovery': cartography and exploration 60
3.3.2 Geography and the expansion of European power 62
3.3.3 Geography at the dawn of the 17th century 64

*Chapter 4 The emergence of geography as a formal academic discipline* 66
4.1 From Varenius to Kant: the reappearance of formal geography 66
4.1.1 Varenius: general and special geography 66
4.1.2 Kant's *Physische Geographie* 70
4.2 Humboldt, Ritter and the foundation of modern geography 74
4.2.1 Alexander von Humboldt and the unity of the cosmos 74
4.2.2 Carl Ritter and the combination of teleology with empirical observation 77
4.3 Institutionalized geography: societies and universities in the age of empire 79
4.3.1 Germany 79
4.3.2 France 80
4.3.3 Britain 83
4.3.4 Geography in the United States of America 85
4.3.5 Imperialism and the anarchist alternative 87
4.4 People, the environment and regional geography 90
4.4.1 Darwin's influence on geography 91
4.4.2 'Man' and the environment 91
4.4.3 The divisions between physical and human geography 95
4.4.4 The region as an object of geographical synthesis and enquiry 98
4.5 Geography in an institutional context 104

*Chapter 5 From region to process: the emergence of geography as an empirical–analytic science* 106
5.1 The demise of regional geography 107
5.1.1 Geographers at war 107
5.1.2 Problems of definition 109
5.1.3 Uniqueness and generality: regions and systems 109
5.1.4 Geographical science and the art of geography 115
5.2 Models, systems and process: the implicit adoption of logical positivism 116
5.2.1 Process and form in physical geography 116
5.2.2 Theoretical approaches to a systematic human geography 119
5.2.3 Systems and theoretical geography 122
5.3 Explanation, relevance and the social origins of concern 129
5.3.1 Explanation in geography 130

5.3.2 Boston 1971 and the relevance debate 131
5.3.3 The failures of logical positivism 134

*Chapter 6 Geography and historical–hermeneutic science: the quest for understanding* 136
6.1 Geography: the magpie discipline 137
6.1.1 Geography, human behaviour and space 137
6.1.2 Geography and the social sciences in France 138
6.1.3 Geography and the social sciences in Britain and the USA 139
6.2 Behavioural geography and the demise of rational economic man 140
6.2.1 Geographers and the behavioural environment 141
6.2.2 Behaviour within the framework of logical positivism 142
6.2.3 Time geography 144
6.3 Humanist perspectives 145
6.3.1 Phenomenology and the understanding of essence 146
6.3.2 Existentialism: individuality and being 148
6.3.3 Idealism and historical experience 149
6.4 The historical–hermeneutic alternative 150
6.4.1 The context and practice of humanistic geography 150
6.4.2 Humanistic geography as an historical–hermeneutic science 152
6.4.3 Physical science and human experience 153
6.4.4 Structure, constraint and the social context 156

*Chapter 7 Critical science and society: the geographer's interest* 158
7.1 The social context: geography in recession 159
7.1.1 Capitalist society in the 1970s and 1980s: power, recession and science 159
7.1.2 Geography and the production of knowledge 161
7.1.3 The origins of radical geography 162
7.2 Radical geography and a structuralist alternative 164
7.2.1 Marxist geography 164
7.2.2 The place of radical geography 166
7.2.3 Structuralist alternatives 169
7.2.4 Space, time and structuration 171
7.3 Realism and postmodernism 175
7.3.1 Reality and realism 175
7.3.2 Concrete buildings and postmodern alternatives 177
7.4 Geography as critical science: the environmental conscience of society 180
7.4.1 Institutions, career profiles and the funding of science 181
7.4.2 Towards a critical geography 182

*Chapter 8 The place of geography* 186
8.1 Geographers and the environment 188
8.1.1 Applied physical geography 189
8.1.2 Landscapes and the domination of nature 191
8.2 Space(–)time and geography 194

8.2.1 The social production of space 194
8.2.2 Relativity and quantum theory 196
8.2.3 Interpreting space 203
8.3 The theory and practice of geography 205
8.3.1 Geographical education 205
8.3.2 The choice of research 207
8.3.3 Geographers in society 208
8.4 The future of geography: peoples, environments and places 209

*Glossary* 212

*Bibliography* 218

*Index* 261

# Preface

The central aim of this book is to provide a readable and yet challenging account of the emergence of geography as an academic discipline. It is written primarily for undergraduate students, as an introduction to the large, and highly complex, literature on the subject that has emerged in recent years, and it concentrates primarily on geography in the English-speaking world. However, it has three further interconnected aims. First, it seeks to trace the development of geography back to its formal roots in classical antiquity, when many of the topics of debate in 20th-century geographical thought had their origins. Second, it interprets the changes that have taken place in geographical practice within the context of Jürgen Habermas's critical theory, and it thus places considerable emphasis on the social construction of knowledge and on the uses to which different types of knowledge are put. Third, it concludes that the separation of geography into two distinct parts, the one focusing on the physical world and the other addressing issues of human concern, has meant that geographers have had remarkably little to say about some of the most pressing environmental issues of the late 20th century.

The structure of the book is broadly chronological. The first chapter examines the contemporary image of geography and the forces shaping academic disciplines in general. This is followed by a theoretical chapter which provides the framework through which the remainder of the book explores the development of geography as an academic discipline. The next two chapters trace the evolution of geography from classical antiquity to the end of the 19th century, and these are followed by three chapters interpreting 20th-century geography through its practice as an empirical-analytic science, a historical–hermeneutic science and a critical science. The final chapter returns to questions raised at the beginning of the book and addresses a range of theoretical and practical issues of contemporary significance for geographers. There is also a short glossary which provides notes on some of the terms with which readers new to the subject may be unfamiliar.

In any such endeavour, the use of language is of fundamental importance. It is through language that we either succeed or fail to communicate with each other. The way in which geographers and philosophers have written about their ideas is therefore part and parcel of their practical discourse. Consequently, the present endeavour seeks, as often as possible, to let authors speak for themselves through their own written words. In trying to cover the whole span of geographical enquiry, this book can only open doors; it cannot examine the detailed contents of each room of geographical experience. The journey is thus essentially a personal one. However, as Ludwig Wittgenstein (1967: viii) commented in the preface to *Philosophical investigations*, 'I should not like my writing to spare other people the trouble of thinking. But, if possible, to stimulate someone to thoughts of their own.'

The shape of this book owes much to those who have shared their ideas with me over the last eighteen years, both as teachers and as students. At Cambridge, my early geographical explorations were heavily influenced by five people: Dick Chorley, who taught me much more than just what I remember about the gunfight at the OK Corral; H. C. Darby, whose use of language and breadth of discourse opened up a whole new realm of geographical enquiry; Derek Gregory, who introduced me not only to the work of Jürgen Habermas, but also to a much wider field of social theory; Jim Lewis, who convinced me of the lasting relevance of Marx's political economy; and David Stoddart, who almost persuaded me to become a physical geographer. At Durham, between 1976 and 1981, the outline of the book first took shape, initially through a series of postgraduate discussion seminars, and then through the opportunity to share my ideas with undergraduates through the framework of the Geographical Philosophical and Methodological Discussion Group; Eric Berthoud provided the contact with Neuchâtel, and ensured European enlightenment. In 1981 I joined the Department of Geography at Bedford College in the University of London, and had the opportunity further to formulate a critical approach to geographical enquiry through teaching a Masters' course on Geography: Theory and Practice. This emphasized only too clearly the difficulties encountered by many students attempting to reflect on the changes in 20th-century geographical practice. The subsequent process of merger with Royal Holloway College during the mid-1980s, to form Royal Holloway and Bedford New College, provided first-hand experience of the institutional changes associated with the restructuring of higher education. I owe a considerable debt of thanks to my colleagues in the department for all they have taught me about geography and for their patience in responding to my often obscure questions.

The following people have kindly found time to read and comment on draft chapters of the book: Felix Driver, David Gilbert, Peter Haggett, David Lambert, Roger Lee, Gunnar Olsson, Rob Potter, Robert Sack, Don Thompson and Pam Unwin. I have learnt much from their perceptive comments, and the final text is greatly improved as a result of their critical appraisal. I am also particularly grateful to the British Acadamy for a grant to enable me to visit geography departments in California in the spring of 1990 and to Justin Jacyno for drawing figures 4.1 and 8.1. Finally, if it had not been for Vanessa Lawrence at

Longman this book would never have been finished. Not only did she encourage its formal beginnings, but she also nurtured its progress and retained a remarkable degree of patience as the delays in completing it built up.

Tim Unwin
Virginia Water
15th November 1991

# *Acknowledgements*

We are grateful to the following for permission to reproduce copyright figures and tables:

The Association of American Geographers for Fig. 4.1; and Heinemann Educational and the author, J Habermas for extracts (Habermas, 1978).

CHAPTER 1

# *Geography: the social construction of a discipline*

> Geography should be encouraged to seize the central fortress, ejecting both pure science and that grossly over-promoted intellectual exercise called mathematics. Geography should stand alone with its one educational equal, the study of the human spirit in English language and literature. Geography is queen of the sciences, parent to chemistry, geology, physics and biology, parent also to history and economics. Without a clear grounding in the known characteristics of the earth, the physical sciences are mere game-playing, the social sciences mere ideology.
>
> (Leader in *The Times*, 7th June 1990: 13)

Geography is one of the oldest forms of intellectual enquiry, and yet there is little agreement among professional geographers as to what the discipline actually is, or even what it should be. Moreover, what has been practised as geography has changed substantially over the last two millennia. In recent decades the pace of this change has accelerated dramatically. One result of this is that the public image of what it is that geographers spend their time doing is often at considerable variance from reality. This book is about the causes of such changes, and the way in which an academic discipline is integrated into the society of which it is a part. In particular it addresses how specific societies produce knowledge. It suggests that what a society accepts as being truthful statements are the result of a series of interactions between social, political, economic and ideological interests. More formally, it is designed as a historically oriented reflection on the emergence of contemporary geography, which seeks to reveal the underlying connections that exist between knowledge, power and human interest.

In contrast to the fervent espousal of the importance of geography expressed within the quotation which opens this chapter, the discipline is not widely seen as forming an important element of the world's educational systems. Moreover, even in Britain where geography has for long been one of the most popular subjects at the secondary level, professional geographers play a very limited role in influencing political decisions. This is surprising given the wealth of research by geographers on subjects such as environmental monitoring, economic restructuring and climatic change, which are all currently seen as being politically sensitive. There would therefore seem to be some kind of discontinuity between the views society holds about the discipline as a subject to be taught at school and society's response to the way in which the discipline is practised by people claiming to be professional geographers.

In the last resort, academic disciplines exist not only because their practitioners believe in their validity, but also because the societies of which they are a part believe in their utility. Both teaching and research are expensive, and, particularly during times of economic recession, the content of both therefore reflects a process of negotiation between academics and the society in which they live. However, the public image of geography should also bear some resemblance to what it is that people who call themselves geographers do. It is at this juncture that the present book takes its departure, first by examining what the public image of the subject is, and then by looking in more detail at the ways in which disciplines are defined.

## 1.1 Geography in the public domain

In *The geographer at work*, one of the few books designed to introduce geography to the wider public domain, Peter Gould (1985: xiv–xv) suggests that 'Most people have little idea what modern geography is all about.' His book begins with a description of a cocktail party at which the following conversation took place (Gould, 1985: 4):

> 'And what do *you* do?' she said.
> 'Oh,' I said, grateful for the usual filler, 'I'm a geographer.'
> And even as I said it, I felt the safe ground turning into the familiar quagmire. She did not have to ask the next question, but she did anyway.
> 'A geographer?'
> 'Er . . . yes, a geographer,' said with that quietly enthusiastic confidence that trips so easily from the tongues of doctors, engineers, airline pilots, truckers, sailors and tramps . . .
> 'Oh really, a geographer . . . and what *do* geographers do?'

He continues: 'It has happened many times, and it seldom gets better. That awful feeling of desperate foolishness when you, a professional geographer, find yourself incapable of explaining simply and shortly to others what you really do' (Gould, 1985: 4).

This account is typical of the experiences of many professional geographers, and well illustrates that the public understanding of what it is that geographers do is extremely limited. Such a situation cannot, though, be blamed on the public in general; geographers themselves have frequently been very poor at actually explaining and justifying their role in society. Indeed, many people teaching and undertaking research in geography departments when faced with the cocktail party question noted above, quickly cover their tracks, with statements such as 'Well, I'm really a soil scientist' or, 'Actually, I'm a consultant on development issues.' It is not easy to establish precisely why this is, but it is probably partly because such geographers realize that the public perception of the discipline is so far removed from their everyday practice that to say simply that they were geographers would appear relatively meaningless. It may also be because geography is such a wide ranging discipline, covering research on topics as diverse as mountain forming processes and the medieval wine trade, that the single word geography conveys little idea of precisely what sort of research is undertaken by geographers. One cannot, though, escape the conclusion that it may also be because many people working in geography departments remain unhappy with the idea that there is indeed something unique and worth while about their own discipline.

The image of geography held by most people is usually derived from their school education. Geography in Britain is thus still widely seen as being about 'capes and bays', and in the United States of America as being concerned with 'states and capitals'. This is well illustrated in a sequence of cartoons by Garry Troudeau (Fig. 1.1), depicting a teacher, clearly a dedicated enthusiast for geography as 'one of the most *basic* disciplines', asking his pupils to point out the locations of a number of places in different parts of the world (*The Guardian*, 23–27 August 1988). The students, in response, show little knowledge, even though the teacher points out that the reason he asked them to identify Nicaragua was that 'we've been involved in a war there for the past eight years'. This cartoon is particularly significant, because even if geography really is about 'capes and bays' or 'states and capitals', it suggests that the subject is evidently failing to provide students with a body of knowledge that they consider to be useful and worth remembering.

In the political sphere, geographers have played a very limited role in influencing decision making and in advising governments at national, let alone international, levels. Likewise, although the media coverage of geography is beginning to increase, geographers are still rarely interviewed on news or current affairs programmes. In contrast, economists are widely asked for advice concerning the handling of the economy, and botanists are turned to for the development of new high-yielding varieties of crop to increase global food production. This situation was well illustrated several years ago during a visit to the headquarters of the Food and Agriculture Organization (FAO) of the United Nations in Rome. The opportunity arose to discuss the potential contribution of research by geographers on the problems and implementation of agricultural development in the poorer countries of the world, and time and again the conversation would come around to the point that geographers had few immediately apparent skills or attributes which were perceived to be of use. Instead the FAO is keen to employ forestry experts, biochemists with specific skills in the development of pesticides, and economists to undertake cost benefit analyses of the effects of introducing certain types of innovation. This is not to deny that some

**Fig. 1.1** Doonesbury by Garry Trudeau (*The Guardian*, August 1988). Universal Press Syndicate © 1988 G.B. Trudeau.

geographers have played a significant part in influencing policy decisions (for example Hall, 1963, 1980, 1988), but it is to suggest that their role is less than that of practitioners in some other disciplines. In part this reflects the widespread separation between the realms of political decision making and academia in general, but it also illustrates that there is no immediate vocational niche for geographers.

Another way of examining public attitudes concerning the value of geography is to look at employers' opinions of geography graduates. A survey undertaken in the mid-1980s (Unwin, 1986), for example, suggested that while no employers saw a geography degree as being a distinct handicap, at least half of the respondents saw it as offering no particular advantages. Of the employers who did see a geographical training as being useful, the majority thought that it was the computing and statistical skills provided in geography degrees that were of most importance. Another more alarming conclusion from this survey was that most employers had little idea about the sorts of teaching and research that were being undertaken within geography departments in Britain during the 1980s. Although this may also be true of other disciplines, given that employers tend to be more concerned with the personal attributes of applicants than with their academic achievements (Unwin, 1986), this once again reflects the poor performance of geographers in developing public awareness about their discipline.

In most countries, questions of public awareness and accountability are central to debates over the future shape of higher education. Although inertia and vested interests make it difficult to change the organizational framework of disciplines, recent experiences in Britain at least, suggest that governments can, through their role as paymasters, have a very significant influence on the type of teaching and research practised in institutions of higher education (for an Australian comparison see Powell, 1990). If specific disciplines are not seen to be providing either useful graduates or useful research, then, particularly during times of financial stringency, the amount of funding that they receive is likely to be reduced. This raises at least two fundamental issues. First, it assumes that it is indeed possible to distinguish between particular types of knowledge that are, and are not, useful. Secondly, though, it also requires the formulation of clear criteria by which disciplines can be identified.

## 1.2 *The construction of disciplines*

Despite the institutional factors giving rise to the apparent immutability of contemporary disciplinary boundaries, there is nothing absolute or sacred about them; all disciplines have been created and argued over by people, and there is no single criterion upon which such boundaries can be agreed. In general terms, disciplines have been identified and justified in four main ways.

First, it has been argued quite simply that a discipline is the collective activity of its practitioners; geography can thus be seen as being whatever it is that geographers choose to do. Bird (1989: 214), for example, advocates that 'geography is what geographers have done; geography is what geographers strive to accomplish'. It is this type of usage, referring to activity within the discipline, that Johnston (1986a) calls *academic*. Such a definition emphasizes that disciplines are social phenomena that reflect the institutional and political frameworks within which they emerge. Geography departments exist in universities and polytechnics, and the staff within them have to compete for resource allocations in order to maintain their collective existence. Accordingly, geographers must continue to attract undergraduate and postgraduate students, by teaching something called geography that is seen as having some interest or utility to those who study it. They must do research that their financial sponsors deem to be useful. Changes are brought about by the activities of influential groups of people within the discipline. Success is achieved through the propagation of a beneficial image of the discipline within society; failure results from an inability to produce useful products.

A second way in which attempts have been made to identify individual disciplines has been through reference to particular objects of study or subject matter. It is this usage that Johnston (1986a) terms the *vernacular*. Such definitions suggest that there are certain objects that are geographical, and that are not, for example, social or geological. This implies that some specific order exists in the world of phenomena, within which practitioners of any discipline simply need to identify their niche. Competition then exists between disciplines for particular objects, with successful disciplines expanding at the edges and swallowing up less successful ones. In general this is the system of definition and justification that is most often resorted to (Holt-Jensen, 1988), and can be exemplified by much geographical work in the first half of the 20th century on the region. Geographers such as Fenneman (1919) thus saw the region as being their particular object of study, arguing that its use would serve to prevent geography from being absorbed by other sciences.

Thirdly, disciplines have also been described in terms of their methodology or techniques. Mentions of historical methods (Bloch, 1954; Norton, 1984; Driver, 1988) and geographical or geomorphological techniques (Ebdon, 1977; Silk, 1979; Clark, Gregory and Gurnell, 1987a; Goudie, 1990) are thus widespread, and many geography undergraduate degree programmes have courses that incorporate phrases such as 'geographical methods and techniques'. Again, as with subject matter definitions, such justifications attempt to delimit disciplinary boundaries with reference to a unique set of technical tools, that can be learnt and applied to different phenomena. Disciplines thus expand through the creation of new types of technique, or through the poaching and development of methods from other disciplines. A classic current example of this practice has been the rapid development of so-called geographical information systems, and the spate of lectureships in this subfield of the discipline that were advertised in Britain during the late 1980s (Chrisman *et al.*, 1989; Maguire, Goodchild and Rhind, 1991).

Both subject matter and methodological definitions tend to imply a static and unchanging view of the academic world, built upon the justification that there really are methods and techniques that can create specific single disciplines. One thus becomes a geographer by learning a particular set of skills and subjects that comprise some kind of geographical truth. A fourth way of defining disciplines attempts to avoid such a replicative stance, by focusing on the sorts of questions that disciplines ask, and the ways in which these questions are framed. Although such definitions again seek to divide up the realm of academia into distinct cells, disciplines defined on the basis of the questions that their practitioners ask can no longer remain stagnant and unchanging.

For the majority of people, the content of any particular discipline does not, though, depend on a well-formulated theoretical debate, but rather on their practical experiences of it in the classroom. It is therefore particularly important to examine the connections between geography as it is practised at different levels within the educational system. This is significant not only for understanding the wider public image of the discipline, but also for examining the type of teaching and research activity within it.

## 1.3 Geographical education

Gritzner (1986: 252) has suggested that,

> No building or field of knowledge is stronger than its foundation; the 'temple' of academic geography rests upon the geographic knowledge and skills possessed by the products of the nation's elementary and secondary schools and the students who, both quantitatively and qualitatively, ultimately enroll in our own courses.

Such experiences, though, vary greatly, not only between countries, but also within a single country depending on the type of syllabus followed. Indeed, the amount of geography studied at the primary and secondary levels can vary from almost none, as in much of the United States of America, to about one-third of the curriculum for those doing A-level geography in England and Wales. In most European countries, the required amount of geography varies somewhere between these two extremes. Broadly speaking, it is possible to categorize educational systems into those that maintain a breadth of subjects throughout the primary and secondary curriculum, as in much of Europe, and those that allow considerable specialization, as currently in England and Wales where those in the last two years of their secondary education can specialize in as few as two or three subjects. The contrasts between the role of secondary geography in the United States of America and in England provide a good indication of the different problems faced by undergraduates reading for geography degrees in the two countries.

### *1.3.1 Geography in the secondary curriculum in North America*

Within the United States of America, the content of the school curriculum varies not only by state, but also by local school district. Despite this potential for diversity, there are only very few pupils who study a specific high school subject called geography. Indeed, as Hill and LaPrairie (1989: 2) point out 'school geography has almost disappeared as a separate subject'. The geography that is taught at the secondary level is usually included in a combined social studies course, or in general courses on history, more often than not taught by people without a geographical background. As Holcomb and Tiefenbacher (1989: 161) have thus noted,

> Geography's place in the school curriculum has been slipping for many years, despite an effort in the 1960s to reintroduce the scientific inquiry method through the High School Geography Project funded by the National Science Foundation. In most school curricula, geography was merged into a more general 'social studies' and frequently subsumed into American history.

Moreover, they go on to note that 'the closing of geography departments in such prestigious universities as Michigan, Chicago and Columbia, and the virtual absence of geography in the Ivy League, caused a perceived vulnerability of the discipline in American academia' (Holcomb and Tiefenbacher, 1989: 161).

The effect on the higher education sector of this dearth of geographical education at the primary and secondary levels is threefold. First, freshmen, and even second-year sophomore students, have only a very limited knowledge of either the content or the skills associated with geography. This in turn means that higher-education lower-division geography courses must frequently begin with somewhat elementary content based courses, covering the fundamentals of such subjects as the physical environment and cultural geography. Trimble (1986: 271) thus bemoans 'the lack of intellectual preparation of entering college students as it affects the learning of physical geography'. Secondly, though, the lack of geography in high school means that most secondary students have little idea in advance of the content of higher education courses in geography. Although Hill and LaPrairie (1989) suggest that other disciplines such as psychology are readily able to overcome such a handicap, it can nevertheless have serious repercussions for undergraduate recruitment, with many potential undergraduates opting for subjects with which they have some direct familiarity, rather than for the unknown. This is related to a third effect, alluded to by Holcomb and Tiefenbacher (1989), which is that the relatively low academic status of geography means that many of the most intellectually able students opt for disciplines which are seen by society at large as being more demanding and thus more prestigious. These influences are all closely related in a complex structure of causation, since the level at which undergraduate geography courses commence has a direct influence on the abilities with which students graduate, and thus with the overall academic status of a discipline.

During the 1980s it was increasingly argued that one way of reversing the declining status of university geography in the United States of America was to revitalize it in the secondary curriculum. Unfortunately, as Jumper (1986: 254) has pointed out, 'the rewards system in higher education places little value upon faculty efforts directed towards the pre-College level'. Despite this, curriculum reform in states, such as South Dakota (Gritzner, 1986), Tennessee (Jumper, 1986) and Texas (Boehm and Kracht, 1986) has led to a considerably improved status for high school geography. Among the earliest and most important of such reforms was that initiated in California and commented upon by Stutz (1985) and Salter (1986, 1987).

The Californian evidence illustrates the ways in which a few individuals played a key role in driving through educational reforms to the advantage of geography. The subject had disappeared from the local school curriculum in 1962, but twenty years later, in the autumn of 1982, articles in a San Diego newspaper generated considerable 'awareness in the community of the value and need for more geographic education' (Stutz, 1985: 391). Meanwhile, a new superintendent of the San Diego Unified School District outlined plans to introduce an additional year of social studies in the local school curriculum. Stutz saw this as an ideal opportunity to include an element of geography, and through a process of lobbying and committee work gradually evolved a new local curriculum in San Diego that did indeed include geography. At the state level in 1983, Senate Bill 813 provided both funds and impetus for educational reform in California as a whole (Salter, 1986), and a group of geographers at the University of California, Los Angeles (UCLA) also saw this as an opportunity to expand the profile of the discipline. Senate Bill 813 required the creation of Model Curriculum Standards which were designed to act as state-wide curriculum guidelines, and it also specified that the following three courses in the domain of history–social sciences should be taught in all California public schools in grades 9–12: (a) United States history and geography, (b) world history, culture and geography, and (c) American government, civics and economics. This led to the formal configuration of the California Geographic Alliance based at UCLA, including both university and school geographers who pooled resources in order to develop the relevant Model Curriculum Standards (Salter, 1986). For many teachers, the incorporation of geography into the new courses was initially considered to be little more than the inclusion of knowledge about states and capitals, but as Salter (1986: 13) has stressed, 'The Model Curriculum Standards were intended to provide a dynamic example of geography that is more than place-names.' The geographical goals identified in the Model Curriculum Standards for the course on world history, culture and geography thus included the following geographical goals:

1. 'To understand and employ the tools of geography, such as the effective use of maps, aerial photos, and geographical models';

2. 'To comprehend the physical and cultural characteristics of various regions of the world';
3. 'To appreciate the complex interactions of peoples with their environments'; and
4. 'To evaluate the effects of the human alteration of the physical environment' (Salter, 1986: 11).

While these Californian initiatives did lead to an increase in the amount of geography within the curriculum, the discipline nevertheless remained within a framework dominated by history or social studies. Geography was by no means elevated to the position it holds, for example, in the English curriculum. Based on the Californian example, though, a number of other Geographical Alliances, bringing together geographers with elementary, secondary and tertiary experiences, have been established in the United States of America, and Fuller (1989: 480) notes that within only six years the Alliance network had 'spread to approximately 30 states, 21 of which are currently funded by matching grants from the National Geographic Society'. At the same time, the Association of American Geographers (AAG) and the National Council for Geographic Education (NCGE) produced a publication entitled *Guidelines for Geographic Education in the Elementary and Secondary Schools* in 1984, and subsequently developed the Geographic Education National Implemetation Project, designed to enhance the position of geographical education in schools throughout the United States of America (Jumper, 1986). According to Bednarz (1989: 486) this has led to a veritable 'renaissance for geographic education'.[1]

Alongside the development of Geographical Alliances, Salter also obtained the support of the mayor of Los Angeles to declare a city-wide Geography Awareness Day in 1985. This initiative was subsequently followed by geographers in other cities, and eventually in 1987 the first National Geography Awareness Week in the United States was held from 15 to 21 November. As Holcomb and Tiefenbacher (1989: 159) note, 'The proclamation, passed by Congress and signed into law by President Reagan, noted the national ignorance of, and the need for, the discipline of geography, and called upon the people of the United States to observe the week with "appropriate ceremonies and activities".' Much of the effort and financial support behind the Geography Awareness Week was generated by the National Geographic Society, and as a media event it was undoubtedly a success. However, as Holcomb and Tiefenbacher (1989: 165) stress, it did little to enhance the academic reputation of geography:

> By emphasizing trivial pursuit geography with its one-right-answer quiz questions and the oft repeated dummy stats, the public image of geography as a suitable occupation for eight-year-olds may have been reinforced. In little of the media material is there much impression of geography as a field worthy of a place in higher education.

In contrast to its tenuous position in the United States of America, geography is much better established in the school curriculum in

Canada (Robinson, 1986). In Wolforth's (1986: 18) words, it has for a long time

> been an independent high school subject in most Canadian provinces, is usually taught by well-trained specialists, and has a solid content that academic geographers would recognize as respectable, if not rigorous. It is supported by an enviable array of generally well-written textbooks, and imaginative teaching materials.

However, even in Canada there are problems: 'in no province is it a mandatory subject throughout the high school years', and 'as an optional subject, it is often thought to appeal to the academically weaker students' (Wolforth, 1986: 18). These comments raise two central issues. First, it is important that geography at the primary and secondary levels is taught by qualified geographers, rather than for example by historians, as it is in many high schools in the United States. Secondly, though, Wolforth stresses a crucial difficulty faced by geography as a discipline, in that it is widely considered to be academically weak. Both issues require further elaboration, and are examined in greater detail in the ensuing chapters.

### *1.3.2 Geography in the educational system of England and Wales*

Of all the countries in the world geography is probably in the strongest position in the primary and secondary curriculum in England and Wales (Storm, 1989). Prior to the late 1980s geography was well established in the school system, and was widely taken both at O-level, the main academic exam taken by 15- and 16-year-olds before its replacement in 1988 by the GCSE examinations, and at A-level, the final exam taken by secondary pupils usually at the age of 17 or 18. The only subjects that were more popular at O-level in 1980 were English language, mathematics, English literature and biology, although at A-level in 1982 it ranked as the ninth most popular subject after mathematics, English, physics, chemistry, biology, economics, general studies and history (Lee, 1985). Despite geography's relative success during the 1970s and 1980s, many geography teachers still complained about 'the government's apparent neglect of the value of maintaining a significant geographical element in the curriculum' (Bunce, 1986: 325). Much of this uncertainty came in the wake of a Prime Ministerial speech by James Callaghan in 1976 which initiated the so-called 'Great Debate' about the future of the country's educational system and which was eventually to lead to the Education Reform Act of 1988, described by Walford (1989: 161) as possibly 'the most significant educational reform of the century'.

Early government thinking on educational reform, as outlined in reports by the Department of Education and Science (1980, 1981) did not bode well for the future of geography. However, in 1985 the then Secretary of State for Education, Sir Keith Joseph, addressed a special conference of the Geographical Association, the main body representing geography teachers at the primary and secondary level in Britain, at the end of which he posed seven questions concerning the role of geographical education. In response to these, the Geographical Association prepared a document entitled *A case for geography* (Bailey and Binns, 1987), which presented a powerful argument for the inclusion of the discipline of geography at all levels in the primary and secondary curriculum. By the time of the publication of this document, a new Secretary of State, Kenneth Baker, had taken office and in August 1986 the Geographical Association sent him a bridging letter designed to maintain a dialogue between those involved in geographical education and the government. This letter suggested that

> 1) geography was a fundamental part of every child's education (and that it was currently weak in primary schools); 2) that geography should not be seen solely as a humanities subject; 3) that geography had powerful integrating qualities in itself 'with strong links in both the sciences and the arts'; 4) that direct observation and investigation of the environment was an important part of geographical work; 5) that geography should be concerned with controversial issues and problems in the latter years at school (Walford, 1989: 162).

These points succinctly capture the views of the Geographical Association concerning the future directions in which its officers wanted the discipline to move, with a central emphasis being placed on its integrative nature at the boundary between the physical and the human world, concentrating on relations and interactions between people and the environment.

In July 1987 the Department of Education and Science and the Welsh Office (1987) published a consultation document on its proposals for the implementation of a National Curriculum for all pupils aged between 5 and 16 commencing in the autumn of 1989. This designated maths, English and science as core subjects which would occupy the majority of curriculum time at primary level, and between 30 and 40 per cent of curriculum time during secondary schooling. A modern foreign language, technology, history, geography, art, music and physical education were designated as foundation subjects which should also be followed by all pupils during compulsory schooling. There was, though, considerable confusion about the exact role that geography would play in this, since in the allocation of curriculum time in years 4 and 5 it specified that 'History/Geography or History *or* Geography' should occupy 10 per cent of the curriculum time (Department of Education and Science, Welsh Office, 1987: 7). This clearly left open the possibility that some kind of combined humanities course in history and geography might be offered, and the non-alphabetic preference given to history also suggested that as far as the government was concerned it was to be the dominant partner.

Following a successful period of lobbying by geographers, 'When the Education Reform Bill appeared early in 1988, geography and history appeared as separate entities, with even the hybrid formulations of the Consultation Document dropped' (Walford, 1989: 163). Subject working groups were then set up, whose task it was to specify precise attainment targets and programmes of study. The Science and Mathematics Working Groups were the first to be established, and delays in formalizing the others meant that the Geography Working Group was not actually formed until after the Final Report of the Science Working Group had been published. Meanwhile, the Geographical Association, whose machinery for dealing with the National Curriculum was now well oiled, had once again set up its own working group to prepare a document designed to inform the debate concerning the sort of geography that children should study in the National Curriculum (Bailey, 1989; Daugherty, 1989). In particular, this emphasized the importance of enquiry-based learning and field work, and argued that the key skills that a geographical education could provide included graphicacy and map skills. In terms of content, the report emphasized that the curriculum should provide 'general locational knowledge; the understanding of processes affecting the physical, social and economic landscape; the competent handling of maps, diagrams, pictures', adding that 'Knowledge and understanding of the geography of Britain, of Europe, and some grasp of general world geography should be accommodated' (Daugherty, 1989: 31).

The composition of the Subject Working Groups indicated the wide-ranging fields of interest that were to be reflected in the establishment of educational attainments for the different subject areas. What is particularly significant is that those actively involved in geographical research in institutions of higher education had remarkably little input into the Geography Working Group's proposals. Thus, of the twelve people in the Geography Working Group, only one of them was a practising academic in a department of geography in an institution of higher education; others included the chairman of a travel company, a regional secretary of the Country Landowners' Association, and a Member of the Countryside Commission. The Geography Working Group's Interim Report was eventually sent to yet another new Secretary of State for Education and Science, John MacGregor, at the end of October 1989, and this formed the basis for his proposals which were formally published in June 1990. In these the nature of geography is majestically defined in the following terms:

(a) Geography explores the relationship between the earth and its peoples through the study of *place, space,* and *environment*. Geographers ask the questions *where* and *what*; also *how* and *why*.
(b) The study of *place* seeks to describe and understand not only the location of the physical and human features of the Earth, but also the processes, systems, and inter-relationships that create or influence those features.
(c) The study of *space* seeks to explore the relationships between places and patterns of activity arising from the use people make of the physical settings where they live and work.
(d) The study of the *environment* embraces both its physical and human dimensions. Thus it addresses the resources, sometimes

scarce and fragile, that the Earth provides and on which all life depends; the impact on those resources of human activities; and the wider social, economic, political and cultural consequences of the interrelationship between the two' (Department of Education and Science and the Welsh Office, 1990: 6).

Eventually, following further amendment by the Secretary of State, a Draft Order for Geography was published in January 1991, designed to provide the basis for all future geographical education in England and Wales. This divides the subject into five Attainment Targets (geographical skills, knowledge and understanding of places, physical geography, human geography and environmental geography) within each of which pupils are to be assessed against up to ten different levels of attainment.

This process of negotiation and amendment provides an excellent example of political intervention in the educational arena. While the Secretary of State's formal proposals of June 1990 represented a considerable change from the Working Group's Interim Report of October 1989, there were substantial further changes apparent in the eventual Draft Order of 1991. These reduced the Working Group's initial eight attainment targets to the five noted above, but more importantly also reflected a change in the type of education proposed. The Secretary of State, for example, recommended that 'the attainment targets should put more emphasis on knowledge and understanding of aspects of geography and less on assessment of skills which, however desirable, are not particular to geography' (Department of Education and Science, 1991: Annex B, 1). In particular he recommended that the concept of 'geographical skills' should exclude the use of secondary sources and enquiry skills. This clearly reflects an attempt to define geography in terms of specific skills and types of knowledge which are unique, and in so doing it denies the possibility that there may be important analytical skills or knowledge of relevance to more than a single discipline.

Moreover, the type of education proposed by the government is further well illustrated by the following three statements made by the Secretary of State:

> The Secretary of State recognizes that geography lessons will sometimes deal with conflicting points of view on important geographical issues. However, he considers that the main emphasis in the statutory requirements should be on teaching a knowledge and understanding of geography rather than on the study of people's attitudes and opinions (Department of Education and Science, 1991: Annex B, 2).
>
> Some statements of attainment in Attainment Target 4 (human geography) have been amended so that they concentrate on geographical knowledge and understanding rather than on political or economic issues (Department of Education and Science, 1991: Annex B, 2).
>
> Those statements of attainment in Attainment Target 5 (environmental geography) which dealt with viewpoints and attitudes rather

than knowledge and understanding have been removed (Department of Education and Science, 1991: Annex B, 3).

Taken together these statements reflect a highly technicist view of geographical education, and, as the ensuing chapters illustrate, they run counter to much of the most exciting geographical research that has been undertaken in institutions of higher education over the past twenty years. Whatever the outcome, though, the general debate that has surrounded the introduction of a National Curriculum has nevertheless raised the public profile of geography, and has at least brought the subject to a position where it can find an ardent supporter in the leader writer of *The Times*.[2]

### *1.3.3 Geography from primary to higher education*

The above examples of changes in geographical education in North America and England and Wales have focused mainly on the secondary curriculum, because it is at this level that the majority of people are exposed to the only geographical learning experiences that they ever encounter. The main function of school geography should therefore be to reflect the needs of those for whom it is the only geography that they will ever learn. However, this creates substantial problems if such a geographical education fails to provide potential undergraduates with sufficient insight into the sorts of research undertaken by professional geographers, and the types of courses involved in geography degree programmes (Unwin, 1989). It is, after all, the geographical research and teaching undertaken in institutions of higher education that identify the characteristics of geography as a formal academic discipline. Whereas in the past, many secondary geography syllabuses were largely determined by the requirements of higher education, the balance, at least in Britain, has now changed substantially (Kirby and Lambert, 1978). New syllabuses, such as the Geography 16–19 Project (Naish and Rawling, 1990), are being introduced specifically to provide pupils with geographical skills and knowledge suitable, not so much as a training for higher education, but rather for their day-to-day lived experience of society. This has not always been welcomed by those in higher education, many of whom see the changes that have taken place in secondary education as having led to a diminution in the knowledge levels of first-year undergraduates in their specific field. In particular, the Geography 16–19 Project has been widely, if not necessarily appropriately, blamed for a demise in the teaching of the scientific skills necessary for potential undergraduates satisfactorily to comprehend courses on physical geography in institutions of higher education (Bailey, 1989).

The issue of the type of geography that is taught at different levels in the education system is inherently tied up with the way in which the discipline is defined. For example, if such a definition is based on the collective actions of academic geographers, then it could be argued that geography at the primary and secondary levels should provide an introduction to the corpus of such material. However, if content- or skills-based definitions are used, as for example in the National Curriculum in England and Wales, then the issues concern the way in which these are divided up between the different educational levels, with different content or skill levels being seen as appropriate to different stages of education. This can be illustrated by a consideration of ways in which the geography of Vietnam might be studied by North American pupils at various educational stages. At the primary level it might be appropriate just to know where the country is, and to be able to describe its various characteristics. At the secondary level, aspects of the social and economic structures of the country could be explained, with particular reference to the military involvement of the United States during the 1960s. At university level, these issues could be integrated into a theoretical framework concerning the international political and economic system, and the way in which this influences contemporary socio-economic change. Doctoral research might then focus on the environmental influence of different types of rice cultivation in lowland areas of the country.

The clear differences between public perceptions of geography, based on people's experience of the subject at school, and the professional practice of the discipline, also necessitate a brief examination of broader issues concerning the role of education in society (Bloom, 1956; Butcher, 1968; Rowntree, 1987). At its most basic level, the central role of education is to provide people with the means of everyday survival. However, it is socially determined, and in most states it is also controlled, to varying degrees, by central government. This control is designed to ensure that people are socialized within the dominant ideology of the society in which they live. Consequently, education is usually designed to promote norms of behaviour, to 'improve' society, to further economic growth, and to counter any tendency towards anarchy or social critique. Such education is about conformity rather than change; it is concerned with the passing on of accepted knowledge (Hall, 1989). Most education systems therefore stifle enquiry and critique. They propagate an image that there is something called the truth, and that students merely have to remember this hallowed object in order to regurgitate it in examinations. These are the initiation rituals of modern society; once passed they enable young people to enter the adult world of social conformity.

Anyone who has experienced young children growing up will recognize that by the age of 2, most have started to ask a seemingly endless series of questions: 'What's Mummy doing?' 'Why is Daddy doing that?' 'Where are we going?' 'How does it work?' and the questions are often repeated time and time again. Such children have an innate sense of exploration, and a quest for understanding. However, only too rapidly parents find themselves forcing children to conform to the rules and regulations of the household, in order to remain sane. Although it may be fun, it is, for example, highly inconvenient for children continually to empty the bookshelves and cupboards of everything in them, and to cover the walls with wax crayon drawings. From a very early age, children are therefore encouraged to conform to the norms of the social group in which they live.

The difference in intellectual approach between a young child and many first-year undergraduates is enormous. Whereas many 2-year-olds are incessantly asking questions, most undergraduates sit passively in lectures, apparently eager to grasp the passed-on wisdom of their professors. The ability to recollect such knowledge is then tested in many universities, particularly in North America, through the use of multiple choice examinations, checked by computer; there really is one right answer, and four wrong ones to each question. Even when essay questions are set, students frequently go to their tutors saying 'How do I answer this question?', as if there is a correct way of so doing. The process of socialization is complete. The pursuit of a creative, critical knowedge is dead.

In contrast to such a bleak caricature, this book argues that higher education geography should essentially be a creative and alive experience (Gold *et al.*, 1991). Its central role should not be to encourage the acquisition and regurgitation of widely accepted knowledge, but rather to provide opportunities for the development of knowledge that has taken on a critical stance.

## 1.4 Geographical self-reflection

Little research has been undertaken on the particular types of courses favoured by geography undergraduates, but what evidence there is suggests that courses on the history and practice of the discipline are widely disliked, being matched only in unpopularity by those on statistical techniques (Unwin, 1989). One reason for this is that much of the material on the institutional and disciplinary development of geography is new, with almost nothing being taught on it at the secondary level. At school, geography therefore is in a very real sense only the subject matter that is taught. Pupils have very little opportunity to consider how the subject has developed; geography quite simply *is* the syllabus. Another reason why such courses are unpopular is that much of the theoretical and philosophical subject matter is inherently complex and difficult to comprehend. Very often this is made worse by the obfuscating language used by social theorists and historians of science, which seems to be designed much more to obscure than to illuminate (Billinge, 1983). As the following extract indicates, initiates have to learn a whole new language game before they can even begin to get to grips with the meanings underlying the text:

> Alternatively, the 'illusion of transparency' dematerializes space into pure ideation and representation, an intuitive way of thinking that equally prevents us from seeing the social construction of affective geographies, the concretization of social relations embedded in spatiality, an interpretation of space as a 'concrete abstraction', a social

hieroglyphic similar to Marx's conceptualization of the commodity form (Soja, 1989: 7).

It is important that geographers should take heed of Wittgenstein's (1961: 3) assertion that 'what can be said at all can be said clearly, and what we cannot talk about we must pass over in silence'.

The widespread dislike of courses on the historical development of geography is, however, also closely related to the role of education within society. Many undergraduates, who have been socialized to see education as the acquisition of skills and knowledge that are seen as being useful by society, enter higher education in order to be better placed later to receive the rewards of such knowledge. For such people, courses that encourage them to question the validity of the knowledge are at the best meaningless, and at the worst positively damaging to their future careers. But, for those who seek to change society, however minutely, it is essential that they grasp some understanding of the ways in which knowledge and power are related, and thus how academic disciplines create the knowledge that society sees as being useful.

Despite the unpopularity of such courses, geographers continue to consider them to be important, and support them with a range of introductory texts on the subject (K. Gregory, 1985; Clark, Gregory and Gurnell, 1987a; Holt-Jensen, 1988; Bird, 1989; Gregory and Walford, 1989; Kobayashi and Mackenzie, 1989; Peet and Thrift, 1989a; Cloke, Philo and Sadler, 1991; Johnston, 1991a). There are at least three main grounds upon which this importance is usually justified. The first is that the role of disciplines, and indeed science as a whole, has changed throughout history, and it is consequently argued that in order to understand the present value of geography it is essential to have some understanding of its past. As Billinge, Gregory and Martin (1984b: 20) have commented, 'the separations between past and present geographies are thus the very conditions of critical intelligibility; they allow us to make sense of our collective biographies'. Secondly, though, knowledge about the organization of academic disciplines can have a strategic significance. If disciplines can be vehicles for social change, then knowledge and understanding of the way in which they have developed can play an important part in influencing future social change. Thirdly, the questions asked by geographers have also varied, and if it is accepted that the philosophical positions adopted by individuals constrain their practice, then a consideration of the ways in which particular disciplines have emerged is essential for the comprehension of the limits of possible understanding.

While accepting all of these justifications for reflecting on the historical context of geography, this book seeks to fulfil three more specific aims. First, it argues that an understanding of the way in which geography was conceived prior to the 19th century is important if we are to grasp the full complexities of many of the issues of current debate within the discipline. To this end, Chapters 3 and 4 address the development of geographical enquiry in classical antiquity, the medieval period and the philosophically highly formative years of the 17th and

18th centuries. In the conclusion it is suggested that contemporary geographical enquiry has much to benefit from a re-engagement with some of these surprisingly neglected traditions. Secondly, the last twenty years have seen the discipline become increasingly divided both practically and theoretically into two separate areas of enquiry, those of physical and human geography. Advocates of the former have sought their intellectual nurture from the perceived strengths and stability of the physical sciences, whereas those of the latter have tried to draw human geography ever closer into the mainstream of thought in the social sciences and humanities. One result of this has been that geographers have had remarkably little to say in the public domain about critical issues of environmental concern, which once formed an important part of the subject. The conclusion argues that practitioners on both sides of the discipline could usefully benefit from a revitalized dialogue, seeking to reintegrate the physical and human worlds into some kind of conceptual unity. Finally, the book advocates a particular conceptualization of the role of scientific activity within society. This owes much to the critical theory of the Frankfurt School, and it is therefore to an examination of the underlying propositions of this school of thought that the next chapter now turns.

1. More recently, with the signing by President Clinton of the *Goals 2000: Educate America Act* into law on 1 April 1994, geography has become one of the nine core subjects in which all students must demonstrate competence at grades four, eight and twelve by the year 2000, and it has led to the formulation by the Association of American Geographers, the American Geographical Society, the National Council for Geographic Education and the National Geographic society of a set of national standards which will be published late in 1994 as *Geography for Life: the National Geography Standards.*

2. Since 1991 there have continued to be changes to the proposed National Curriculum for England and Wales. In 1992 the Secretary of State determined that Geography as a single subject was no longer to be compulsory at Key Stage 4 (over the age of 14), and then in 1993 Sir Ron Dearing was invited to undertake a major review of the National Curriculum in order to slim it down and improve its administration. With respect to geography, the draft proposals for the new National Curriculum published in May 1994 argued for four main changes: a substantial reduction in content; a restructuring at each Key Stage in order to clarify geographical knowledge, understanding and skills; the replacement of the five existing attainment targets with a single new one called *Geography*; and the use of level descriptions rather than the previous statements of attainment. The main components of the existing Order, namely skills, places and themes, the latter divided into physical, human and environmental geography, have nevertheless been retained. It is also worth noting that a subject core for GCE A and AS examinations was published by the School Curriculum and Assessment Authority in December 1993.

CHAPTER 2

# *The place of theory*

Pfuhl was one of those hopelessly, immutably conceited men, obstinately sure of themselves as only Germans are, because only Germans could base their self-confidence on an abstract idea – on science, that is the supposed possession of absolute truth. A Frenchman's conceit springs from his belief that mentally and physically he is irresistibly fascinating to both men and women. The Englishman's self-assurance comes from being a citizen of the best organized kingdom in the world, and because as an Englishman he always knows what is the correct thing to do. An Italian is conceited because he is excitable and easily forgets himself and other people. A Russian is conceited because he knows nothing and does not want to know anything, since he does not believe that it is possible to know anything completely. A conceited German is the worst of them all, the most stubborn and unattractive, because he imagines that he possesses the truth in science – a thing of his own invention, but which for him is absolute truth. Pfuhl was evidently of this breed.

Leo N. Tolstoy (1869), *War and Peace*, Book Three, part 1, 10, (Harmondsworth, Penguin, 1978: 757)

## *2.1 Science and society*

In the above quotation, Tolstoy captures both the essential definition of science, and also its key problem. Science, at the broadest level, is concerned with the pursuit of truth, but, as Tolstoy reminds us in his account of Pfuhl, it is also an abstract idea. It is an invention of the human mind, which once created takes on the status of absolute truth. This means that we need to have some understanding both of what truth is, and also of the method by which it is pursued (Russell, 1961; Popper, 1968; Harvey, 1969; Kuhn, 1970; Habermas, 1978). This section of the chapter addresses three main aspects of the relationship between truth and its pursuit. First, it concerns the distinctions between science and ideology, and more specifically between the claims of *empiricism*, according to which reason is subordinated to the senses, and *metaphysics*, which is concerned with questions about the essential being of things which science is unable to solve (Scruton, 1981). Second, it addresses questions of *epistemology*, or theories of knowledge, and third it is concerned with the ways in which knowledge is *used* by societies.

### 2.1.1 Definitions of science

Memories of school bring back readily accessible definitions of science: it is studied in laboratories; it involves experiments; and one's answers are wrong because they never quite agree with the truth in the answer book. Science consists of subjects such as physics, chemistry and biology, in contrast to the arts or humanities, such as English, classics and painting, which are somehow less precise and more to do with imagination and creativity. Science is concerned with the production of generalizations which explain particular phenomena. More formally, Braithwaite (1960: 1) has suggested that the aim of scientific explanation is thus 'to establish general laws covering the behaviour of the empirical events or objects with which the science in question is concerned'. According to Popper (1968: 27) the scientist 'puts forward statements, or systems of statements, and tests them step by step. In the field of the empirical sciences, more particularly, he constructs hypotheses, or systems of theories, and tests them against experience by observation and experiment.'

In this model, the sciences and the arts are distinguished both by the objects they study, and also by the methods they pursue. However, geography, combining an interest in both the physical and human worlds, has traditionally been seen as fitting neatly within neither camp. This ambiguity, for example, is clearly expressed in cases where students doing exactly the same higher education degree can graduate with either a BSc or a BA qualification. If geography is neither a science nor an art, it is often seen as a bridging discipline, one that brings together the arts and the sciences, involving an interaction between the subjective human world and the objective natural world. This view is, for example, clearly expressed in the National Curriculum proposals for geography in England and Wales, where it is argued that the core elements of the subject 'create a bridge between the humanities and the physical sciences' (Department of Education and Science, Welsh Office, 1990: 6).

The widely accepted view of science, which sees it as being 'a highly logical, ordered activity which tries to understand the world as it is independently of ourselves' has commonly been described as *rationalist* in approach (Haines-Young and Petch, 1986: 24). Such science is concerned with the formulation of hypotheses derived from observations, their subsequent testing, and their eventual elevation into theories and laws. Central to its practice is the development of theory, which for most researchers can be seen as 'the sum-total of propositions about a subject, the propositions being so linked with each other that a few are basic and the rest derive from these' (Horkheimer, 1972: 188).

As Horkheimer (1972: 188) goes on to say, 'The real validity of the theory depends on the derived propositions being consonant with the actual facts. If experience and theory contradict, one of the two must be re-examined. Either the scientist has failed to observe correctly or something is wrong with the principles of the theory.' Two arguments are central to such a view of science: that facts exist independently of the observer, and that they can be identified through observation and experience.

Such a conceptualization of science has its origins in the 17th century, and in particular with the writings of René Descartes (1596–1650) (Russell, 1961; Horkheimer, 1972). In his *Discourse on method* (1637), Descartes examined what has come to be called the method of Cartesian doubt, in which in order to establish a basis for his philosophy he made himself doubt everything that could be doubted. One thing remained that he could not doubt, and that was that he himself existed. In Descartes's (1968: 53–54) own words,

> while I decided thus to think that everything was false, it followed necessarily that I who thought must be something; and observing that this truth: *I think, therefore I am*, was so certain and so evident that all the most extravagant suppositions of the sceptics were not capable of shaking it, I judged that I could accept it without scruple as the first principle of the philosophy I was seeking.

As Russell (1961: 548) has pointed out, one effect of this emphasis on the mind as being of more importance than matter was that there is 'in all philosophy derived from Descartes, a tendency to subjectivism, and to regarding matter as something only knowable, if at all by inference from what is known of mind'. From this foundation, however, Descartes (1968: 41) sought to conduct his 'thoughts in an orderly way, beginning with the simplest objects and the easiest to know, in order to climb gradually, as by degrees, as far as the knowledge of the most complex, and even supposing some order among those objects which do not precede each other naturally'. For Descartes, theory was thus developed in a hierarchical way, starting with simple observations and building up more and more complex and comprehensive derivations (Horkheimer, 1972: 189).

To clarify these issues, it is essential to distinguish between two different ways in which theory and observation are related, the one *inductive* following the arguments of Bacon (1561–1626), and the other essentially *deductive* derived from the ideas of Leibniz (1646–1716) and more recently Popper (1968, 1976), (Haines-Young and Petch, 1986). Most empirical scientists have generally followed Bacon's inductive method, by which *universal* statements, such as hypotheses and theories, are inferred from *singular* statements, such as the results of experiments or observations. As the hallmark of science, induction is thus seen as the criterion of demarcation between science and non-science (Magee, 1973). This was generally the method adopted by philosophers such as John Locke (1632–1704), David Hume (1711–76) and later John Stuart Mill (1806–73). However, there are two fundamental problems with the inductive method. The first, as summarized by Popper (1968: 27), is that

> it is far from obvious, from a logical point of view, that we are justified in inferring universal statements from singular ones, no matter how numerous; for any conclusion drawn in this way may always turn out to be false: no matter how many instances of white swans we may have observed, this does not justify the conclusion that *all* swans are white.

This led Popper to argue that the fundamental demarcation criterion for admitting a system to be scientific should not be *verifiability*, but rather *falsifiability* (Popper, 1968: 40). According to Popper, induction based upon verifiability is unreliable, because however many experiments are undertaken to test a given hypothesis, there is no logical reason why the next experiment should not produce a different answer. In contrast, he argues that the scientific method should proceed through the establishment of scientific systems which can be refuted by experience, and are thus falsifiable. Secondly, though, induction also suffers from the problem that 'Usually some hypothesis is a necessary preliminary to the collection of facts, since the selection of facts demands some way of determining relevance' (Russell, 1961: 529). In other words, our preconceived hypotheses closely influence our choice of data, and it is therefore impossible to follow a methodology that is purely inductive. The alternative to induction is deduction, in which singular statements are derived from universal ones. The main development of this method was worked out by Leibniz, building in part on the ideas of Descartes (Russell, 1961). However, it too is problematic, since all of our theories or hypotheses can themselves be seen as having been influenced to some extent by our own experience. Consequently, it is impossible to follow either a purely inductive or a purely deductive method, and some way out of this impasse must be sought.

One solution is to reject the whole rationalist approach to science altogether, and to argue instead that there is no such thing as a successful rational scientific method. Foremost among those who advocate such an approach is Feyerabend (1975, 1978), who, like Tolstoy, sees science as only one among a number of competing ideologies, all of which 'must be seen in perspective. One must not take them too seriously. One must read them like fairytales which have lots of interesting things to say but which also contain wicked lies, or like ethical prescriptions which may be useful rules of thumb but which are deadly when followed to the letter' (Feyerabend, 1981: 156; see also Newton-Smith, 1981). Feyerabend essentially attacks the rational view of science on two grounds. First, he argues that the history of science has shown that all scientific rules have been violated at some time or another, and second he suggests that scientific knowledge is no more superior than any other kind of knowledge (Zelinsky, 1974; Haines-Young and Petch, 1986). In essence, because all scientific rules or laws are sooner or later superseded, he suggests that they should all be treated with caution. More importantly, Feyerabend draws attention to the fact that knowledge can be classified into different types, and that the usual grounds upon which it is argued that scientific knowledge is somehow superior do not bear up to investigation.

Science can thus be thought of and described in a variety of different ways. Indeed science has meant many contrasting things at different periods in the past. It is therefore important to identify some kind of overall framework in which to view such changes. Three of the most important of such overviews are those presented by Kuhn (1962, 1970), Foucault (1972, 1980) and Habermas (1978).

### *2.1.2 Kuhn: paradigms and scientific revolutions*

Some of the most pervasive ideas to have influenced recent accounts of the history of geography (Johnston, 1979, 1983, 1987, 1991a; Holt-Jensen, 1981, 1988), have been those of Thomas Kuhn, a theoretical physicist, in his book *The structure of scientific revolutions*, first published in 1962. Observing that 'the practice of astronomy, physics, chemistry, or biology normally fails to evoke the controversies over fundamentals that today often seem endemic among, say, psychologists or sociologists', Kuhn (1970: viii) sought to explain why this difference existed. In so doing, he coined the term 'paradigms', which he took 'to be universally recognized scientific achievements that for a time provide model problems and solutions to a community of practitioners' (Kuhn, 1970: viii). He goes on to argue that 'A paradigm is what the members of a scientific community share, *and*, conversely, a scientific community consists of men who share a paradigm' (Kuhn, 1970: 176). In essence, according to Kuhn, scientific communities therefore consist of groups of researchers using commonly agreed methods, to derive solutions to unsolved problems defined by the existing framework. Paradigms thus define both the problems and the methods by which they are solved, with the solution of one problem generating the questions posed in the next (Barnes, 1982).

Kuhn (1970: 10) also introduced another term closely related to the paradigm, namely *normal science*, which he took to mean 'research firmly based upon one or more past scientific achievements, achievements that some particular scientific community acknowledges for a time as supplying the foundation for its further practice'. Achievements which attract adherents away from other methods of scientific enquiry, and yet which are sufficiently open-ended to leave enough new problems to be resolved are thus paradigms. Young scientists become socialized into a paradigm by reading the textbooks produced by practitioners of this normal science, and then participate as puzzle-solvers in its propagation by resolving some of its problems. As Kuhn (1970: 35) himself observed, however, 'Perhaps the most striking feature of the normal research problems . . . is how little they aim to produce major novelties, conceptual or phenomenal.'

The practice of science changes, according to Kuhn (1970: 92), through periodic revolutions, which he defines as 'non-cumulative developmental episodes in which an older paradigm is replaced in whole or in part by an incompatible new one'. These 'scientific revolutions are inaugurated by a growing sense, . . . often restricted to a narrow subdivision of the scientific community, that an existing paradigm has ceased to function adequately in the exploration of an aspect of nature to which the paradigm itself had previously led the way' (Kuhn, 1970: 92). Furthermore, such revolutions can apply both to major paradigm changes, such as that attributed to Copernicus, and also to much smaller ones, such as that associated with the discovery of X-rays. In essence, the choice between competing paradigms is 'a choice between incompatible modes of community life' (Kuhn, 1970: 94). Periods of scientific revolution commence when a few scientists are worried by anomalies which cannot be accounted for by normal science, and begin to undertake *extraordinary research* outside the bounds of the accepted paradigm. These scientists are nearly always young or new to the paradigm which they eventually overthrow, and in time they present a new paradigm to the scientific community for judgement. If this new paradigm is accepted, then a scientific revolution is seen to have taken place.

Such concepts were first introduced to geographers by Haggett and Chorley in 1967, and by the late 1970s they had become 'common currency in geographical writings' (Stoddart, 1981: 70; see also Harvey, 1969: 16–18; Berry, 1973; Johnston, 1979; Buttimer, 1981). However, as Stoddart (1981: 72) points out, 'This ready acceptance of Kuhn's vocabulary has occurred without any close attention to Kuhn's own statements or to the critical literature on them in the history and philosophy of science.' Three central criticisms have been made of Kuhn's arguments. The first, elucidated by Masterman (1970), is that in his initial formulation Kuhn used the term 'paradigm' in no less than twenty-one different ways. This criticism led Kuhn (1970, 1977) to reformulate some of his arguments, and in particular to pay greater attention to the role of scientific communities and rather less to the paradigms themselves. However, the diversity of ways in which Kuhn initially used the term paradigm, was mirrored in geography by 'The confusion implicit in Haggett and Chorley's initial usage and the plasticity with which the concept has subsequently been applied' (Stoddart, 1981: 73).

A second criticism of Kuhn's argument is that scientific revolutions do not take place as rapidly as he suggested, but rather that there can be long periods when different paradigms compete for acceptance (Masterman, 1970; Lakatos, 1978; Mulkay, 1978). Lakatos (1978) thus suggests that within any research programme there are tendencies both towards its reinforcement and towards its degeneration. While a programme is in decline, there will be other competing research programmes, and there may be a lengthy period of criticism before a new programme is adopted. In this context, Stoddart (1981) has pointed out that geography has frequently been characterized by numerous competing approaches. Thus, even though the ideas of Davis (1909) were widely adopted by geomorphologists in the first half of the 20th century (Chorley, Beckinsale and Dunn, 1973), there were nevertheless other alternative approaches to the practice of geomorphology, such as those proposed by Penck (1924) and Hettner (1921).

Thirdly, Popper (1968, 1970) has argued against both Lakatos and Kuhn, and suggested that science should be critical, characterized by the design of experiments specifically intended to falsify, rather than to verify, previous theories. At the core of this critique is Popper's suggestion that science advances through the testing rather than the replacement of ideas. While Barnes (1982) has argued that Popper's views are essentially normative, reflecting one vision of what science *should be* rather than what it *is*, Stoddart (1981) nevertheless uses this insight to suggest that one reason why the adoption of the paradigm concept was so popular in geography was that its concentration on the replacement of ideas could also be extended to apply to the replacement of practitioners. Thus, those who advocated the paradigm concept could see themselves as heroes, replacing an older generation of fools and knaves. In Stoddart's (1981: 78) words, 'the concept of revolution bolsters the heroic self-image of those who see themselves as innovators and who use the term paradigm in a polemical manner'.

### 2.1.3 *Foucault: truth instead of knowledge; power instead of ideology*

The above criticisms of Kuhn's ideas have largely been formulated within the natural sciences, and as Johnston (1987: 22) has pointed out, 'In all cases, there is no questioning the basic world-view, which (implicitly) is that science is the study of an empirical world, in which subject and object can be separated, and that scientific progress is measured by the volume of successful predictions.' Like Pfuhl, such scientists equate science with truth, but fail to examine the criteria by which both science and truth are socially produced. A fundamentally different approach to understanding the history of science has been proposed by the French social theorist Michel Foucault (1972, 1980), one of whose central concerns has been to examine 'the political status of science and the ideological functions which it could serve' (Foucault, 1980: 109). More specifically, he has argued 'that in certain empirical forms of knowledge like biology, political economy, psychiatry, medicine etc., the rhythm of transformation doesn't follow the smooth, continuist schemas of development which are normally accepted' (Foucault, 1980: 111–12). Instead, he suggests that such sciences undergo periods of rapid change in terms of their discourse and forms of knowledge. The importance of these changes, however, is not so much their rapidity, but rather that they reflect 'modifications in the rules of formation of statements which are accepted as scientifically true' (Foucault, 1980: 112). Central to his arguments is his assertion that there is thus a fundamental connection between power, knowledge and truth.

Foucault (1980: 131) argues that '"Truth" is linked in a circular relation with systems of power which produce and sustain it, and to effects of power which it induces and extend it.' Thus, 'each society has its régime of truth, its "general politics" of truth: that is, the types of discourse which it accepts and makes function as true' (Foucault, 1980: 131). For Foucault, therefore, truth is a relative concept, depending on the power relations within the societies which produce it. In a very real sense, Foucault is thus concerned with the political economy of truth, which he sees in capitalist society as being characterized by five traits: it

> is centred on the form of scientific discourse and the institutions which produce it; it is subject to constant economic and political incitement . . . ; it is the object, under diverse forms, of immense diffusion and consumption . . . ; it is produced and transmitted under the control, dominant if not exclusive, of a few great political and economic apparatuses . . . ; lastly, it is the issue of a whole political debate and social confrontation (Foucault, 1980: 131).

For Foucault (1980: 131), therefore, 'It is necessary to think of the political problems of intellectuals not in terms of "science" and "ideology", but in terms of "truth" and "power".'

Foucault firmly locates science within the societies which produce it, and of which it is a part. In order to interpret changes in science he examines the variations in the relationship between words and things throughout the course of modern history (Foucault, 1966). In particular, he focuses on the various world-views, or structures of thought, that people have held, which he calls *epistemes* (Foucault, 1972: 19). In the 16th century, Foucault thus argues that reality was all on one plane, and that both words and things were perceived at the same level. However, he suggests that this Renaissance episteme, or system of thought, was overthrown during the first half of the 17th century by a classical episteme, which established a break between things and their representations. The problem for science at this period was therefore that it had to find a language which reflected the apparent ordering of the world (Claval, 1981). At the beginning of the 19th century Foucault suggests that interest shifted once again, this time towards an understanding of function rather than appearances. In particular, in this modern episteme, the sciences concerned with people found themselves in the new position of being both observer and observed. This came to be of particular importance for the development of the natural sciences, linguistics and political economy, which were in turn concerned with questions of life, language and work (Claval, 1981).

Foucault's analysis of power and truth is not, though, without its critics. Johnston (1991a) thus argues that Foucault fails satisfactorily to explain the processes through which one episteme is replaced by another. He thus concentrates on the way in which particular societies create their own world views, rather than on periods of competition between such views. Moreover, his insistence on the existence of a 16th century episteme, a classical episteme during the 17th – century and a modern episteme emerging in the 19th century, denies the possibility that there were other systems of thought in existence during these periods. A second difficulty with Foucault's project is that it 'contains an element of the arbitrary' (Claval, 1981: 238). In his analysis of the modern episteme, he thus focuses particular attention on the life sciences, linguistics and political economy, which he sees as being 'true sciences because they have been able to create a perfectly definable object, which is by nature objective; life, work, language in its material manifestations' (Claval, 1981: 237). This relegates other disciplines, and particularly many of the social sciences, to a secondary position. Furthermore, Foucault fails satisfactorily to consider the developments in sciences such as physics and chemistry, for which his analysis seems much less appropriate. Thirdly, Foucault's insistence on the relativity of truth has brought him into conflict with those pursuing the quest for a single perspective that will provide a definitive view of society (Boyne, 1991). Not only does Foucault reject any form of absolutism, but he also tends to reject the possibility of a vision of totality. Foucault's arguments are thus, in this respect, diametrically opposed to those of Giddens and to Althusser. As Poster (1984: 39) notes,

> Foucault, rejecting the category of totality in general and the Marxist version of it in particular refuses to limit himself to an analysis of the working class. The category discourse/practice is thus not inserted into a totalized theory but floats like a hawk over the social historical process, ready to swoop down upon any topic that seems appropriate.

Despite these problems, Foucault's analysis is important because it suggests that people use knowledge in order to achieve power; even if there is such a thing as absolute truth, the relative truth that societies create is a truth that is designed to reflect and reinforce the power relations within that society.

### 2.1.4 *Habermas: power, knowledge and truth*

Foucault's concerns with life, language and work, and with power, knowledge and truth, are also reflected in the writings of Jürgen Habermas (1974, 1976, 1978, 1984, 1987a). However, as Poster (1984) has illustrated, Habermas approaches these issues in a substantially different way. While Foucault challenges the links between reason and democracy, Habermas still retains a belief in the 'value of individual autonomy through reason' (Poster, 1984: 32). Habermas also advocates the justification of theory on transcendental grounds, whereas Foucault rejects any attempt to create an overarching systematic theory and concentrates instead on a historical approach which does not privilege any particular form of discourse.

Building on a long tradition of German philosophy, particularly associated with the Frankfurt School (Tar, 1977; Held, 1980), Habermas provides a further critique of the philosophy of science, which is of particular relevance to an understanding of recent developments within geography. The home of the Frankfurt School was the Institut für Sozialforschung, or Institute of Social Research, established in Germany in 1923. Its first director, Carl Grünberg, captured the central purpose of the School in his 1924 inaugural address when he 'emphasized his opposition to the trend in German universities toward teaching at the expense of research and toward the production of "mandarins" only capable of serving the existing balance of power and resources' (Held, 1980: 30). The changing organization of higher education in the capitalist states of the world in the 1990s, makes it evident that such sentiments are as of much relevance today as they were when Grünberg first pronouced them. Following Grünberg's retirement in 1929, a diverse group of scholars from many different disciplines including Walter Benjamin, Theodore Adorno and Herbert Marcuse, continued the work that Grünberg had set in motion, under the new directorship of Max Horkheimer. In particular they were determined that their social theories should have profound political implications. Deeply opposed to the rise of Nazi power, the Institute was transferred to Geneva in 1933, and then to New York in 1935. Eventually, in 1953 it was re-established in Frankfurt, although several of its key members, most notably Marcuse, remained in the United States.

Geuss (1981: 1) has summarized the two central foundations of the critical theory established by the Frankfurt School as follows:

> The members of the Frankfurt School think that Freud, too, was a conceptual revolutionary in more or less the sense in which Marx was, and that the theories of Marx and Freud exhibit such strong similarities in their essential epistemic structure that from a philosophical point of view they don't represent two different kinds of theory, but merely two instances of a single new type.

He goes on to claim that there are three essential distinguishing characteristics of such critical theories: they provide guides for human action by producing enlightenment in those who hold them and by being inherently emancipatory; they have cognitive content, and are forms of knowledge in themselves; and they are fundamentally different from theories in the natural sciences, in that whereas the former claim to be objective, critical theories are reflective (Geuss, 1981: 1–2).

Although the work of Habermas 'should not simply be regarded as the outcome of a path of progressive development that begins with the earliest writings of Horkheimer and Adorno' (Held, 1980: 249), it is his arguments that have become most associated with a new generation of critical theory (McCarthy, 1978, Roderick, 1986; Outhwaite, 1987). Central to much of Habermas's earlier work was his concern to develop a critique of scientism as reflected in the theory and practice of the natural sciences. In the preface to *Knowledge and human interests*, he thus argues that he is 'undertaking a historically oriented attempt to reconstruct the prehistory of modern positivism with the systematic intention of analyzing the connections between knowledge and human interests' (Habermas, 1978: vii). This powerful analysis has much to offer our understanding of the relationships between science and society, and indeed the position of geography within the academic division of labour.

**Table 2.1** Habermas's theory of cognitive interests

| Form of science | Knowledge-constitutive interest | Social medium | Expression through |
|---|---|---|---|
| Empirical–analytic | Technical | Work | Material production |
| Historical–hermeneutic | Practical | Language | Communication |
| Critical | Emancipatory | Power | Relations of domination and constraint |

*Source*: Derived from Habermas (1978).

## 2.2 *Science, knowledge and interest*

One of the key positions from which Habermas develops his arguments is that all 'knowledge is historically and socially rooted and interest bound' (Roderick, 1986: 51). In particular he develops a theory of what he terms cognitive interests, or knowledge-constitutive interests, in order to explain the connections between knowledge and human action. These cognitive interests arise from human participation in nature. Habermas (1978: 47) thus argues that 'From the level of pragmatic, everyday knowledge to modern natural science, the knowledge of nature derives from man's primary coming to grips with nature; at the same time it reacts back upon the system of social labor and stimulates its development.' Knowledge is therefore both derived from human involvement in nature, and also influences such action in the form of social labour. Despite such a claim, Habermas does not subsequently really develop a detailed analysis of the relationships between people and nature, either through a consideration of the ways in which nature acts as a force in people's lives, or through the ways in which societies change nature. It is here therefore that an engagement between the traditions of geography and critical theory can be seen as having much potential interest.

Habermas's critical theory suggests that there are three types of knowledge-constitutive interests: knowledge which enables the human species to control objects in nature and thus produce the necessities of its material existence; knowledge which enables people to communicate; and derived from these two types, knowledge which enables the species to act rationally, to be self-determining and self-reflective (Table 2.1). These three knowledge-constitutive interests he terms *technical*, *practical* and *emancipatory*. Moreover, he also argues that each of these interests is developed in a particular social medium: the technical through *work*, which is essential for material production; the practical through *language* which is essential for communication; and the emancipatory through *power*, expressed through relations of domination and constraint (Habermas, 1978: 313). Furthermore, he then suggests that 'There are three categories of processes for which a specific connection between logical–methodological rules and knowledge-constitutive interests can be demonstrated' (Habermas, 1978: 308). Thus, 'The approach of the empirical–analytic sciences incorporates a *technical* cognitive interest; that of the historical–hermeneutic sciences incorporates a *practical* one; and the approach of critically oriented sciences incorporates the *emancipatory* cognitive interest' (Habermas, 1978: 308).

### 2.2.1 *Empirical-analytic science*

Habermas's (1978) basic aim in *Knowledge and human interests* was to develop a critique of the role of science in society. He suggested that prior to the 19th century philosophy was centrally concerned with a number of different theories of knowledge, or epistemologies, but that since then the rise of scientism has meant that science cannot be properly comprehended by philosophy. '"Scientism" means science's belief in itself: that is the conviction that we can no longer understand science as one form of possible knowledge, but rather must identify knowledge with science' (Habermas, 1978: 4). Thus Habermas argues that once science had become equated with knowledge, it became impossible for a philosophy concerned with the grounds upon which theories of knowledge were possible to comprehend science. This fundamental change was brought about by the emergence in the 19th century of a new type of philosophy known as positivism, which strengthened 'science's belief in its exclusive validity after the fact, instead of to reflect on it' (Habermas, 1978: 4).

Positivism, or positive philosophy, was developed during the 19th century in the writings of Auguste Comte (1798–1857), most notably in his *Cours de philosophie positive* (1830–42) and his *Système de politique positive* (1848–54). It was founded on the belief that phenomena of the human social world are no different from those of the natural inorganic and organic world and that they can therefore be investigated through similar means, yielding equally reliable results (Simon, 1963). However, Comte's positivism was also to be a kind of new world religion, which would provide general rules for the benefit and improvement of society (Kolakowski, 1972; Thompson, 1976). At the heart of Comte's philosophy was a particular conceptualization of the way in which the human mind has developed, and also of the ways in which science was classified. Comte claimed to have discovered a fundamental law 'that each of our leading conceptions – each branch of knowledge – passes successively through three different theoretical conditions: the theological, or fictitious; the metaphysical, or abstract; and the scientific, or positive' (translated from *Cours de philosophie positive* in Thompson, 1976: 39). In the first stage, the human mind is seen as supposing that all phenomena are caused by supernatural beings; in the second, transitional, stage they are explained with reference to abstract forces; and in the last, the human mind searches for laws, and explanation is seen as 'the establishment of a connection between single phenomena and some general facts, the number of which continually diminishes with the progress of science' (translated from *Cours de philosophie positive* in

Thompson, 1976: 40). Sciences can also be classified according to this Law of the Three Stages, and Comte argued that different kinds of knowledge passed through the stages at different rates, with the most general, simple and independent sciences, such as astronomy, arriving first at the positive stage, to be followed by physics, chemistry, physiology and eventually social physics, the most individual, complex and interdependent. Comte saw his own task to be that of bringing social physics, or sociology, to this its final positive stage.

Underlying all of these laws, though, was the central tenet of Comte's positivism which did away entirely with human subjectivity, and argued that social phenomena were subject to laws and methods of enquiry, which directly paralleled those of the natural sciences. In methodological terms, Comte's sociology was based upon five central rules, all of which were covered by the term positive (Kolakowski, 1972; Gregory, 1978; Habermas, 1978). He thus uses the term positive to refer to the *actual* rather than the imaginary, the *certain* in contrast to the undecided, the *exact* rather than the imprecise, the *useful* instead of the vain, and the *relative* in contrast to the absolute. His concern with the actual is based on the preponderance of observation over imagination; the former is the realm of science, the latter of metaphysics. This in turn is based on the central rule of empiricist schools of thought, namely *phenomenalism*, which asserts that there is no distinction between essence and phenomenon, and therefore that we can only record that which is directly experienced (Kolakowski, 1972). However, Comte was keen to point out that observation did not negate the need for theory. In his own words

> The next great hindrance to the use of observation is the empiricism which is introduced into it by those who, in the name of impartiality, would interdict the use of any theory whatever. No logical dogma could be more thoroughly irreconcilable with the spirit of the positive philosophy, or with its special character in regard to the study of social phenomena than this. No real observation of any kind of phenomena is possible, except in so far as it is first directed, and finally interpreted, by some theory (translated from *Cours de philosophie positive* in Thompson, 1976: 102).

Closely linked to this is his emphasis on *certainty*, which is achieved by a 'common experience of reality' (Gregory, 1978: 26), and which requires all scientists to adopt a unity of method (Kolakowski, 1972). His demand for precision was based on the establishment of theories which could be tested, and in this he saw himself as being descended in the rationalist tradition from Descartes. The testing of theories, and their development into laws, in turn precluded value judgements and normative statements from the realm of science because they could not be tested. Furthermore, such laws had specific utility, with the combination of science and technology providing the means by which society could be improved. Finally, Comte emphasized that this methodology, involving observation, experiment and comparison (Thompson, 1976: 101–15) was relative and unfinished. However, as Habermas (1978: 77–8) has emphasized,

'The knowledge of laws, checked by experience, methodologically arrived at, and convertible into technically exploitable predictions, is relative knowledge insofar as it can no longer pretend to know what there is in its essence, that is absolutely.'

Thompson (1976: 21) argues that one of Comte's greatest achievements was to establish a methodology based on observation, experiment and comparison at the heart of the emerging discipline of sociology, in contrast to the 'previous tendencies for social theorizing to be theologically or metaphysically speculative and incapable of verification by empirical observation'. He goes on to claim that

> It was his lasting achievement to have staked out a claim for a social science which, both in the definition of its subject matter and its proper method of study, would respect the position of humanity as an integral part of the world of nature and yet unique in that world. . . . Comte's sociology encourages us to believe that – on the basis of knowledge we are able to establish – enlightened and informed social action can hasten the movement towards a more just and harmonious state (Thompson, 1976: 33).

In his later life, Comte was increasingly concerned with the practical implementation of his ideas in the form of a new secular religion, in part influenced by his earlier contacts with Saint-Simon. He was also seriously affected by the death of Clothilde de Vaux, and as his ideas took on a manic proselytizing zeal, they became less and less accepted by the wider scientific community. Nevertheless, his earlier work had profound implications on the writings of sociologists such as Émile Durkheim, John Stuart Mill and Herbert Spencer. In particular, Mill's treatment of the inductive method in his *System of logic* (1843) owed much to his reading of, and correspondence with, Comte.

Although Comte's own influence dwindled, the seeds of positivism had been sown, and by the beginning of the 20th century a new form of the philosophy emerged in Vienna, initially under the leadership of Ernst Mach (1838–1916), an Austrian physicist who held the chair in philosophy at the University of Vienna from 1895 to 1901 (Blackmore, 1972). Following the end of the First World War in 1918, a group of scholars who had been closely influenced by Mach's writing, including Moritz Schlick (1882–1936) and Rudolf Carnap (1891–1970), established a regular Thursday discussion session, members of which soon came to be known as the Vienna Circle. Simon (1963) has suggested that Comte's positivism was at best a minor influence on this group, but what they shared with him was a fundamental aversion to metaphysics. Unlike Comte, Mach did not believe in a classification of science, but rather he sought to unify all sciences through the application to them of the methods of physics. The philosophy which emerged from this group has become known as logical positivism, and owed much to the legal positivism of the early 20th century which claimed that the body of laws enacted by a sovereign state were valid in themselves and could not be overruled by any such concept as natural laws (Simon, 1963).

During the 1920s members of the Circle were strongly influenced by the publication in 1921 of Ludwig Wittgenstein's *Tractatus logico-philosophicus*, which had developed a scheme for a logically perfect language. One of the notions developed by Wittgenstein (1961: 7) was that '1.11 The world is the totality of facts, not things', and members of the Vienna Circle interpreted these *facts* in the light of Mach's own writings concerning *hard data*. This misinterpretation eventually led to a rift between Wittgenstein and the Vienna Circle, but not before its members had also misinterpreted his ideas concerning the distinction between *analytic* and *synthetic* propositions. Kant (1724–1804), the founder of German Idealism, had first drawn this distinction, noting that analytic propositions, in which the predicate is part of the subject, as in 'a tall man is a man', could be distinguished from all other propositions, which can be termed synthetic. Wittgenstein (1961: 127) took this argument further by arguing that analytic propositions are either tautologous or self-contradictory, and that such 'logical propositions cannot be confirmed by experience any more than they can be refuted by it'. Members of the Vienna Circle, particularly Carnap (1935), then argued that such propositions constitute the domain of formal sciences, whereas the truth of all synthetic statements can only be determined by empirical verification in the domain of factual sciences.

Among the clearest expressions of the methodological implications of these ideas for the factual sciences is that by Hempel (1965: 232), who asserted with respect to the context of history that the scientific explanation of an event E consists of:

1. A set of statements asserting the occurrence of certain events $C_1, \ldots C_n$ at certain times and places,
2. A set of universal hypotheses, such that
   (a) the statements of both groups are reasonably well confirmed by empirical evidence,
   (b) from the two groups of statements the sentence asserting the occurrence of event E can be logically deduced.

Thus, explanation of past events and prediction of future ones can be achieved from a set of determining conditions and a set of general laws. The problem is to establish general laws from theories, and this is usually achieved by positing hypotheses, which when applied to initial conditions, can be seen to produce events. These in their turn can be compared with those events revealed by empirical observation. This therefore returns the argument once again to the problem of induction, because no matter how many experiments are undertaken to verify a theory and establish a law, there is no *logical* reason why the next experiment should produce the same result.

Habermas (1978) develops his critique of empirical–analytic sciences at this point, focusing on the claim that positivism disavows reflection on its own epistemological foundations (Hesse, 1982). Thus he asserts that 'The glory of the sciences is their unswerving application of their methods without reflecting on knowledge-constitutive interests' (Habermas, 1978: 315). Elsewhere, he argues that 'by making a dogma of the sciences' belief in themselves, positivism assumes the prohibitive function of protecting scientific enquiry from epistemological self-reflection. Positivism is philosophical only insofar as is necessary for the immunization of the sciences against philosophy' (Habermas, 1978: 67). Furthermore, he suggests that 'The positivistic attitude conceals the problems of world constitution. *The meaning of knowledge itself becomes irrational* – in the name of rigorous knowledge' (Habermas, 1978: 68–9). Positivist science is thus fundamentally geared to the production of technically useful knowledge, and is concerned with the prediction and control of processes that have been objectified (Roderick, 1986).

### 2.2.2 *Historical–hermeneutic science*

The second type of science on which Habermas (1978) focuses attention is what he terms historical–hermeneutic science. In origin, hermeneutics was concerned with the establishment of an authentic version of biblical texts, but by the 19th century it had developed into a powerful alternative to the empirical–analytic science of positivism. Whereas positivist social science sought *explanation*, hermeneutics sought *understanding* (Bauman, 1978). In recognizing that human actions have purpose, hermeneutics attempted to retrieve an understanding of intentions. Habermas develops his critique of hermeneutics first through an analysis of the work of Dilthey (1833–1911), and then through a critique of Husserl's phenomenology (Gadamer, 1975; Held, 1980; Roderick, 1986; Outhwaite, 1987).

Dilthey (1958) suggests that there is a central difference between objects that can be understood, and those which can only be studied externally. Moreover, this difference lies in the nature of the objects themselves rather than in any human intervention. It is here that, following in the tradition of Hegel, Dilthey returns to the central concern of German ideology with *der Geist* (the mind or spirit), by arguing that the potential objects of our understanding are nothing less than expressions of the universal Spirit. In Bauman's (1978: 36) words, 'It is *because* they are expressions of the Spirit that we *can* understand them.' However, herein lies one of the central problems of hermeneutics, because any understanding of an individual object must therefore be related to a pre-existing understanding of the universal Spirit, which can thus only be incomplete. Hermeneutic understanding can itself therefore only be inconclusive and partial.

According to Habermas (1978: 141), Dilthey's (1958) central aim was to demonstrate that the cultural sciences (*Geisteswissenschaften*) had different foundations from the natural sciences. In contrast to the natural sciences, 'the position of the subject in the cultural sciences is distinguished by unrestricted experience' (Habermas, 1978: 143). Consequently, in his interpretation of Dilthey, Habermas argues that in order to understand socio-cultural phenomena it is essential to go beyond the restrictive conditions of experimental situations. In the cultural sciences, Habermas (1978: 143) thus suggests that

> the experiencing subject is given free access to reality. The perceptual responses of all prescientifically accumulated experiences are called

> into play. The larger part played by the receptive faculties in the subject exposed to the entire breadth of experience has its counterpart in a lesser degree of objectivation. Reality seems to open itself up to experience from within.

This last sentence, drawing attention to the internal experience of reality, rather than simply the external objectivization of logical positivism, illustrates one of the fundamental differences between the two types of science. Hermeneutics involves understanding on the part of thinking, living individuals, whereas logical positivism is concerned with explaining a reality which is seen as lying outside the interpretative influence of individuals.

In order to communicate shared cultural values, Dilthey (1958) argued that individuals needed to employ symbols which are essentially grounded in ordinary language. In Habermas's (1978: 157) words 'Language is the medium in which meanings are shared.' However, such a concern both with shared meaning and with individual interpretation, leads back to the central problem of the relationship between the universal and the particular. As Habermas (1978: 157) has noted, 'Hermeneutic understanding must employ *inevitably general* categories to grasp an *inalienably individual* meaning.' The problem of hermeneutics thus becomes to grasp and represent the meanings of individuated life structures through inevitably general categories.

It is at this juncture that Habermas (1978) points to a fundamental inconsistency in Dilthey's arguments, because Dilthey suggests that the practice of hermeneutics has to proceed through the objectivity of scientific knowledge. Dilthey (1958) draws attention to the conflicting tendencies in hermeneutics between a practical relation to life and scientific objectivity. As Habermas (1978: 179) points out, 'He would like to free hermeneutic understanding from the interest structure in which it is embedded on the transcendental level and shift it to the contemplative dimension according to the ideal of pure description.' Although Dilthey advanced hermeneutics from a naïve form of empathy to a methodology of self-reflection, in which the mind is seen as 'externalizing itself in objectivations and at the same time returning to itself in the reflection of its externalizations' (Habermas, 1978: 147), he failed to overcome problems concerned with the contemplative nature of truth. Dilthey never completely abandoned the empathy model of understanding, and Habermas (1978) suggests that his model of re-experiencing other people's subjectivity is essentially the equivalent of observation. As Habermas (1978: 181) has cogently argued 'He who puts himself into the place of another's subjectivity and reproduces his experiences extinguishes the specificity of his own identity just like the observer of an experiment.'

Habermas (1978: 179) thus suggests that Dilthey is 'caught in a covert positivism'. In essence, Dilthey's hermeneutics was concerned with gaining individual understanding of historical contexts through a form of empathy involving the transference of the self into a particular other expression of life. For Habermas, the central problem of this approach is that it fails to question the basis of understanding on which it is established: 'Dilthey remains in the last analysis so much subject to the force of positivism that he leaves off the self-reflection of the cultural sciences just at the point where the practical cognitive interest is comprehended as the foundation of possible hermeneutic knowledge and not as its corruption' (Habermas, 1978: 179). It is thus to *self-reflection*, and particularly to the insights of Freud that Habermas turns for his development of a critical science.

In his Frankfurt inaugural address in 1965 Habermas (1978: 301–17) also developed a critique of Husserl's hermeneutics, which combined the German traditions of Dilthey with 'the French–Cartesian legacy of rationalism' (Bauman, 1978: 19). According to Bauman (1978: 19–20) one of Husserl's main contributions was to show that meaning could be understood as a result of 'the possibility of freeing the meaning from its tradition-bound context, instead of meeting it there, in its "natural" habitat. . . . This can be done by a phenomenological contemplation of "pure meanings" as disclosed by the experience of phenomena laid bare of their historical–structural guise'. For Husserl, it was essential for consciousness to separate itself from its social and historical roots, and establish itself through a form of continual reduction, akin to Cartesian doubt, as some kind of an absolute.

Edmund Husserl (1859–1938) was one of a group of German philosophers, others including Moritz Geiger and Max Scheler, who published a series of studies under the general title of *Jahrbuch für Philosophie und Phänomenologische Forschung* in the second and third decades of the 20th century (Mercer and Powell, 1972). These formed the basis of a philosophical movement, known as phenomenology, which, although quite focused in its initial propositions, has subsequently devolved into a number of quite different variants. This makes any attempt at generalizing about its aims and methods fraught with difficulty (Kockelmans, 1967a, b; Relph, 1970, 1981; Billinge, 1977; Jackson, 1981; Pickles, 1985). The nearest 'approach to a phenomenological platform ever formulated' according to Spiegelberg (1960: 5) is found in the following statement published by the editors at the head of the *Jahrbuch*:

> It is not a system that the editors share. What unites them is the common conviction that it is only by a return to the primary sources of direct intuition and to insights into essential structures derived from them that we shall be able to put to use the great traditions of philosophy with their concepts and problems (Mercer and Powell, 1972: 9).

A fundamental characteristic of phenomenology, with its emphasis on direct intuition, is thus that it is once again diametrically opposed to the empirical sciences based on positivism. According to Pickles (1985: 3) 'Phenomenology seeks precisely to disclose the world as it shows itself *before* scientific enquiry, as that which is pre-given and presupposed by the sciences. It seeks to disclose the original way of being prior to its objectification by the empirical sciences.'

The term phenomenology was not, however, invented by Husserl, having been used in a variety of ways long before the 20th century. Hegel (1977) had thus incorporated it into the title of his *Phenomenology of Spirit* in 1807, and it had been used in the mid-18th century by another German philosopher Johann Lambert (Spiegelberg, 1960; Mercer and Powell, 1972). A central concern of recent phenomenology, though, has been its 'rejection of the traditional metaphysical assumption of the separation of subject and object as the description of the fundamental state of affairs' (Pickles, 1985:17). This distinction was made clear in the writings of Immanuel Kant (1724–1804), the founder of German Idealism, who differentiated between things as they are perceived (phenomena) and things as they are (*noumena*) (Russell, 1961). In his *Critique of pure reason*, Kant argued that our perceptions (phenomena) are caused by two things: *sensations*, the product of objects, and the *form* of the phenomena caused by our subjective apparatus. For Kant, the form is *a priori*, independent of experience and not connected to the context of our perceptions. Phenomenology seeks to redefine this relationship between object and experience; for phenomenologists there is no objective reality external to, and separate from, the mind.

Five basic propositions underlie Husserl's philosophy:

1. 'That *experiences* are the main object of philosophical enquiry';
2. That 'language reflects the structure of experience';
3. That 'No absolute criterion of precision exists: rather, it is a function of both subject matter and context';
4. 'That we do not necessarily have to define a term precisely before we start analysing both it and the corresponding experience'; and
5. 'That philosophy should be concerned with the search for the absolutely presuppositionless' (Mercer and Powell, 1972: 9–10).

These propositions have profound methodological implications, contrasting markedly with those of the objective sciences. At their heart lies Husserl's oft quoted phrase 'Back to the things themselves'. According to Kockelmans (1965: 18), this expression 'indicates that in philosophy one should renounce all principles and ideas that are insufficiently explained or incorrectly founded, all arbitrary ways of thinking and all prejudices, and be guided only by the things themselves'. At the heart of phenomenology is thus the pursuit of meaning, and the revelation of essences through the search for pure consciousness.

Habermas (1978) develops his critique of Husserl in three stages. First, he notes that Husserl's phenomenology is directed against the objectivism of the sciences, and that it 'discloses the products of a meaning-generative subjectivity' (Habermas, 1978: 304). Second, he suggests that Husserl tries to remove this subjectivity by creating an objective self-understanding, and third he argues that Husserl then identifies this transcendental self-reflection with traditional theory. However, for Habermas (1978: 305) Husserl 'errs because he does not discern the connection of positivism which he justifiably criticizes, with the ontology from which he unconsciously borrows the traditional concept of theory'. According to Habermas, Husserl claimed that phenomenology sought to bring to consciousness the relationships between knowledge and the interests of the life-world, and because it brought this to consciousness it was free from such interests and was thus a form of pure theory. Habermas (1978: 306) refuted this:

> the error is clear. Theory in the sense of the classical tradition only had an impact on life because it was thought to have discovered in the cosmic order an ideal world structure, including the prototype for the order of the human world. Only as cosmology was *theoria* also capable of orienting human action. Thus Husserl cannot expect self-formative processes to originate in a phenomenology that, as transcendental philosophy, purifies the classical theory of its cosmological contents, conserving something like the theoretical attitude only in an abstract manner. Theory had educational and cultural implications not because it had freed knowledge from interest. To the contrary, it did so because it derived *pseudonormative power* from *the concealment of its actual interest*.

For Habermas, Husserl, while criticizing the objectivism of the natural sciences, himself falls prey to another form of objectivism traditionally associated with theory. Hermeneutics could thus provide no substantial grounding for human action.

However, Husserl was acutely aware of the difficulties of his enterprise, and in particular of the methodological implications of his separation of the life-world from the phenomenological world of meaning. As Bauman (1978: 127) has noted, he 'spent the last part of his life haunted by the realization that his solution to the problem of understanding was evidently ethereal. He tried hard to build a bridge from the phenomenologically reduced, back to the "life" world, over the gap between the two which he himself had dug' (for suggested solutions to this dilemma see Heidegger, 1959; Schutz, 1962, 1967).

### 2.2.3 *Towards a critical science*

One of Habermas's central concerns in *Knowledge and human interests* was to develop an understanding of the connections between knowledge and action. As the above section has illustrated, he argued that historical–hermeneutic sciences failed to achieve this because they did not satisfactorily examine the ways in which the worlds of meaning and action are integrated. In searching for a framework in which such an interpretation could be achieved, Habermas turned to psychoanalysis and the writings of Freud. These had two striking features of relevance to his enquiry. First, they offered a new and action oriented conception of self-reflection which overcame the problems associated with Dilthey's empathy, and second they were able to claim legitimation as a rigorous form of science. As Habermas (1978: 214) has argued, 'The birth of psychoanalysis opens up the possibility of arriving at the dimension that positivism closed off, and of doing so in a methodological manner that arises out of the logic of enquiry.'

The central task of psychoanalysis is to enable a subject to understand past events, and thereby to resolve the anxieties which give rise to particular forms of disturbed behaviour in the present. Freud (1953–74) approached this through the interpretation of dreams, and specifically through a concern for 'those connections of symbols in which a subject deceives itself about itself' (Habermas, 1978: 218). It was in this focus on understanding self-deceipt that Freud provided Habermas with a way of connecting knowledge and action. Underlying disturbed behaviour, Freud suggested that there are repressed meaning structures, which a subject splits off and detaches from everyday usage. Access to these structures can, however, be gained through dreams, and subjects can thus be helped to overcome their anxieties through an understanding of the meaning of the dreams in the realm of consciousness. Analysis is self-reflection, 'because the critical overcoming of blocks to consciousness and the penetration of false objectivations initiates the appropriation of a lost portion of life history; it thus reverses the process of splitting-off' (Habermas, 1978: 233). Moreover, Habermas (1978: 233–6) suggests that there are three other aspects of psychoanalysis that also demonstrate its claims to self-reflection. First, it is driven by interest in self-knowledge; second, the patient must regard illness as part of the self, and thus assume responsibility for it; and third, psychoanalysis can only be practised by someone who has already undergone analysis, in order to become free of the very illnesses that it is the practitioner's intent to treat. This concern with self-reflection is central to Habermas's formulation of a critical social science. In his words, 'The methodological framework that determines the meaning of the validity of critical propositions of this category is established by the concept of *self-reflection*. The latter releases the subject from dependence on hypostasized powers. Self-reflection is determined by an emancipatory cognitive interest' (Habermas, 1978: 310). For Habermas (1978: 310) both the critique of ideology and psychoanalysis share in common the characteristic that they 'take into account that information about lawlike connections sets off a process of reflection in the consciousness of those whom the laws are about'.

Self-reflection therefore provides the central methodological core of Habermas's conception of a critical science. For an individual, such self-reflection involves one part of the self splitting off from the remainder, so that it can attempt to understand the whole, and thus be in a position to help itself. For society at large, it involves people figuratively stepping outside that society, to understand the structural distortions inherent within it, so that they can then reveal them to the remainder of the population. This process is closely related to a second central concern of Habermas's critical theory, namely *emancipation*. This is only achievable through self-reflection. In essence, for Habermas there is no difference between knowledge and the emancipatory interests of knowledge; the fully knowledgeable person is the fully emancipated one. To be emancipated, to be free from the adverse structural constraints of society, is to understand them. Habermas (1976) suggests that under advanced capitalism knowledge is systematically distorted, giving rise to crisis tendencies in the economic, administrative, legitimation and socio-cultural systems.

He thus argues that:

> It is a consequence of the fundamental contradictions of the capitalist system that, other factors being equal, either
> – the economic system does not produce the requisite quantity of consumable values, or;
> – the administrative system does not produce the requisite quantity of rational decisions, or;
> – the legitimation system does not provide the requisite quantity of generalized motivations, or;
> – the socio-cultural system does not generate the requisite quantity of action-motivating meaning (Habermas, 1976: 49).

The role of the critical scientist is thus to reveal how systematically distorted communication takes place, and in so doing to provide society with the means to solve these crises to the emancipated benefit of its constituent population.

This is associated with a third key feature of Habermas's critical theory, which involves a particular conception of the relationship between *theory and practice*. Like Tolstoy, in the quotation with which this chapter begins, Habermas considers that modern science has divorced itself from the means of understanding its social context. In essence, modern science has disavowed the fundamentally critical nature of classical science:

> The conception of theory as a process of cultivation of the person has become apocryphal. Today it appears to us that the mimetic conformity of the soul to the proportions of the universe, which seemed accessible to contemplation, had only taken theoretical knowledge into the service of the internalization of norms and thus estranged it from its legitimate task (Habermas, 1978: 304).

Not even Marx escapes Habermas's critique with respect to the conceptualization of the relationship between theory and practice. Habermas (1974, 1978) thus argues that Marx, by reducing social practice to labour, failed fully to comprehend the interconnectivity between theory and practice, and the central importance of critique therein: 'By equating critique with natural science, he disavowed it. Materialist scientism only reconfirms what absolute idealism had already accomplished: the elimination of epistemology in favor of unchained universal "scientific knowledge" – but this time of scientific materialism instead of absolute knowledge' (Habermas, 1978: 63). It is thus through his critique of Dilthey and Husserl that Habermas seeks an alternative framework for examining the relationship between theory and practice, and this found its expression in his theory of cognitive interests discussed above. For Habermas (1978: 62–3)

> ultimately a radical critique of knowledge can be carried out only in the form of a reconstruction of the history of the species, and that conversely social theory, from the viewpoint of the self-constitution of the species in the medium of social labor and class struggle, is possible only as the self-reflection of the knowing subject.

The fullest expression of Habermas's (1984, 1987a) development of a critical theory is to be found in his theory of communicative action. This returns to the central role played by language in communication and the reinforcement of social life, and it is designed 'to make possible a conceptualization of the social-life context that is tailored to the paradoxes of modernity' (Habermas, 1987a: xl). His theory is based around three interlinked topics: a concern with the rationality of communication; a two-tier conceptualization of society involving the interaction of life-world and system; and a particular view of modernity, which explains social pathologies through the way in which autonomous, formally organized systems of action subordinate communicatively structured domains of life. In essence, he is seeking to explain social disorder through an understanding of the way in which communication is structured. In more detail, Habermas (1984, 1987a) suggests that there are three worlds with which people interact when they speak: an *objective* world, which is the sum of all entities about which it is possible to make true statements; a *social* world, which is the sum of all relations legitimately regulated between people; and a *subjective* world, made up of the sum of the individual experiences of the speaker. Moreover, in making any statement he suggests that an actor must raise three validity claims: that the statement being made is truthful; that the speech act conforms to the existing normative context; and that the speaker means what is said. For Habermas (1984), it is the actors who seek consensus and measure the validity claims of a speech act in the context of the three worlds with which they exist. In brief, he seeks to combine considerations of life-world, reproduction processes, communicative action and the structure of speech acts. As McCarthy (1984: xxv) has summarized, Habermas's theory is based on the argument that

> to the different structural components of the lifeworld (culture, society, personality) there correspond reproduction processes (cultural reproduction, social integration, socialization) based on the different aspects of communicative action (understanding, coordination, sociation), which are rooted in the structural components of speech acts (propositional, illocutionary, expressive).

## 2.3 Theory, practice and geographical interest

While Habermas's theory of communicative action has had little direct influence as yet on the practice of geography, some of his earlier work has informed geographical reflection on the post-industrial city (Ley, 1980), on regional science (Lewis and Melville, 1978) and on systems theory (Gregory, 1980). More significantly, though, his distinction of three different types of science has been widely accepted as providing a useful basis within which to examine the changes that have taken place in geography since the 1960s (Gregory, 1978; Jackson and Smith, 1984; Johnston, 1991a: 30–4). Habermas's distinction between empirical–analytic and historical–hermeneutic sciences is therefore used as the organizing framework for Chapters 5 and 6. However, it is with his more general theoretical principles that most of this book's emphasis is concerned. In particular it focuses on the historical relationships between geographical theory and practice, with the particular interests served by geographical enquiry, and with the possibility of developing a type of geography which seeks to be self-reflective and emancipatory.

Habermas's conceptualization of these issues is not, though, unproblematic, and his approach has been criticized on at least six broad fronts (Thompson and Held, 1982a):

1. That there is a tension between his theoretical and practical projects with respect to the Marxian heritage;
2. That the conceptualization of critical theory is ambiguous, conflating Kant's programme of transcendental philosophy with the Young Hegelians' concept of negation;
3. That his analysis of the concept of reason seems to preclude the possibility 'of encountering a nature-in-itself which would set limits to the human interest in technological control' (Thompson and Held, 1982b: 18);
4. That in his examination of the relationships between reason and history there are problems in attempting to link a universal ethics back to historically specific situations;
5. That his theories of action, language and science lack clarity; and
6. That difficulties in his attempts to link the concepts of life-world and system give rise to obscurities in his theory of legitimation crisis (for a response to these criticisms see Habermas, 1982).

Although undoubted difficulties remain with his programme, there are nevertheless two ways in which Habermas's arguments can be seen to be of very direct relevance to many of the issues which have continually challenged geographers. First, his critical theory is directed centrally at understanding the place of people in nature. Knowledge has a twofold identity: it both derives from and influences human involvement in nature; but such knowledge must also be socially communicated. In his analysis of these technical and practical interests Habermas addresses key questions that have been at the heart of geography since classical antiquity. Nevertheless, while such arguments are implicit in much of Habermas's work he rarely examines the involvement of people in nature explicitly. Consequently, it is important to integrate Habermas's ideas with a more detailed understanding of the interactions between people and the environments in which they live. Second, Habermas has drawn attention to the existence of very different practices in both the natural sciences and the cultural sciences. This distinction closely parallels the increasingly apparent division of geography into two parts, focusing on the physical world and the human world. While most 'physical geographers' have anchored themselves in the certainty of empirical–analytic science, under the beguiling guise of the so-called scientific method, the majority of human geographers have thrown in their credentials with a social science dominated by historical–hermeneutic and structuralist conceptions of science. The grounds for communication between human and physical geographers have therefore become increasingly attenuated. The arguments of critical theory in providing a critique of both types of science, are therefore of particular relevance for those embarking on a reintegration of the physical and human parts of the discipline, and seeking to reopen the paths of communication between the two.

CHAPTER 3

# *Geography and society: classical context and a world of discovery*

> What is most striking in conceptions of nature, even mythological ones, is the yearning for purpose and order; perhaps these notions of order are, basically, analogies derived from the orderliness and purposiveness in many outward manifestations of human activity: order and purpose in roads, in the grid of village streets and even winding lanes, in a garden or a pasture, in the plan of a dwelling and its relation to another.
>
> Glacken (1967: 3)

This chapter provides an overview of geographical writing prior to the 17th century, in order to set the context for a discussion of more recent changes in the discipline in the subsequent chapters. Few aspects of geographical debate are really new, and the origins of many issues of current dispute can be found in the writings of earlier scholars. Although geography was not widely established as an academic discipline in universities until the latter part of the 19th century, this formalization of the discipline was built on an ancient tradition of geographical writing. Moreover, just as there is no uniformity of agreement today as to what constitutes geography, so too was there disagreement in the past. Indeed, many works which might be termed 'geographical' in the light of the recent work of geographers were never envisaged as such by their authors. There is therefore a fundamental problem of definition in determining what should be included in a discussion of early geographical writing.

Three main types of geographical literature can be distinguished: first, there are the writings of scholars who called themselves geographers for all or part of their lives; second, there are the works of authors who specifically referred to geography, or geographical ideas, in their writings, but who did not claim to be geographers; and third, there are the works which have subsequently been used as sources of inspiration by geographers but which at the time that they were written made no claim to be geographical. The last of these alternatives implies the possibility that later generations of geographers can somehow claim earlier knowledge as being geographical. For all geography there is therefore a pre-geography, a body of knowledge which lays the foundation for the creation of a formal geography.

Geographical knowledge, though, does not just exist; it is created by societies. Hence it is also useful to distinguish between formal and informal geographies. While individuals can produce their own informal geographies, formal geography is only that which is accepted and recorded as such by the society within which it is produced. For it to be accepted, such knowledge must be deemed to have some utility, and to be worthy of preservation. However, although formal geographies might exist in pre-literate societies, if such geographies remain unrecorded their reconstruction is problematic. Consequently, the development of geography in any society can be seen as progressing from an unrecorded informal pre-geography, through a formal pre-geography to a formal geography.

## *3.1 Greek and Roman geography*

### *3.1.1 The origins of classical geography*

The word 'geography' is derived from the Greek *ἡ γεωγραφία*, combining the words *γῆ*, meaning 'earth', and *γράφω*, meaning 'writing' or 'describing'. In literal terms, geography therefore means earth-writing, or writing about the earth. However, the production of books specifically called geographies was a relatively late phenomenon, and until the 1st century BC most accounts of the earth were written in works that did not explicitly claim to be geographical. Before the advent of writing there was an informal pre-geography created as a result of the explorations and discoveries of prehistoric peoples. This led to the establishment of a body of knowledge about the world, which gradually formed the basis of the formal pre-geography of the classical world. In early Greek writing three such traditions can be identified:

1. A topographical tradition concerned with descriptions of the earth and the people living on it;
2. A mathematical and astronomical tradition concerned with measurement of the earth (Dreyer, 1953; Dicks, 1970; Neugebauer, 1983);
3. A theological tradition, concerned with answering questions about the very reason for human existence on the earth (Bunbury, 1879; Thomson, 1948; Glacken, 1967).

The topographical tradition of geographical writing was derived directly from the travels of pre-literate peoples. As commercial exchange became firmly established it was essential for people to have knowledge of sea and land routes, and as explorers discovered new places it was important for knowledge about them to be disseminated within their societies. From its earliest days, therefore, geographical knowledge can be seen to have served a utilitarian purpose, and maps sketched on rocks and bones survive in the archaeological record from as early as

13,000 BC (Dilke, 1985; Harvey, 1980). With the advent of writing such knowledge became formalized, and many of the earliest surviving poems and literary works provide considerable details about different parts of the then known world. Homer's *Iliad* and *Odyssey* (9th century BC) are thus widely considered (Bunbury, 1879; Thomson, 1948) to be among the first surviving geographical works. These accounts of the epic conflict between the Achaeans and the Trojans, and the subsequent wanderings of Odysseus, provided Homer with a marvellous backdrop against which he could develop detailed descriptions of the peoples he described, and the places to which they travelled. The subsequent exploration and colonization of much of the Mediterranean by the Greek city states in the 8th and 7th centuries BC led to a vast expansion of knowledge, much of which was written down in treatises which have not survived. It is also from this period that the first serious attempts at representing the known world cartographically were attempted, with Anaximander of Miletus (611–547 BC) widely being recognized as the author of the first map of the world's surface (Thomson, 1948). The first major description of the earth about which we have any real knowledge was the *Periodus*, or *Description of the earth*, completed by Hecataeus of Miletus (c. 550–476 BC), and divided into two books entitled *Europe* and *Asia*. Unfortunately only about 300 fragments of these works survive, and although Hecataeus is most usually considered to be a historian (Pearson, 1939), Thomson (1948: 47) suggest that in his work 'there are hints of intelligent curiosity about climate and customs, flora and fauna, so that it deserves to be called a general geography'.

All of this earlier topographical writing culminated in the first great prose work of European literature, *The histories* of Herodotus (*c*. 485–425 BC) (1954). In this, the result of what he termed his enquiries into history, Herodotus (1954: 13) aimed 'to do two things: to preserve the memory of the past by putting on record the astonishing achievements both of our own and of the Asiatic peoples; secondly, and more particularly, to show how the two races came into conflict'. In so doing, he provided a wide-ranging account of the whole of the known world, based on his own experiences, on existing inscriptions and literary works, and on oral accounts. Although many of his stories now seem fantastic, much of Herodotus's account has been substantiated by other literary and archaeological evidence. In Book Four, for example, although he had little knowledge about the interior of Africa, which he termed Libya, he was aware that it was 'washed on all sides by the sea except where it joins Asia' (Herodotus, 1954: 254). Likewise, in Book One he noted with surprise how Babylon was supplied with wine by boats made of hide, which were taken to pieces on arrival and the hide then carried by donkey overland back to Armenia since 'It is quite impossible to paddle the boats upstream because of the strength of the current' (Herodotus, 1954: 92). A century after the death of Herodotus, Alexander the Great (356–323 BC) vastly expanded the horizons of the Greek world with his eastern conquests of the Persian Empire. Although Herodotus had written much about Persia, these conquests provided the opportunity for a considerable increase in the accuracy of Greek topographical writings about the region. These found their fruition in the subsequent writings of Dicaearchus, a pupil of Aristotle, and of Eratosthenes (*c*. 276–194 BC).

At the same time as a topographical awareness of the earth's surface was growing, there also developed an interest in its measurement and in astronomy. To the Greeks of Homer's age, the earth was envisaged as a circular plane surrounded on all sides by an ocean. However, by the 6th century, a growing interest in astronomy was leading to a reappraisal of this view. Thales of Miletus (*fl*. 580 BC), one of the founders of the Ionian school of philosophy, for example, is widely regarded as having understood the causes of solar eclipses (Dreyer, 1953). Thus Herodotus (1954: 42) commented that Thales had predicted an eclipse during the battle between the Medes and the Lydians, and Plutarch (*c*. AD 46–*c*.120; *De placit. philosoph*. ii. 24) claimed that he had attributed solar eclipses to the interposition of the moon between the earth and the sun. However, there is no direct indication that he considered that the earth was a sphere (Thomson, 1948). According to Plutarch (*De placit. philosoph*. iii. 10), Anaximander is thought to have conceived of the earth as a cylinder, but it is only with Pythagoras (*fl*. 6th century BC) that the concept of the earth as a sphere began to gain credence. Since he left no writings, it is once again only in later commentaries, particularly through the works of Aristotle (384–322 BC) (*De Coelo*, ii. 13), that his ideas have survived (Dicks, 1970). In essence, Pythagoras is claimed to have argued that the earth, along with the other visible heavenly bodies, revolved around a central fire, which occupied the mid-point of the universe. Since the sphere and the circle are the most perfect forms, according to Pythagoras, then the heavenly bodies, including the earth, must all be spheres revolving in circles. It was thus by somewhat mystical and philosophical, rather than strictly empirical, reasoning that it was argued that the earth was round.

The description of the earth as a sphere by Plato (*c*. 427–347 BC) in *The republic* (Plato, 1974) and in *Timaeus* (Plato, 1971: 44) was to have a marked significance for the future understanding of its shape (Dreyer, 1953; Neugebauer, 1983). Indeed, Thomson (1948: 114) has argued that 'it was Plato's adoption that gave the globe a wider currency'. However, by placing the spherical earth at the centre of the universe, with the sun, moon and planets orbiting around it, Plato's account was to hamper deeper astronomical understanding for centuries to come. In contrast to Plato's concern with the unseen world of ideal forms, Aristotle's emphasis on the collection of empirical facts was eventually to lead to a greater concern with the measurement of the earth's circumference. These ideas were taken up by his pupils, notably Dicaearchus, and subsequently reached their fruition in the research undertaken at Alexandria, particularly under Eratosthenes, who is generally acknowledged to have made the first accurate measurement of the globe (Dreyer, 1953; Neugebauer, 1983), and who has been described by Bunbury (1879: I. 615) as 'the parent of scientific geography'. Eratosthenes, moreover, also developed a system of latitude and longitude, and published a three volume treatise entitled *Geogra-*

*phica*, which, although it does not survive in its original form, was one of the main sources used by Strabo and Ptolemy.

These topographical and astronomical accounts were also closely associated with the third, theological, tradition concerning the origins of the earth and the reasons for human existence on it. Indeed, they can also be seen as combining together to give rise to an interest in astrology, the use of astronomy to predict human and natural events on the earth. Glacken (1967: 5) has suggested that three basic ideas underlay much early mythology concerning the relationships between people and the environment: that the earth had an order and purpose designed by a deity; that the environment had an influence on people; and that people were able to modify the environment. Gradually the Ionian school of philosophers began to develop a particular kind of cosmology, or theory of the composition of the universe. Thales of Miletus had suggested that everything was made of water, but this theory was subsequently revised by Anaximander who envisaged a primal infinite substance from which the many substances of the world, including water, were derived (Kahn, 1960; Russell, 1961). At the same time, Anaximander also developed an argument that 'Order is characterized by a struggle of opposites' (Glacken, 1967: 9). Eventually, in the 5th century BC the influential theory of the four roots or elements, was formulated by Empedocles of Acragas (Agrigentum) in southern Sicily (*c.* 492–432 BC). His major contribution to science was to identify air as a separate substance. This then enabled him to suggest that there were four primitive independent elements, earth, air, fire and water, which could be combined in different proportions to produce the many different substances found in the world. These elements were held together by love, and separated by hate, thus representing a further refinement of Anaximander's theory of opposites. It was not until the 4th century that Aristotle added a fifth element, ether (*On the heavens*). For Aristotle, the spherical earth was at the centre of the universe. Everything below the moon was considered to be made of the four elements, subject to change and decay, whereas everything above the moon, was made of the fifth element, ether, which was indestructible.

Meanwhile, alongside this emerging cosmology, there developed a range of ideas concerning the place of people within the global order. These differed markedly, and as Glacken (1967: 13) has pointed out 'The single most important generalization to be made about the attitudes toward nature held by the peoples of the classical world is that these varied greatly throughout the long span of history.' Underlying the classical vision, though, lay much older religious concepts, the most important of which were those associated with fertility, death and rebirth. At the heart of this mythology lay the worship of a female, mother earth, later to be formalized as the great deity of the primitive Greeks, Gaea (Guirand, 1968). Even before the origins of agriculture, human survival depended on the annual cycle of the plants, and on the fertility of nature. The fertility of the natural world was the same as that of the human world, and human ritual sought to propitiate a bountiful mother earth, frequently through the sacrifice of male animals. This concept of a female earth was not, though, universal, and in Egyptian mythology it is salient to note that the earth-god known as Geb was male, in contrast to the female goddess of the sky, Nut or Hathor (Viaud, 1968).

The idea of a planned world subject to divine intervention is extremely ancient, and hints of it can be found in the early myths of Egypt, and the peoples of the Tigris–Euphrates basin. By the 5th century, though, a firm teleology, or doctrine concerning design and purpose in nature, had emerged at Athens. Thus Xenophon (c. 435–354 BC) in his *Memorabilia* describes Socrates (before 469–399 BC) as using three main types of evidence to demonstrate 'divine providence: the proof of physiology, of the cosmic order, and of the earth as a fit environment' (Glacken, 1967: 42). Plato developed some of these ideas in his *Timaeus*, where he provides an account of the framer of the universe, creating the ordered cosmos consisting of fire, earth, air and water out of a state of disorder (*Timaeus* 29–30). In particular, Plato suggested that god was good, and therefore that he designed his ordered creation so that 'all things should be good, and so far as possible nothing be imperfect' (Plato, 1971: 42). Aristotle appears to have abandoned the idea of a creator god, and instead envisaged an internal logic in nature, which drove it forward of necessity (*Physics*); nature acts for a purpose, but for a purpose that is unconscious. Such arguments, particularly those of Socrates, had a major influence on the Stoic school of philosophy, established by Zeno in the early part of the 3rd century BC, and through them on much subsequent European philosophy. Zeno believed that the whole of nature was determined by rigorous laws, designed for the benefit of humanity. Russell (1961: 262) thus argues that for the Stoics 'Everything has a purpose connected with human beings. Some animals are good to eat, some afford tests of courage; even bed bugs are useful, since they help us to wake in the morning and not lie in bed too long.' Such views did not, however, go unquestioned, and found their most powerful opposition in the school of philosophy established by Epicurus (*c.* 341–270 BC). The Epicureans, including Cicero (106–43 BC) and Lucretius (*c.* 99–55 BC), pursued pleasure as the main good, and conceiving that fear of death and fear of the gods were the main evils, they were thus forced to separate the gods from nature. Given that not everyone living on the earth is fully content, they argued that nature cannot have been designed purely for human beings.

### *3.1.2 The advent of classical formal geography: Strabo and Ptolemy*

By the end of the Hellenistic age, and the emergence of Rome as the dominant Mediterranean power, geography was considered to encompass not only topographical descriptions of the world, but also questions of astronomy and teleology; right at the heart of classical geography lay profound philosophical questions concerning the place of people in nature. Its subject matter was closely similar, if not necessarily interchangeable, with that of physics, τα φυσικά literally meaning 'natural things', and geometry, ή γεωμετρία, meaning 'earth-measuring'. Earth-writing (geography) thus included work on the realm of natural things, the measurement of the earth, and theological matters concerning the role of divine power in forming the earth.

Specific works on geography, however, remained few in number. While authors such as Polybius (*c.* 205–*c.* 123 BC) and Posidonius (*c.* 135–51 BC) included much topographical description in their works, they described them as histories rather than geographies. What geographical works that were written have generally failed to survive. It is thus with the *Geography* of Strabo (*c.* 60 BC–post AD 21), a Greek born at Amasia in Pontus, that a formal geography can finally be considered to have found its surviving expression. Significantly, Strabo did not consider himself to be purely a geographer, and if his forty-seven book history had survived he might well have been better known as a historian. Nevertheless Strabo's work does enable us, in Bunbury's (1879: II. 209) words, 'for the first time to obtain a complete and satisfactory view of the state of geographical science'.

More importantly, Strabo also provides a justification for the enterprise of writing geography. He begins the first of the seventeen books of his *Geography* by arguing that 'The science of Geography, which I now propose to investigate, is, I think, quite as much as any other science, a concern of the philosopher' (Strabo, 1949: 3). He asserts this on the grounds:

1. That those who had previously discussed the subject, such as Homer and Anaximander, were philosophers;
2. That 'wide learning, which alone makes it possible to undertake a work on geography, is possessed solely by the man who has investigated things both human and divine – knowledge of which, they say, constitutes philosophy' (Strabo, 1949: 3);
3. That the utility of geography 'presupposes in the geographer the same philosopher, the man who busies himself with the investigation of the art of life, that is, of happiness' (Strabo, 1949: 5).

Nowhere does Strabo provide a clear and straightforward definition of geography, but within Book One he claims that it requires encyclopaedic learning and also that 'All those who undertake to describe the distinguishing features of countries devote special attention to astronomy and geometry' (Book I. I. 12–13) (Strabo, 1949: 25). Moreover to this he adds terrestrial history, which he defines as the history of animals, plants and everything produced by land or sea (Book I. I. 16), and also a considerable emphasis on the political development of states (Book I. I. 18). In addition to these empirical aspects, he claims that geography 'involves theory of no mean value, the theory of the arts of mathematics, and of natural sciences, as well as the theory which lies in the fields of history and myths' (Book I. I. 19) (Strabo, 1949: 39).

It is in Strabo's account of the utility of geography, though, that we can gain a real insight into the role that the discipline played in Greek and Roman society. Geography's central task was seen to be political.

Strabo thus argues that 'the greater part of geography subserves the needs of states' and 'that geography as a whole has a direct bearing upon the activities of commanders' (Book I. I. 16) (Strabo, 1949: 31). Indeed, he asserts that 'the greatest generals are without exception men who are able to hold sway over land and sea, and to unite nations under one government and political administration' (Book I. I. 16) (Strabo, 1949: 31). The role of geography was thus to provide the information that enabled rulers to conquer more territory, and to retain power over the lands that they ruled; 'the description which geography gives is of importance to these men who are concerned as to whether this or that is so or otherwise, and whether known or unknown. For thus they can manage their various affairs in a more satisfactory manner, if they know how large a country is, how it lies, and what are the peculiarities either of sky or soil' (Book I. I. 16) (Strabo, 1949: 33). Geography is, though, not only important in great undertakings, and Strabo uses the illustration of hunting to emphasize its importance for matters of smaller concern (Book I. I. 17): 'A hunter will be more successful in the chase if he knows the character and extent of the forest; and again, only one who knows a region can advantageously pitch camp there, or set an ambush, or direct a march' (Strabo, 1949: 35).

In terms of content, Strabo freely admits that he has made use of the works of his predecessors, notably Eratosthenes, Polybius and Aristotle, and indeed it is primarily through Strabo's writing that there remains any detailed knowledge of many earlier geographical works at all. He assumes that the earth is spherical, lying in the centre of the universe, and divided into five zones; its inhabited portion is described as a large island, surrounded on all sides by the ocean. The bulk of his *Geography*, though, consists mainly of a topographical description of the known world and after the first two introductory books, eight books are allocated to Europe, six to Asia, and one to Africa.

The next major description of the known world after Strabo's *Geography* was the *Natural history* of Pliny (AD 23–79) (Pliny, 1855–57). Although this did not specifically claim to be a work of geography, it aimed to present an overview of everything that was then known of the physical constitution of the universe. Pliny's work has been subjected to considerable criticism, largely on the grounds of its lack of structure and poor scientific understanding. Indeed Bunbury (1879: 374) has argued that when Pliny's geographical writings are compared 'with the writings of Eratosthenes and Strabo, we are struck with the almost total absence of any scientific comprehension of his subject, or of those general views which, however imperfectly developed, were certainly present to the minds of the Greek geographers'. The problems with Pliny's *Natural history* derive in part from its design and method of compilation. In his preface (Book I) Pliny thus claimed that he had included 20,000 topics gained from the perusal of 2,000 volumes written by some 100 authors (Pliny, 1855: I. 7), and that these were put together at interrupted intervals. It is therefore not surprising that there is a certain amount of confusion and disorder in his work. Book II of the

*Natural history* is 'An account of the world and the elements', which sets the globe within its astronomical setting. This is followed by four books on geography which Pliny describes as containing 'an account of the situation of the different countries, the inhabitants, the seas, towns, harbours, mountains, rivers, and dimensions, and the various tribes, some of which still exist and others have disappeared' (Book I) (Pliny, 1855: I. 11). By making a break between his astronomical and geographical observations, Pliny therefore appears to be adopting merely a topographical definition of geography. However, it is interesting to note that by discussing tribes which have disappeared he is able to include a historical dimension in his geographical writing. In the remaining books, Pliny describes man and his inventions, animals, birds, plants, medicines and minerals, and once again by allocating separate chapters to these subjects he can be seen as excluding them from his explicit account of geography.

In complete contrast to the works of Strabo and Pliny is the *Geography* of Claudius Ptolemaeus (*c.* AD 90–168) known to subsequent generations as Ptolemy (Ptolemy, 1966; Rylands, 1893). Above all else, Ptolemy was an astronomer, who worked at Alexandria, and before turning his attention to geography he had written a compendium of astronomy, subsequently known by the Arabic name *Almagest*. In this, he assumed that the globe is at rest in the centre of the universe, and among other information he provided a detailed table of thirty-three latitudes up to the Arctic Circle (Thomson, 1948; Dreyer, 1953). In the opening chapter of his *Geography*, Ptolemy defines geography as 'an imitative delineation of that part of the earth comprehended within our knowledge as a whole, with its parts roughly (lit. generally) appended' (Rylands, 1893: 18), and he is at pains to contrast this with chorography, which he sees as the detailed description of selected regions of the earth. Geography deals with the whole, whereas chorography considers only the part. Moreover, for Ptolemy the mathematical sciences are central to geography, whereas chorography which is merely concerned with topographic description has no need of them. Having explained how to construct a globe, together with its parallels and meridians, Ptolemy then provided details of two projections, one conical and the other globular, which can be used to project the known world onto a plane surface. Once the theoretical outline was completed, he then devoted the bulk of his *Geography* to a division of the continents into regions and to a listing of the latitudes and longitudes of all of their major features. This vast compendium provided an atlas which was to form the basic account of the European world until the Portuguese and Spanish explorations of the 15th and 16th centuries. Unfortunately, despite its apparent precision, it was based on fundamental inaccuracies. As Bunbury (1879: II. 553) has commented, 'the means at his command did not enable him to carry his ideas into execution; the substance did not correspond to the form; and the specious edifice that he reared served by its external symmetry to conceal the imperfect character of its foundations and the rottenness of its materials'. A fundamental inaccuracy entered his work because Ptolemy relied for his calculations of longitude on the maximum circumference of the earth as calculated by Posidonius at 180,000 stades. Ptolemy's adoption of this figure is somewhat surprising because Eratosthenes had earlier calculated it much more accurately at 252,000 stades. There are fundamental problems in equating the Roman stade (*stadion*) into modern measurements, but that used by Ptolemy is usually (Thomson, 1948: 161; Dreyer, 1953: 175; Holt-Jensen, 1988: 12) seen as being equivalent to 157.5 metres. If this figure is used, the difference between the figures of Posidonius and Ptolemy amounts to some 11,340 kilometres. Ptolemy's measurements thus underestimated the true circumference of the earth by some 11,729 kilometres, and this was one of the reasons that medieval explorers using Ptolemy's calculations vastly underestimated the size of the globe.

### 3.1.3 The Greek and Roman concept of geography

Most commentators on the Greek and Roman contributions to geography (Bunbury, 1879; Thomson, 1948) contrast the innovative, dynamic and original contributions of the Greeks with the utilitarian, replicative and stagnant geographies of the Romans. Gould (1985: 13) has thus commented that 'Unfortunately, the marvellous intellectual and artistic efflorescence of the Classical and Alexandrian Greek worlds could not last. The Roman spirit was of a different order, and fundamentally inimical to the Greek tradition of probing enquiry, the sort of inquiry that was prepared to go wherever the questions might lead.' It is from the geographies of Strabo and Ptolemy, both Greeks writing at the height of the Roman Empire, that the remnants of classical geography were to survive through the early medieval period. Their geographies represented the marked contrasts that existed within the discipline at that time. Strabo's geography was of the type decried by Ptolemy as mere chorography. However, the basic questions tackled by geographers ever since were firmly considered by the geographers of the Greek and Roman world.

Three central issues lay at the heart of classical geography. First there was a concern with topographical description, or what Ptolemy called chorography. This involved the detailed description of places, accounting for the genesis of human occupance therein, and thus including an element of historical understanding. Second, and related to this was a mathematical and astronomical concern with measurement, represented in its purest form, for example, by Ptolemy's *Geography*. This involved not only the measurement of the earth's dimensions, but also considered the world's position in the universe, and its relationship with the stars and planets. Third, though, geographical considerations were also philosophical ones, concerning the origins of nature. In particular, geographers sought to understand the place of humanity within the natural world.

All of these considerations were tempered and conditioned, though, by an underlying appraisal of the role that geography played within society. This was made explicit in Strabo's account of the utility of geography, but the use of knowledge about different places and peoples was intimately tied up with the exercise of political control throughout the Greek and Roman worlds. This was well illustrated in the close relationships between geography and conquest, both in the eastern campaigns of Alexander and also in the subsequent domination of Europe and the Mediterranean by the Romans. The Persian conquests of Alexander, for example, which took him as far east as the Indus, were in part enabled by earlier geographical writings, but also led to a considerable subsequent expansion in geographical knowledge as expressed in the writings of Dicaearchus and Eratosthenes.

## *3.2 Chinese and Islamic geography*

### *3.2.1 Chinese geography: a separate tradition*

With the collapse of the western Roman Empire in 476, European geography entered a period of dark stagnation. A few copies of Greek and Roman geographical texts survived, but the Germanic tribes that overran the vestiges of the Roman world had little use for such works. Indeed, many had already been lost, particularly at Alexandria, where the fire of 47 BC had destroyed some 400,000 manuscripts in the great Library, and the disturbances of AD 391 had led to the loss of perhaps 300,000 more works in the Temple of Serapis (Cornell and Matthews, 1982; Gould, 1985). Parallel to, but totally separate from, the Greek and Roman world a completely different culture of science had meanwhile evolved to the east in China. It was here that the subsequent focus of global intellectual and scientific activity was to be encountered, particularly under the Tang (AD 618–907) and Southern Sung (1127–1279) dynasties, the latter of which was described so magnificently by the Venetian Marco Polo. Once again, the emergence of a tradition of geographical writing at this time can be seen in part to have been influenced by military conquests and the need for the emperors to have a sound knowledge of their lands in order to retain their positions of power. Moreover, the development of new surveying and cartographic skills enabled the Chinese to produce maps of a quality far surpassing anything being produced in medieval Europe.

The earliest Chinese geographical document is reputed to be the Yü Kung (Tribute of Yü) chapter within the *Shu Ching* (Historical Classic) dating from the 5th century BC. This provides an inventory of the Chou Empire mainly in terms of its physical geography, and lists 'the traditional nine provinces, their kinds of soils, their characteristic products, and the waterways running through them' (Needham and Wang Ling, 1970: 500). Other ancient travellers' guides, such as the *Shan Hai Ching* much of which dates from the 4th century BC, can also be considered to be geographical, but most include mythological and magical elements together with details of semi-human races and peoples. Needham and Wang Ling (1970) suggest that there were five main types of Chinese geography:

1. Anthropological geographies, known as *Chih Kung Thu* (Illustrations of the Tribute-Bearing Peoples), dating from the mid-6th century AD;
2. Descriptions of the folk customs of the countries to the south of China (*Fêng Thu Chi*) and descriptions of unfamiliar regions (*I Wu Chih*) both dating from the 2nd century AD;
3. Hydrographic books and coastal descriptions, such as the *Shui Ching* (Waterways Classic);
4. Local topographies or gazetteers, such as the *Hua Yang Kuo Chih* (Historical Geography of Szechuan) which were mostly written from the 4th century AD onwards;
5. Geographical encyclopaedias compiled from the Chin dynasty (3rd and 4th centuries AD) onwards, in a style similar to that of Strabo.

Chinese geography was also closely tied to astronomy and cartography. Astronomy played a central role in Chinese science because of its religious concern with cosmic unity, and its links with astrology. Moreover, knowledge of astronomy and the compilation of agrarian calendars were also a way in which the state could control the productive capacity of the population. Traditionally, there was an ancient belief in China that the heavens were round and the earth square, but by the 2nd century AD three main schools of cosmology and astronomy had emerged (Needham and Wang Ling, 1970): the Kai Thien theory which envisaged the heavens as a hemisphere covering the earth which was shaped like an upturned bowl; the Hun Thien school, which corresponded with the Greek view of heavenly spheres revolving around the globe; and the Hsüan Yeh teaching which envisaged an infinite space in which the heavenly bodies floated freely about.

Among the most important contributors to Chinese astronomy and cartography were Chang Hêng (AD 78–139) and Phei Hsiu (AD 224–271). Although maps from as early as the 3rd century BC are recorded in China, it was these two scholars who first developed a scientific method of cartography based on a rectangular co-ordinate system. The corpus of cartographic work in the ensuing centuries was then brought together by Chu Ssu-Pên (1273–1337), who used it to summarize the wealth of new information which had become available as a result of the Mongol unification of Asia. His map of China prepared between 1311 and 1320 is a remarkable achievement, and remained a basic work of reference for over two centuries. Although Chu Ssu-Pên was cautious about depicting lands distant from China, it is evident that he had a level of knowledge surpassing anything current in Europe at that time. He thus recognized that Africa was a southward pointing triangle, whereas on contemporary European and Arabic maps it was always represented as pointing eastwards (Needham and Wang Ling, 1970).

### 3.2.2 *The Islamic contribution to geographical understanding*

Situated between Europe and China, the explosion of energy that accompanied the emergence of Islamic power in the 7th and 8th centuries AD also saw the creation of another great tradition of geographical scholarship. Although the first waves of Islamic conquest led to the destruction of many ancient archives, such as those that remained at Alexandria, the subsequent emergence of centres of scholarship at the cities of Cairo, Damascus, Baghdad and Granada, for example, formed the focus for much geographical writing. Not only were surviving Greek and Roman texts translated into Arabic, but with their trading contacts to the east the Arabs were also open to cultural and scientific influences from China.

One fundamental factor influencing Islamic geographical writing was the religious requirement for the faithful to make the pilgrimage to the holy places in the vicinity of Mecca at least once in their lives (Robinson, 1982). This gave rise to numerous travel guides, which provided pilgrims with descriptions both of the journeys and of the holy places of Arabia themselves. Moreover, this requirement to travel brought scholars from all over the Arab world into contact with each other, and provided a forum for considerable intellectual debate. Many of the works of early Islamic geographers, such as Ibn Khurradādhbih (*fl.* 850), al-Ya'qūbī (*fl.* 900) and Ibn Ḥawḳal (*fl.* 953) (Baker, 1937), had been based on compilations of previous writers, but with al-Muqaddasī (*c.* 945–*c.* 988) a new emphasis was introduced. He had travelled widely himself and is generally acknowledged to have been among the first to insist that all of the information he presented should be based on his own experiences (Scholten, 1980; Holt-Jensen, 1988). Few of these early Arabic geographers however had any influence on Christian Europe.

In contrast, al-Idrīsī (Abū 'Abd Allāh Muḥammad B. Muḥammad B. 'Abd Allāh B. Idrīs al-'Āli Bi-Amr Allāh, called Al-Sharīf Al-Idrīsī) (1099–1180) (*Encyclopedia of Islam*, 1971), who completed his descriptive geography entitled *Kitāb Nuzhat al-mushtāḳ fi 'Khtirāḳ al-āfāḳ* in 1154, was to have much greater influence because of his connections with the court of Roger II of Sicily. This text had been written to accompany a large silver planisphere which he had previously constructed for Roger II, and in working for one of the most powerful Christian kings of the time he was also able to gain information on Europe that had been unavailable to previous Islamic geographers. In his preface, al-Idrīsī outlines his task as being to describe the towns and territories shown on the planisphere, the nature of the agriculture and settlements, and the extent of the seas, mountains, rivers and plains (Jaubert, 1975: xxi). Moreover, he goes on to include details of the main crops, the arts and crafts, the import and export trade, and the customs, curiosities and religions of the people living within each of the seven climates into which he divides the world. This is a remarkable undertaking, and there are few other surviving accounts of Europe in the first half of the 12th century to match it. The best known Arabic travel writer was Ibn Baṭṭūṭa (Shams al-Dīn Abū 'Abd Allāh Muḥammad B. 'Abd Allāh B. Muḥammad B. Ibrāhīm B. Muḥammad B. Ibrāhīm B. Yūsuf al-Lawātī al-Ṭandjī) who was born at Tangier in 1304 and who died in Morocco in 1368–69 or 1377 (*Encyclopedia of Islam*, 1971). His *Tuḥfat al-nuẓẓār fī gharā'ib al-amṣār wa-'adjā'ib al-asfār* was completed in 1357 and was of major significance for its descriptions of India, Anatolia and West Africa. Moreover, during his wide travels, Ibn Baṭṭūṭa was at one time appointed by the ruler of Delhi as ambassador to the court of China, and as a result he also visited the Maldive Islands and Ceylon on his journey to the east (Sykes, 1934; Baker, 1937).

Ibn K͟haldūn (Walī Al-Dīn ʿAbd al-Raḥmān B. Muḥammad B. Muḥammad B. Abī Bakr Muḥammad B. al-Ḥasan) (1332–1406), a contemporary of Ibn Battuta, is also sometimes considered to have been a geographer (Holt-Jensen, 1988), but he is best 'regarded as a historian, a philosopher of history, and a proto-sociologist' (Morgan, 1988: 202). Despite this, his great cyclical view of world history, the *Muḳaddimah*, provides a good overview of the state of Arabic thinking concerning geography in the 14th century (Ibn K͟haldūn, 1967). Although the main focus of Ibn K͟haldūn's writing concerned the processes of state formation and decline, he developed his ideas through a consideration of the physical environment which he saw as forcing people to live together in social and political groups. Central to his argument was the view that states develop through a natural sequence of growth, maturity, decline and fall, because group solidarity is inevitably eroded by the process of civilization. Ibn K͟haldūn's geographical account is mainly included in Chapter 1, which explicitly acknowledges his debt to Ptolemy and al-Idrīsī. Indeed his map and subsequent topographical description of different parts of the world is based almost entirely on that of al-Idrīsī. Within his second preparatory discussion he describes the earth as having a spherical shape and being enveloped by water. He again uses the sevenfold division of the cultivated area of the world, noting that the northernmost and southernmost zones are least populous and civilized. For Ibn K͟haldūn (1967: I. 104) 'civilization has its seat between the third and the sixth zones'. In developing this argument he explicitly accounts for the distribution of civilization by the harshness of the climatic regime and physical environment at the two extremes; in the south it is too hot, and in the north too cold. Moreover, elsewhere in his fourth preparatory discussion he also argues that climate has a significant direct influence on human character: people living in hotter coastal and southern regions are seen as being joyful whereas those living in cold hilly and mountainous regions are described as sad and gloomy (Ibn K͟haldūn, 1967: I. 174). Such ideas were to have a lasting influence on geographical thought, and survived through their incorporation into environmental determinism well into the 20th century.

As well as topographical accounts and cartographical descriptions of the world Islamic scholars also added considerably to the body of astronomical writing, although there is much debate concerning their original contributions in this field (Mieli, 1938; Dreyer, 1953; Needham and Wang Ling, 1970). Again, religion played a significant role in determining the importance given to astronomy, both because of the need to determine lunar movements for the calendar, and also for the identification of the direction to which the faithful should turn for prayer. It seems that Arab astronomy was based both on ideas introduced from India during the 8th century and also on the rediscovery and translation of earlier Greek texts. In particular, Islamic astronomers sought to revise the works of Ptolemy and Aristotle, but this they were largely unable to do because they failed to recognize that the earth was not at rest. Their main astronomical achievements thus appear to have been in the general development of a tradition of Greek scholarship which had disappeared in Europe in the face of Christian dogmatism, and also in the construction of observatories such as those at Damascus, Baghdad and Cairo.

The Chinese and Islamic contributions to geography in the first millennium AD represented fundamental advances on anything happening in Christian Europe. The Chinese had developed a culture of science in which knowledge was used extensively in the support of the Emperor's wishes. Great advances were made in cartographic representations of the Empire, and topographical accounts provided surveys both of the human and physical characteristics of the regions under the Emperor's sway. Islamic science was, in contrast, closely influenced by religious requirements. This was particularly so with respect to astronomy, but topographic accounts and cartographic skills were also largely related to the uniting of the Islamic world, and an interpretation of divine instruction.

## 3.3 *The resurgence of European geography*

Following the collapse of the Roman Empire, the dominance of Christianity in Europe led to a rejection of much of the earlier scientific scholarship of the Greek world. If Roman culture had led to a stagnation in the inquiring, critical spirit of the Greek world, then many Christians sought completely to destroy it. In Dreyer's (1953: 207) words, 'A narrow-minded literal interpretation of every syllable in the Scriptures was insisted on by the leaders of the Church, and anything which could not be reconciled therewith was rejected with horror and scorn.' Although the earliest Christians do not appear to have been antipathetic towards science, the elevation of Christianity to the position of the state religion of Rome under Constantine in the 4th century led to the propagation of doctrines that were inimical to critical scientific enquiry. In particular Lactantius (*fl.* 4th century) in his *Divinarum institutionum*, declared science to be 'both foolish and false' (Kimble, 1938: 14), and sought to ridicule the belief that the world was spherical (Dreyer, 1953).

However, not all Christians rejected earlier scholarship, and within the growing number of monasteries that began to flourish throughout the Christian world, surviving ancient manuscripts were studied and transcribed. Thus the anonymous Geographer of Ravenna, writing in the mid-7th century, although a devout Christian, made extensive use of the works of pagan scholars, and the English monk Bede writing at the beginning of the 8th century copied large extracts from Pliny's work in his *De natura rerum* and his *De temporum ratione*. Bede, for example, appears to have accepted the spherical form of the earth, and the annual motion of the sun and planets, even though he also acknowledged the

Christian belief that there was water surrounding the heavens (Dreyer, 1953).

The dominance of Christianity also had significant ramifications for medieval cosmology. Of most importance was the Judaeo-Christian teleological belief in God's creation and care for the universe as expressed in Genesis (Glacken, 1967; Doughty, 1981). This view consisted of four central elements. First, the world was created out of chaos by God (Genesis 1). Thus, in Glacken's (1967: 153) words 'The creation is a continuing process requiring the constant care, activity, and solicitude of God.' Second, humanity, made in the image of God (Genesis 1: 27), is set apart from nature. There is therefore a clear divide between the realm of nature and the realm of people. Third, the products of nature exist for humanity, and it is the task of people to cultivate and maintain them; people are here seen as caretakers of the environment (Genesis 2: 15). Fourthly, as a result of the Fall, when Adam and Eve disobeyed God, the idyllic relationship between people and the environment is broken, and thereafter humanity had to toil in a cursed environment (Genesis 3: 17). Once more, in Glacken's (1967: 153) words,

> The story of the Fall became important to the Christian idea of nature because it is the source of the belief, widely held through the seventeenth century, that the fall of man has caused disorder in nature and a decline in its powers, an idea clearly distinct from the classical idea of a senescence in nature based on the organic analogy.

These four elements were to play a fundamental role in shaping the subsequent development of geographical ideas concerning the relationship between people and the environment.

From the 12th century onwards, the resurgence of European scholarship and science was underlain by three important processes: the reconquest of Iberia from the Moors beginning in the late 11th century and leading to the incorporation of elements of Islamic science into Christian European understanding; the Portuguese and Spanish voyages of discovery from the 15th century onwards which opened up whole new areas of the world to European scientific examination; and the expansion of the Ottoman Empire and the fall of Constantinople in 1453 which caused a considerable migration of eastern scholars and knowledge into Europe.

### *3.3.1 A world of 'discovery': cartography and exploration*

By the beginning of the 15th century, with the establishment of the House of Avis as the ruling dynasty of Portugal, European horizons were dramatically altered by the so-called Age of Discovery. Following the economically draining wars with Castile, the Portuguese King João I, and his sons Dom Duarte, Dom Pedro and Dom Henrique, turned to overseas exploration as a way of countering many of the problems that beset the country. In particular, any economic gains would provide urgently needed income, long voyages would keep a restless nobility active outside the country, and foreign adventure would distract attention from internal social problems (Bell, 1974; Diffie and Winius, 1977; Unwin, 1987). From the capture of Ceuta in North Africa in 1415, the Portuguese gradually ventured further around the coast of Africa, until Bartolomeu Dias reached the Cape of Good Hope in 1488. Many of these early voyages were sponsored by Dom Henrique (1394–1460), known in English as Prince Henry the Navigator, who established a school of navigation and cartography on the Sagres Peninsula in the far south-west of Portugal. Dom Henrique's central purpose appears to have been to increase his own financial resources, and although he did play an important role in bringing together sailors, navigators and adventurers to undertake these voyages of 'discovery' he was not alone, and many other voyages were sponsored by the King and nobility. This Portuguese expansion was enabled by two main factors: the culmination of sailing experience derived from previous commercial and fishing voyages in the north Atlantic (Diffie, 1960), and the acquisition of Islamic geographical knowledge.

Towards the end of the 15th century the pace of 'discovery' changed dramatically. In 1492 the Genoese Christopher Columbus, in the service of the Castilian Crown reached San Salvador, Cuba and Haiti (Sykes, 1934; Baker, 1937). India was attained during Vasco da Gama's voyage from 1497 to 1499 (Ravenstein, 1898), and Cabral crossed the Atlantic to Brazil in 1500. Portuguese ships sailed to China in 1513, and in 1519 Fernando de Magalhães set sail on his circumnavigation of the globe, which was at last firmly to shatter any remaining illusion that the earth was flat (Diffie and Winius, 1977). These voyages were both enabled by the rediscovery of classical geographical and cartographic works, but they also rapidly led to a fundamental revision of methods of map making (Crone, 1968).

In 1295 Maximus Planudes (*c.* 1260–1310), a monk at the Chora monastery in Constantinople, had eventually rediscovered a manuscript of Ptolemy's *Geography* (Dilke, 1985). This manuscript had no maps with it, and Planudes therefore began compiling ones to accompany the text. However, during the 14th century knowledge of Ptolemy's work remained in Constantinople, and it was only in 1406 in Florence that the Tuscan Jacopo d'Angelo da Scarperia translated the work into Latin (Dilke, 1985). Manuscript versions of the maps were soon prepared, and from the 1470s printed editions of the text and maps together were produced at Bologna, Rome and Ulm. Revisions, however, were soon necessary. Ptolemy had suggested that the Indian Ocean was enclosed by a piece of land running from Africa to the Malay Peninsula, and it was thus the voyages of Dias and da Gama that definitively proved this to be wrong. Moreover, Columbus recalculated the size of the globe to include Marco Polo's eastern discoveries, and although he reached America rather than Asia, it was these calculations based in part on those of Ptolemy that had given him the incentive to sail westwards.

Ptolemy was not, though, the only influence on medieval cartography, and long before the rediscovery of his works European cartographers were grappling with ways of depicting the world in which they lived (Harvey, 1980). Early medieval Christian world maps were often very simple representations, without any indication of latitude or longitude, designed as much to depict theological as topographical truths. The most common, known as T–O maps, depict the three known continents separated by a T, in the midst of the surrounding ocean O. Similar to such maps, but less rigidly divided, is the Hereford Map designed by Richard of Haldingham in the late 13th century. As with most Christian maps, Jerusalem was in the centre, and Dilke (1985) suggests that the prominence given to places of pilgrimage and the Alpine passes 'suggest that one object of the Hereford Map was to act as a help to intending pilgrims'. Other maps, such as the mid-13th century maps of Britain and the Holy Land by Matthew Paris, appear likewise to have been designed essentially as itineraries (Harvey, 1980).

### 3.3.2 Geography and the expansion of European power

The rediscovery of classical cartography was central to the expansion of European political and economic power in the 15th and 16th centuries. However, although Ptolemy's great work was called a *Geography*, there were few medieval or Renaissance explorers who called themselves geographers. Likewise, although a strong tradition of topographic writing had developed in the wake of works such as the late 12th century *Topographica Hibernica* (Topography of Ireland) written by Giraldus Cambrensis (1951), these made few claims to be works of geography. Thus, although the astronomical, cartographic, cosmological and topographic traditions of classical Greek and Roman geography re-emerged in the medieval period, there was no widely accepted medieval definition of geography, and much so-called medieval geography has merely been given this title by subsequent commentators (Wright, 1925; Taylor, 1930). The nearest word encompassing the essence of later geography was 'cosmography', and during the 15th and 16th centuries a number of treatises bearing this title were published, often modelled on the works of Strabo and Ptolemy. Among the most famous of these are the *Cosmographiae introductio* of Martin Waldseemüller, the *Cosmographicus liber* of Petrus Apianus and the *Cosmographia universalis* of Sebastian Münster (Dickinson, 1969).

The beginning of the 16th century saw a fundamental departure from classical understanding of cosmology and astronomy. Around 1530 Nicolas Copernicus (1473–1543) eventually finished the manuscript of his work *De revolutionibus*, the basic ideas for which had probably first been thought out soon after 1506 (Dreyer, 1953; Babicz, Büttner and Nobis, 1982). In this, Copernicus put forward an entirely new theory of planetary movement, which countered the long accepted arguments of Aristotle and Ptolemy by demonstrating that the earth was not the centre of the universe, but instead revolved around the sun together with the other planets. This had important ramifications for the links between geography and astronomy, because while the earth was still considered to be the centre of the universe it was reasonable to include discussions about that universe in geographical writing about the earth.

Astronomy could thus logically be considered an extension of geography. However, following the acceptance of a heliocentric view of the universe, astronomy threw off the shackles of its earth-bound past, and simultaneously broke its links with geography.

The vast expansion of cartographic and topographic knowledge in the 16th century opened up whole new avenues of commercial and political action. With the demise of the Iberian powers, and the emergence of the Dutch and English as the most formidable European states by the beginning of the 17th century, the heart of the impetus for further discovery and exploitation shifted northwards. Thus, in referring to the emergence of geographical ideas in 16th-century England, Taylor (1930: v) has noted that, 'Elizabeth's day saw the map and the globe as the necessary furniture of the closet of scholar, merchant, noble and adventurer alike, and the dreams of Empire were formulated which found expression in Drake's achievement and Humfrey Gilbert's splendid failure'. Maps, globes, topographic accounts, and the few explicit geographies that were written, such as the *Geographia* (1540–41) of Roger Barlow, were designed with one purpose mainly in mind: that of enabling English ships to sail the furthest oceans and bring back the products of China, India and the Americas for commercial profit.

These tendencies found their fulfilment in the writings of Richard Hakluyt the Younger (1552–1616) and the Reverend Samuel Purchas (1577–1626). Hakluyt lectured on cosmography at Oxford, and had benefited greatly from his contacts with other cosmographers, particularly through the practical work of the Portuguese and the theoretical schemes of the Flemish school (Taylor, 1934). However, he described his great work, *The principal navigations, voyages, traffiques and discoveries of the English nation* (1589) as a history (Hakluyt, 1903: xxiii, xxx), and in its preface he confines the term 'geographical' to the depiction of places on a map (Hakluyt, 1903: xxx). Indeed, in the preface to the second edition of 1598 Hakluyt (1903: xxxix) makes this distinction more explicit by referring to 'Geographie and Chronologie (which I may call the Sunne and the Moone, the right eye and the left of all history)'. Here geography is used to refer each incident or voyage to its correct place, whereas chronology relates it to its due time. Hakluyt's central task was to bring together all previous accounts of the voyages of English people abroad and to weld them into a memorial of the achievements of the English nation. The first edition confined itself to English voyages of discovery, colonization and commerce, but in the three volume second edition of 1598–1600 he also included material on naval engagements, particularly the defeat of the Spanish Armada, and some details of voyages by foreigners where they provided additional information on parts of the world not elucidated by English authorities. Hakluyt's intentions in writing his *Principal navigations* were to provide a eulogistic account of the achievements of the English and to preserve documents which would otherwise have been 'buried in perpetuall oblivion' (Hakluyt, 1903: xxxii). However, underlying these he was eager to promote continued English overseas expansion, particularly in the newly established colonies of Virginia, and in his later life he developed this practical intent, serving as an 'adviser on Virginia and East India matters' (Taylor, 1934: 33).

Hakluyt's role as a collector of travel accounts was taken up by Samuel Purchas, who had been educated in divinity at St John's College, Cambridge. Remarkably, unlike Hakluyt, he never travelled outside England, but his publications brought together a wealth of material concerning the world as it was known in the 17th century. His first book, *Purchas his pilgrimage or relations of the world and the religions observed in all ages and places discovered, from the Creation to this present* was published in 1613, and sought to combine his theological and geographical interests by providing a synthesis of the different religious practices of the world's peoples. This work was immediately popular, and was even read by King James (Taylor, 1934). As a result he met a number of influential people, such as Sir Walter Ralegh, who furnished him with yet further manuscripts and books. In particular, Hakluyt gave him access to his own collection of manuscripts, and these were an important source for his later work, *Hakluytus posthumus or Purchas his pilgrimes contayning a history of the world in sea voyages and lande travells by Englishmen and others*, which appeared in two volumes in 1625, nine years after Hakluyt's death. While these volumes bring together a mass of secondary sources concerning voyages and travels, they do not claim strictly speaking to be geography. As in his earlier work, Purchas was seeking to incorporate geographical works into a historical account. Like Hakluyt, his scholarship was imbued with a practical concern with colonial expansion, and by bringing together the results of explorations and discoveries of the New World he was seeking to inspire further English settlement overseas.

### 3.3.3 *Geography at the dawn of the 17th century*

In 1600 geography was a term of many meanings, conjuring up a range of different images to those who used it. At its simplest it referred to topographical writings about particular places, and thus included descriptions of distant lands, coastlines and harbours. In effect this type of geography was none other than Ptolemy's chorography, whereas Ptolemy's concept of geography had become subsumed within cosmography, the science of the globe. However, such descriptions required precise elucidation, and since classical antiquity this had meant that they were frequently associated with mathematical and astronomical calculations. Cartography thus formed a central part of geography, providing the detailed basis upon which voyages of 'discovery' could be made and subsequent descriptions of 'new lands' could be written. These links with mathematics and astronomy also brought the subject into a close alliance with astrology, which sought to explain human and environmental phenomena through reference to astronomical phenomena. Indeed Livingstone (1990a, b) has argued that many 16th- and 17th-century writers about geography such as William Cunningham and John Dee (Matley, 1986) were deeply implicated in magical practices (see also Sack, 1976). This view of geography owed much both to the mystical ideas of Pythagoras and to the writings of subsequent generations of Islamic scholars.

Above all geography had retained the practical importance attributed to it by Strabo. It was essential for successful merchants and statesmen alike, providing the basis upon which European nations came to dominate the globe. The rediscovery of Greek and Roman geographical works, enhanced by later Arab writers, had opened up new visions of the world to the Portuguese and Spanish explorers of the 15th and 16th centuries. In turn, their discoveries overthrew long established traditions and images, and provoked a range of new questions for the emergent sciences of the 17th and 18th centuries to answer. In short, as Livingstone (1990a: 8) has argued, 'First-hand encounter with the world – the very stuff of geography – brought an immense cognitive and cultural challenge to tradition.' But geography, as yet, remained poorly defined. There was not the institutional dominated concern with the clear delimitation of disciplinary boundaries that was to be encountered in the 19th century, and anyone who wrote about the earth could claim to be a geographer. It was not until the 17th and 18th centuries that a formal geography to compare with that of Strabo or Ptolemy began to emerge.

CHAPTER 4

# The emergence of geography as a formal academic discipline

> The usefulness of this study is very extensive. It provides a purposeful arrangement of our knowledge, serves our own entertainment, and provides rich material for social conversation.
>
> (Kant, *Physische Geographie*, 1802, translated in May, 1970: 264)

Science and scholarship do not exist separately from their human context, and it is therefore essential to examine the broader intellectual and social framework within which any discipline has emerged. Nevertheless, when attention focuses on the development of a formal geography, produced by people specifically claiming to be geographers, it is also necessary to examine the roles of those individuals who have played a formative part in the development of the subject. The rapid increase in knowledge about the world gained by European people in the wake of the 'discoveries' of the 15th and 16th centuries, provided the basis for the gradual emergence of a formal discipline of geography in the 19th century. It is with the complex processes giving rise to the eventual institutionalization of academic geography that this chapter is primarily concerned.

## 4.1 From Varenius to Kant: the reappearance of formal geography

### 4.1.1 Varenius: general and special geography

As the previous chapter has shown, the medieval period saw very few major works explicitly on geography. Instead, the term cosmography was widely used to refer to general descriptions of the world, which at a later date have been considered to be geographical. However, in 1650 a firm attempt was made formally to delimit the nature of geography as something separate from cosmography in the *Geographia generalis* of Bernhard Varenius (1622–50) (Baker, 1955). The first half of the 17th century had seen the emergence of a new kind of empirical and rationalist science in the writings of Francis Bacon, Galileo Galilei and René Descartes, and Bowen (1981) has suggested that the work of Varenius was the first real effort to relate geography to such developments.

In his dedication, Varenius defined geography as being concerned only with the earth, thus clearly separating it from astronomical studies of the heavens. He went on to assert that

> Geography itself falls into two parts: one general, the other special. The former considers the earth in general, explaining its various parts and general affections. The latter, that is, special geography, observing general rules, considers, in the case of individual regions, their site, divisions, boundaries and other matters worth knowing. But those who have so far written on geography have discussed at length special geography alone, practically without exception, and have explained very little relating to general geography, with much that is necessary being neglected and omitted, so that young men, learning special geography, are for the most part ignorant of the bases of this discipline, and geography itself scarcely preserves the title of a science (Varenius, translated in Bowen, 1981: 277–8).

Two important observations can be noted about this critical statement. First, Varenius's division of the subject into general and special geography closely parallels Ptolemy's distinction between geography and chorography. Thus, Varenius goes on to suggest that special geography can be divided into two parts, namely chorography and topography, the former being description of a region of a medium size, and the latter being description of a small tract or place. By criticizing the concentration of most geographical studies on chorography, he emphasizes the point that by the 17th century geography had largely devolved into regional description. Secondly, and more importantly for the future development of the discipline, Varenius was keen to promote geography as a science, and the way he sought to do so was through an emphasis on general geography. He thus envisaged his task as being to bring back general principles, or theories, into the subject.

Varenius divided general geography into three parts: the *absolute* part, concerned with the body of the earth and its constituents, such as land masses and rivers, together with their properties such as shape, movement, and size; the *respective* or *relative* part, pertaining to the effects of celestial phenomena on the earth; and the *comparative* part, which seeks to explain those properties emerging from a comparison of different places of the earth (Baker, 1955; Bowen, 1981). Likewise, he saw special geography as being concerned with three kinds of things in any region, namely terrestrial, celestial and human. However, Varenius recognized that there was a problem in trying to incorporate human considerations into geography defined as a mathematical science (Lange, 1961). Ptolemy had circumvented this difficulty by classifying most things that could not be treated mathematically and cartographically as chorography, but, as Bowen (1981: 82–3) has commented, 'Varenius, by proposing to incorporate in geography a consideration of human learning, intelligence, language and so on, raised that problem again and furthermore brought the science as he envisaged it into conflict with the more recent Cartesian dualism of mind and matter.'

Methodologically, Varenius drew an important distinction between general and special geography. He allied general geography with classical methods of science, and argued that most of its proofs were to be conducted by logical argument or demonstration, in a similar way to those of mathematics and geometry. In contrast, in special geography he suggested that 'practically everything is explained without demonstration (excepting celestial affections, which can be demonstrated) since experience and observation, that is the evidence of the senses, confirm most things; indeed they cannot be proved in any other way' (Varenius, translated in Bowen, 1981: 281). Varenius, though, appears to have been somewhat uneasy about making this distinction, and admitted that there could be overlap between the two methods. This confusion arises in part because at times he referred to geography as a whole, and at other times he distinguished between his two divisions of the subject. Thus, he refers to three principles that geography, as a whole, uses to confirm the truth of its propositions: 'First, the propositions of geometry, arithmetic and trigonometry. Second, the precepts and theorems of astronomy. . . . Third, experience, for the greatest part of geography, particularly special geography, rests solely on the experience and observation of men who have described individual regions.' (Varenius, translated in Bowen, 1981: 281). Although he does not say so explicitly in this quotation, the implication is that geometry, mathematics and astronomy are the principles upon which general geography should be based. Nevertheless, by noting that the greatest part of geography relies on experience, he also implies that experience is important for general geography. This tension seems to have arisen because Varenius was seeking to make geography a mathematical science, concerned with general laws and principles, while at the same time recognizing that the inclusion of people and regional description meant that much of it had to be based on empirical experience. It also reflects an attempt to combine the new inductive methodology of Bacon with the logical mathematical proofs of previous science.

A final important feature to note about the *Geographia generalis* is Varenius's justification for the subject. He thus argued that

> The study of geography is commended by, 1. Its value, for it is in the highest degree suitable for man as an inhabitant of the earth and endowed with reason beyond other animals. 2. It is also pleasant and indeed a worthy recreation to contemplate the regions of the earth and its properties. 3. Its remarkable utility and necessity, since neither theologians, nor medical men, nor lawyers, nor historians, nor other educated persons can do without knowledge of geography if they wish to advance in their studies without hindrance. (Varenius, translated in Bowen, 1981: 282–3).

Interestingly, although there are parallels between this justification and that of Strabo, Varenius does not explicitly follow Strabo by arguing that geography is important for politicians and for the exercise of military objectives. Although this was still almost certainly recognized by Varenius, he concentrates instead on its *value*, the *pleasure* of its contemplation, and its *utility*. In terms of value, geography is seen as being highly important for human occupance of the earth, seeking to bring together both mind and matter. Varenius also accords great significance to the reflective importance of the subject by noting the worthiness of contemplation of the earth, and he commends its utility for the advancement of knowledge. These three themes are closely parallel to Habermas's three knowledge constitutive interests. Thus Varenius's concern with the *value* of geography for human occupance of the earth can be compared to Habermas's *technical* interest, interpreted as knowledge enabling people to control objects in nature. His focus on the *pleasure of contemplation* can be seen as broadly similar to the concern of the historical–hermeneutic sciences with a *practical* interest in communication, and geography's *utility* for the advancement of knowledge could be interpreted as an *emancipatory* interest. While this similarity should not be stretched too far, it does suggest that Varenius had a broad enough perception of the subject to seek to justify it according to three very different types of interest.

Varenius died at only twenty-eight, before he had time to write a detailed treatise on special geography to match that which he had written on general geography. The *Geographia generalis* was translated into several languages, and Isaac Newton (1642–1727), the English mathematician and physicist, even revised two editions of it for his students at Cambridge in 1672 and 1681. However, Varenius's theoretical concerns with a comprehensive global approach to the discipline were gradually overtaken by the increased attention being paid to the empirical sciences, resulting in part from the continued exploration of, and need to describe, the world's surface. As Bowen (1981: 90) has argued,

> General geography as Varenius presented it fell out of favour because it involved a comprehensive approach rather than a specialist one, and was not amenable to the experimental method. Meanwhile special geography, in dealing with man, was considered by the mechanists to be outside the ambit of natural science, so that while many specialist sciences flourished during the next century, in geography the regional studies remained in an ambiguous and even defensive position in relation to the scientific tradition.

Thus in an era of considerable scientific advance, when empirical and experimental methods were to come to the fore, geography's image as a science was badly tarnished. It is a position from which some geographers have ever since been trying to resurrect the subject, seeking to raise it to what they argue is its correct place among the natural sciences.

### 4.1.2 *Kant's* Physische Geographie

The second half of the 17th century saw the flourishing of new scientific activity, represented in Europe by the foundation of the Royal Society in London in 1660 and the establishment of the Paris Academy of Sciences in 1666. Bacon's ideas concerning scientific method had an important early influence on both societies, but these were soon supplemented, particularly in France, by those of Descartes. The fields of science that advanced most rapidly were mathematics, physics and astronomy, heavily directed by the work of Galileo and Newton. Subsequently, they were also strongly influenced by the very differing approaches to empiricism of John Locke (1632–1704) and George Berkeley (1685–1753) (Russell, 1961; Scruton, 1981). Philosophically, the central question that Locke, Berkeley, Descartes and Bacon alike were grappling with was to identify what constituted a basis for the foundation of knowledge. What they sought were grounds upon which it was possible to know anything based upon the evidence available. The responses to this question were broadly divided into two camps: the rationalists, following Descartes, Spinoza (1632–77) and Leibniz (1646–1716), who argued that the key to knowledge lay in rational reflection and that all empirical enquiry was underlain by metaphysical principles; and in contrast those such as Locke, Berkeley and Hume, who advocated empiricism, and confined understanding to the limits of human experience.

At the time of this intellectual ferment in the late 17th and early 18th centuries geography appears to have quietly slid into a backwater. This was the era of the emergence of a revitalized natural history, and just as Pliny's *Natural history* can be seen as covering subject matter of apparent relevance to geography, so too do works such as *The theory of the earth* by Thomas Burnet (c. 1625–1715) and the *Wisdom of God* by John Ray (1627–1705). Indeed Bowen (1981) suggests that these natural histories were of considerable significance both for their influence on later geographical work, and also as attempts to combine the new empirical sciences with the religious teachings of the Church. What geographical writing there was still concentrated on voyages of exploration, and descriptive travelogues.

It was in Germany, in the second half of the 18th century that the next main impetus to geographical practice and theory emerged. At a practical level, Anton Friedrich Büsching (1724–93), professor of philosophy at Göttingen, sought to provide an accurate chorographical and topographical description of the earth's surface in his *Neue Erdbeschreibung*, which appeared in eleven volumes between 1754 and 1792 (Büsching, 1762; Büttner and Jäkel, 1982). To do this he organized his material into two broad categories, the one concentrating on civil and political divisions and the other on natural features (Adickes, 1925). Bowen (1981) suggests that his work was also significant for its inclusion of statistics on population density in his regional descriptions, and for his insistence on personal research. However, of much more importance for the future development of geography at a theoretical level was the work of the philosopher Immanuel Kant (1724–1804), who has been described by May (1970: 3) as 'the outstanding example in Western thought of a professional philosopher concerned with geography'.

Kant is best known for his series of three critiques, the *Critique of pure reason* first published in 1781, the *Critique of practical reason* (1788) and the *Critique of judgement* (1790). In these he sought 'to show that the choice between empiricism and rationalism was unreal, that each philosophy was equally mistaken, and that the only conceivable metaphysics that could commend itself to a reasonable being must be both empiricist and rationalist at once' (Scruton, 1981: 137). As such, he became the founding father of German 19th-century idealism. However, from 1756 he also lectured on geography at Königsberg University for forty years, and various versions of manuscript notes from these courses survive, enabling the development of his views concerning the subject to be traced (Adickes, 1911; May, 1970). In 1757 he produced a short *Outline and prospectus for a course of lectures in physical geography* but it was only in 1802 that an official edition of his views was eventually published under the title *Physische geographie* (Physical Geography).

In his 1757 outline Kant argued that there were three ways of looking at the earth: a mathematical way concerning its form, political doctrine concerned with people and their types of government, and physical geography which considered the natural conditions of the earth and what it contains. Once again, as for Varenius, such a classification caused problems relating to the way in which people were to be considered, since Kant included them in both physical geography and in political doctrine. Later, however, in the introduction to his *Physische geographie* he sought to resolve this by referring to physical geography as being concerned with the world as the object of outer sense, in contrast to anthropology as being to do with the experience of consciousness provided by the inner sense. A further distinction between Kant's earlier and later views on geography was the way he perceived its relationship to history. In his 1757 outline, Kant appears to have considered geography as part of history, but by the publication of the *Physische geographie* he had refined this to the assertion that geography and history together make up all knowledge, with geography being description according to space and history description according to time. Moreover, he went on to argue that geography is the foundation of history 'because occurrences have to refer to something. History is in never relenting process, but things change as well and result at times in a totally different geography. Geography therefore is the substratum' (Kant, translated in May, 1970: 261–2).

By the time of the publication of the *Physische geographie* Kant had also reworked his ideas on the relationship between physical geography and other types of geography, arguing that just as physical geography was the foundation of history, so too was it the foundation of all other possible geographies. He divided these other geographies into five types:

1. 'Mathematical geography, in which the shape, size and motion of the earth, and its relationships to the solar system in which it is situated, are dealt with' (Kant, translated in May, 1970: 263);
2. 'Moral geography, in which the diverse customs and characteristics of people of different regions are told about' (Kant, translated in May, 1970: 263);
3. Political geography, whereby the political organization of a state is seen as being totally dependent on its physical geography;
4. Commercial geography, concerned with trade linking together areas of excess with areas of deprivation; and
5. Theological geography, concerned with the way in which theological principles are transformed due to differences in the land.

Overall, therefore, this classification suggests that Kant envisaged the expressions of human occupance of the earth as being heavily influenced and underlain by physical geography.

Kant's geography also had a profoundly pedagogic value, and one of the main arguments that he used to justify its teaching was that it provided students with a basic framework of knowledge (May, 1970). Under the influence of Rousseau (1712–78) he also considered that it could be used to teach morality and theology, and that it could bring ennobling satisfaction (Adickes, 1925). In essence, he argued that geography could provide a unity of knowledge, enabling people to orient themselves within the world. However, in the quotation that begins this chapter, taken from the introduction to his *Physische geographie*, Kant also notes its utility for entertainment and for social conversation, in a way generally similar to Varenius's emphasis on its use as a pleasant and worthy recreation.

A final feature of Kant's geography that needs to be considered is the place he attributed to it in his broader philosophy of science, and for this it is necessary to return to his conception of rational and empirical knowledge as expressed in the *Critique of pure reason*. Kant argued that scientific empirical knowledge is based on experience and deals with matters of fact; it is *a posteriori*. However, as Scruton (1981: 139) has summarized, he went on to argue that 'it rests upon certain universal maxims and principles, which, because their truth is presupposed at the start of any empirical enquiry, cannot themselves be the outcome of such enquiry. These axioms are, therefore, *a priori*'. Such *a priori* propositions can either be analytic or synthetic. Analytic propositions are true because of the words used to formulate them, and specifically they are statements 'in which the predicate is part of the subject; for instance, "a tall man is a man"' (Russell, 1961: 679). In contrast, synthetic propositions are those that are not analytic, and convey meaning about the empirical world. Such synthetic *a priori* truths for Kant are justifiable only through reflection, and their truth must be necessary truth; they 'form the proper matter of metaphysics' (Scruton, 1981: 140). Kant argued that the facts of geography and of history are *a posteriori* and empirical, derived from sense perception and experience, whereas those of mathematics are *a priori* and rational.

For Kant, space and time are forms of intuition rather than concepts; they are subjective aspects of our perception. Moreover, he argued that they are *a priori* forms, on the basis of both metaphysical and epistemological grounds. In order to comprehend Kant's intuition of space, it is necessary to understand something of the way in which Kant conceived of the process of perception. Kant considered that all our sensations have causes, or what he terms *noumena*, things in themselves. However, it is impossible to know a noumenon; all that we can do is to perceive a *phenomenon* through a synthesis of concept and experience. This has been summarized particularly well by Russell (1961: 685) as follows: 'What appears to us in perception, which he calls a "phenomenon" consists of two parts: that due to the object, which he calls the "sensation", and that due to our subjective apparatus, which, he says, causes the manifold to be ordered in certain relations. This latter part he calls the *form* of the phenomenon.' This form is *a priori* and not dependent on experience. For Kant, the form of outer sense is space, and the form of inner sense is time. Thus, in Scruton's (1981: 143) words, 'This means, roughly, that the idea of experience is inseparable from that of time, and the idea of an experienced *world* is inseparable from that of space.' Kant's metaphysical argument concerning the intuition of space had four elements (Russell, 1961). First, he suggested that space is used to refer sensations to something external, and consequently it cannot be an empirical concept abstracted from external experience. Second, he argued that 'we cannot imagine that there should be no space, although we can imagine that there should be nothing in space' (Russell, 1961: 685). Consequently space is necessarily *a priori*, underlying external perceptions. Third, he considered that there is only one space, and therefore that space is not a general concept of the relationships between things. Fourth, he thought that space was an infinite given magnitude, containing all parts of space within it, unlike the relationship between a concept and examples of that concept. Consequently he considered space to be an intuition (*Anschauung*) rather than a concept.

There are, though, very real problems in combining the ways in which Kant uses the intuitions of space and time in his *Critique of pure reason* and his use of the terms in his *Physische geographie* to refer to the realms of geography and history (Kaminski, 1905; Smith, 1923; May, 1970). As a starting point it is possible to suggest that in the former he was using them in a theoretical sense, whereas in the latter he conceived of them empirically. This enabled him to argue that as empirical sciences 'Geography and history fill up the total span of knowledge; geography namely that of space, but history that of time' (Kant, translated in May, 1970: 261), without necessarily bringing geography and history into his theoretical discussion of the intuitions of space and time in his *Critique of pure reason*. Bringing them together would suggest that geography was concerned with the form of outer sense, whereas history was concerned with the form of inner self. However, this is incompatible with the suggestion that anthropology was to do with the experience of consciousness by the inner sense. This could be resolved by arguing that history and anthropology are one and the same, but this Kant chose not to do. May (1970: 120) has therefore suggested that the dilemma is best resolved by considering Kant's distinction between geography and history to refer to a distinction in outer sense, and that 'the concept "time", as employed in relation to the history of nature, is quite distinct from the concept "time" understood as the form of inner sense'.

Kant's linking of geography with space, and history with time, was to have a profound influence on the future development of the subject (May, 1970), but not until long after his death. Indeed, as with Varenius before him, his theoretical and philosophical work seems to have been largely ignored by geographers during his own lifetime.

## 4.2 Humboldt, Ritter and the foundation of modern geography

While Kant had provided geography with a theoretical justification, his other philosophical interests meant that he did not implement this in practice. Such a task was left to two other Germans, Alexander von Humboldt, born in 1769, and Carl Ritter, born in 1779. These two scholars have almost universally been seen as the founders of modern geography (Hartshorne, 1939, 1958; Dickinson, 1969; Schultz, 1980; Holt-Jensen, 1988). Both men died in 1859, the same year that saw the publication of Darwin's *On the origin of species*. Their task was to bring together the wealth of new empirical information that had been gathered about the world in the wake of European exploration and colonial policies. It was the combination of the new methods of scientific enquiry with such political activity that formed the background for the emergence of geography as a formal academic discipline. This was enabled, according to Stoddart (1986: 34), by two great devices which brought 'the huge diversity of nature revealed by this new wave of exploration within the bounds of reason and of comprehension': classification and the comparative method, the former exemplified in the work of Linnaeus (1707–78), and the latter in that of Reinhold Forster (1729–98) (Hoare, 1976). The achievement of Humboldt and Ritter was 'to seize the technical and conceptual advances of the Pacific voyages and so to organize and order knowledge as to show its coherence and significance, Humboldt ecologically, Ritter historically and regionally' (Stoddart, 1986: 36–37).

### 4.2.1 *Alexander von Humboldt and the unity of the cosmos*

Humboldt's great contribution to geography was as an empirical observer and comparer, who sought to develop an understanding of the world in which people were seen as part of nature (Kellner, 1963; Meyer-Abich, 1967; Meyer-Abich and Hentschel, 1969; Dickinson, 1969; Bowen, 1970, 1981; Botting, 1973; Stoddart, 1986). His method was essentially inductive, seeking 'to modify and extend the Baconian tradition in science' (Bowen, 1981: 215). Humboldt's university education was based at Frankfurt-on-Oder and Göttingen, but he then spent a year at an Academy of Commerce in Hamburg and eight months at the Freiburg Academy of Mines. With this background he began work as a geologist and mining engineer, with a strong interest in botany. However, in 1789 while at Göttingen he had met the geographer George Forster, the son of Reinhold Forster, the naturalist on Cook's second voyage around the world (1772–75) (Hoare, 1982). In 1786, six years before his meeting with Humboldt, George Forster had raised objections to Kant's model of geography, which had drawn a distinction between a history of nature and a description of nature. In reply Kant had reiterated his arguments and suggested that to prevent confusion the term physiography should be used for the description of nature and physiogony for the history of nature (May, 1970). Through his contacts with Forster, it therefore seems highly likely that Humboldt was made aware of this debate and that he began to develop an interest both in geography and also in Kant's writings on the subject.

Humboldt's first major work, *Florae Fribergensis*, was published in 1793, and in this he put forward the view that the geography of plants, the geography of rocks and the geography of animals should form a subject which he called in Latin *Geognosia* and in German *Erdkunde*. This was to form an important basis for his later geographical writings. A year later he joined his brother Wilhelm at Jena, and through this visit came into contact with prominent writers and philosophers concerned with developing and extending Kant's idealist philosophy, such as Goethe (1749–1832), Schiller (1759–1805), Fichte (1762–1814) and Schelling (1775–1854) (Scruton, 1981). In particular, Bowen (1981) has suggested that Humboldt found inspiration at this time in the work of Johann Gottfried von Herder (1744–1803), another member of Goethe's circle, who stressed the need for collecting and analysing data about different parts of the world in order to develop a greater understanding of the earth as the home of the human race (Birkenhauer, 1986). Meanwhile, Humboldt had been travelling widely in Europe, and in 1799 he set off for Spanish America for a period of five years of scientific exploration, during which he gathered a wide range of data and material, along the lines suggested by Herder (Kellner, 1963; Meyer-Abich and Hentschel, 1969). Above all he was concerned with examining 'the influence of the inanimate world on the animal and vegetable kingdom' (Humboldt, 1799 in Hamy, 1905: 18).

On his return in 1804 he went to Paris and for the next thirty years concentrated on publishing the results of his American travels. Through his contacts with French scientists he encountered a scientific and philosophical tradition very different from that of his earlier influences in Germany. In particular, while romanticism was growing in strength in Germany, a mechanistic view of the world, derived in part from Lavoisier's (1743–94) research into chemistry, was becoming predominant in France. One of Humboldt's strengths was his ability to combine both influences, incorporating experimentation and careful observation derived from France with a German philosophical concern with the extension of knowledge for the betterment of the human race. His reforming activities, however, were to cause Humboldt problems when faced with the increased political repression in France and Germany in the second and third decades of the 19th century. When Humboldt, then holding the post of Chamberlain to the Prussian King, was recalled from France in 1826, this was according to Bowen (1981: 241) 'probably in an effort to control his reforming activities'. His recall took him to Berlin, and there in 1827 he began a series of lectures on physical geography. These were to lead eventually to his great, but uncompleted, work *Kosmos*, the first volume of which was published in 1845. This is widely seen as 'one of the greatest scientific works ever published' (Thorne, 1961: 672) and in it he sought to represent the whole material world in the tradition of the *Geographia generalis* of Varenius.

The period during which Humboldt's ideas for his *Kosmos* developed was one of the most formative in scientific and philosophical thought in western Europe. In particular it saw the emergence of Comte's positivism, Hegel's idealism, and Marx's political economy. Although Humboldt espoused empirical research he rejected Comte's formulation of positivism; although inspired by some of Hegel's ideas concerning the historical development of society, he rejected much of the latter's idealist philosophy; and although sympathetic to radical liberalism, he was unprepared to participate in the revolutionary programme of Marx and Engels.

Humboldt's method sought to collect data, group them together, make generalizations, and subsequently to develop a comprehensive view of the world. Hartshorne (1939) has thus characterized his position as using systematic studies to achieve regional masterpieces. However, central to Humboldt's understanding of physical geography was his conception of the relationship between mind and matter, based in part on Kant's ideas. Humboldt's argument in *Kosmos* is that 'Science begins where the mind itself takes hold of matter and attempts to subject the mass of experiences to a rational understanding; it is mind directed towards nature [I, 69]' (translated in Bowen, 1981: 257). It is thus the task of science to understand the world of perception; the way in which people, as part of nature, comprehend nature. In the end, Humboldt's great project was barely begun: the first two volumes (1845, 1847) formed the Prolegomena, followed in 1850 by a volume devoted to astronomy, and in 1858 by a fourth volume on the earth. The fifth volume, unfinished at his death, was on geology and vulcanism, and this was to have been followed by other volumes on the distribution of such things as organic life, plants, animals, human races, and languages. The *Kosmos* can thus be seen in many ways to have followed much earlier cosmographical writings in its bold attempt to provide a science of the whole cosmos. Indeed, by its inclusion of astronomy, in many ways it returned to Greek and Roman concepts of geography. However, by incorporating people as part of nature in his physical geography, Humboldt was seeking an essential unity to the subject which was to disappear soon after his death.

Remarkably, Humboldt's theoretical and methodological work had a surprisingly small influence on geographers in the second half of the 19th century (Hartshorne, 1958; Bowen, 1981). While his descriptions of Spanish America were widely admired, even these were criticized by conservative reactionaries who objected to his strong liberal views on social and political reform. Instead, it was the geographical works of Carl Ritter, who had become the first professor of geography at the University of Berlin in 1820, which were to have the greatest influence over German geography in the closing decades of the 19th century.

### *4.2.2 Carl Ritter and the combination of teleology with empirical observation*

Born ten years after Humboldt, Carl Ritter was subject to many of the same influences as the older man. Yet his career was substantially different. Whereas Humboldt had used a legacy following his mother's death to support his explorations and travels, and had then entered the service of the Prussian King, Ritter's life was largely spent within academic and military institutions. He was educated at Halle University, and in 1798 became tutor to the sons of a Frankfurt banker. This provided him with an income and also the time to undertake his own research. Soon afterwards in 1804 the first part of his two volume work on Europe appeared. The second volume was published in 1807, and during the same year he first met two people who were to have an important influence on his subsequent work, Humboldt and the Swiss educationalist Johann Heinrich Pestalozzi (1746–1827) (Linke, 1981). In 1813 Ritter moved to Göttingen, and here continued with his studies, which led in 1817 to the publication of the first volume of his life's work, *Erdkunde*. Over the next 42 years a further 20 volumes, making a total of 19 parts and some 23,000 pages, of this major, but unfinished, work were produced. *Erdkunde* was subtitled 'a general comparative geography', and together with Humboldt's *Kosmos*, it has been widely seen as one of the two founding works of modern geography (Hartshorne, 1939). The published volumes of *Erdkunde* concentrated on Africa and Asia, and sought to deal with three main types of subject: topical, concerned with the established forms of the continents; formal, to do with the movable features of the continents; and material, concerning the localized distributions of aspects of nature (Dickinson, 1969). The main units used for his analysis were the continents, but these were then subdivided into broad physical regions and then into smaller units 'arrived at from the detailed configuration and content . . . of particular areas' (Dickinson, 1969: 40). Ritter has thus been seen as the father of regional geography, in contrast to Humboldt, the father of modern systematic geography (Hartshorne, 1939; Dickinson, 1969).

This simple distinction between the works of Humboldt and Ritter is, however, too starkly drawn and there were many similarities in their overall approach to the subject. Indeed, both men frequently quoted from each other's works, and Ritter in particular acknowledged 'the leading contribution of the older geographer' (Bowen, 1981: 238). Thus

Ritter's apparent concentration on regional geography was designed to enable subsequent comparisons and generalizations to be made, whereas Humboldt's systematic approach sought eventually to achieve regional 'masterpieces' (Hartshorne, 1939: 258). Both were concerned with the unity of nature, although Humboldt sought that unity in ecological concepts whereas Ritter emphasized the importance of historical and regional coherence (Stoddart, 1986). Likewise, both emphasized the need for accurate empirical analysis as a basis for their inductive approaches, even though Humboldt's focus was on direct observation and experimentation, whereas Ritter tended to base 'his works on observations of other students' (Hartshorne, 1939: 55). Both stressed the importance of a comparative methodology, and both sought to develop geography as an integrative science.

Where they differed most was in their consideration of the place of people in the natural world. For Humboldt, people were part of nature, but Ritter's teleological emphasis, profoundly influenced by the natural theology of his times, saw the earth as being designed by God for the benefit of humanity. Moreover, partly derived from the idealism of Kant, Schelling and Hegel, Ritter saw the history of the continents as being the product of divine will; the world itself was a place where people could learn to know God (Glacken, 1967). In combining an empirical methodology with his theological beliefs, Ritter was thus seeking to present geography as a science which could provide people with a greater understanding of God. Because of this, Ritter's regional geography has widely been seen as primarily human in focus, in contrast to Humboldt's largely physical systematic geography (Dickinson, 1969; Holt-Jensen, 1988). However, each of them continually emphasized the importance of both human and physical elements in their work, and a more appropriate distinction would be to suggest that Humboldt's work was the more scientific and Ritter's the more ideological in concept and approach.

A second important difference between Humboldt and Ritter was in their direct involvement in the practice of education. Ritter had gained two important ideas from his contacts with Pestalozzi: that education should comply with the natural laws upon which human nature depended, and that observation was of central importance in the learning process. While these can be encountered in his written work, more importantly they also found expression in his practical teaching. Thus he became particularly concerned with the introduction of a new type of school geography based on maps, atlases, pictures, and the detailed study of pupils' home areas, which eventually exerted an important influence on the world view of German people during the age of imperialism in the latter part of the 19th century. In 1820 Ritter had been appointed to a teaching post at the Berlin Military Academy and to a professorship in geography at Berlin University. As holder of the only university chair in geography in Germany at that time he was able to exert a considerable influence on the future direction of the discipline. His lectures were popular, with Linke (1981) even noting that Karl Marx apparently attended them in 1838. Among those directly influenced by Ritter were the Swiss Arnold Guyot (1807–84), who obtained a chair at Neuchâtel in 1839 and later went to North America where he became professor of physical geography and geology at Princeton (Hartshorne, 1939), the French geographer Elisée Reclus (1830–1905) (Dunbar, 1981), and the Germans Heinrich Kiepart (1818–99), who succeeded Ritter at Berlin, Karl Neumann (1823–80) who became professor of geography and ancient history at Breslau in 1865, and Johann Wappaeus (1812–79) who became a full professor at Göttingen in 1854 (Dickinson, 1969). Ritter's work at the Military Academy was also of great significance, particularly through its influence on the German Field Marshal Count Helmuth von Moltke (1800–91) (Hartshorne, 1939), who was chief of the general staff in Berlin from 1858 to 1888. Moltke was responsible for the reorganization of the Prussian army, and his strategic skill was displayed to great success in the wars against Denmark (1863–64), Austria (1866) and France (1870–1871) (Thorne, 1961). Ritter's influence on military action and political decision making was thus following in a tradition clearly identified by Strabo. This can also be seen through his work as a member of the Royal Prussian Academy of Sciences, and through the key role he played in establishing the Gesellschaft für Erdkunde (Geographical Society) in Berlin, of which he became the first president in 1828 (Linke, 1981). With the establishment of such societies in other countries of Europe, and the formal introduction of geography as a university subject, the discipline had come of age.

## 4.3 *Institutionalized geography: societies and universities in the age of empire*

The first three decades of the 19th century witnessed a ferment of intellectual activity throughout Europe leading to the establishment of academic societies of many different kinds. Among these were the famous geographical societies of Berlin, Paris and London. However, the creation of university chairs of geography was, with a few notable exceptions, a later phenomenon, taking place after the 1870s. Both the societies and the universities played a fundamental part in shaping the institutional structure of geography (Capel, 1981), and the role that geographers were to play in society; geography was the discipline of exploration, and geographers frequently the servants of imperialism (Driver, 1991).

### *4.3.1 Germany*

At the beginning of the 19th century, according to Richthofen (1928: 18), Berlin 'was a small town with a petty outlook on life and narrow intellectual horizons', and it is perhaps surprising that by the 1820s it had become the centre of German geography. However, the lectures that Humboldt delivered in 1827 and 1828 'provided the impulse which led to the foundation of the Berlin Geographical Society in April 1828 (Lenz, 1978). Ritter, who had held the chair of geography in the city since 1820 was invited to become president of the society, and was its leading figure until his death. However, during this thirty year period the society did not gain popular recognition, and its lack of finance meant that it was unable to sponsor scientific expeditions (Lenz, 1978). It was only following Barth's explorations in Africa in the 1850s, and his subsequent efforts to raise financial support for further expeditions as president of the society from 1863 to 1865, that it really became the centre of German exploration.

Very soon afterwards, and influenced in part by the earlier expansion of geographical teaching at the primary and secondary level (Capel, 1981), academic geography received government approval as a university subject. Chairs of geography were established at Leipzig and Halle in 1871 and 1873, and then in 1874 the Prussian government decided to establish further such chairs at all of the state universities (Dickinson, 1969). This move brought two people who were to have a major influence on the future direction of German and European geography to the fore: Friedrich Ratzel (1844–1904), who obtained the chair at Munich in 1875, and Ferdinand von Richthofen (1833–1905) who became professor at Bonn in 1877. Ratzel came to dominate German human geography, and had close links with ethnographers, whereas von Richthofen had a background primarily in geology. Both men, however, played major roles in the expansion of German interests overseas. Thus von Richthofen had undertaken field research on the geology of China in the late 1860s, and on his return to Germany in 1872 he became an ardent advocate of the strategic importance of a German foothold in the country. Ratzel's work likewise provided some support for German expansionist policies, but in a different way, largely through the biological analogies he applied to political geography, and through his concept of *Lebensraum*, the geographical area within which living organisms develop. In particular, Ratzel believed that states have a natural tendency to expand, unless constrained by stronger neighbours, and this provided a firm basis for the development of German colonies overseas and also for the expansion of German interests in Europe.

Between 1870 and 1900 the Berlin Geographical Society became 'recognized as a focal point in the scientific and cultural life in the city of Berlin and far beyond the state borders' (Lenz, 1978: 222). It had sponsored polar expeditions, and it had played a central role in the expansion of German colonies and interests in Africa (Bader, 1978). Over the same period there had been a dramatic increase in the teaching of geography in German universities. Eleven chairs of geography were thus established between 1870 and 1880, and by 1914 the total number of such chairs in Germany had reached twenty-three (Elkins, 1989).

### 4.3.2 *France*

The earliest institutional development of geography in France preceded that in Germany, but the subject did not become firmly established as an academic discipline until much later. Thus, although the first chair of geography, albeit a joint one with history, at the Sorbonne had been established in 1809 (Broc, 1974), the second Paris chair, in colonial geography, was only founded in 1892 (Dickinson, 1969). The Paris Geographical Society created in 1821 was the earliest surviving such society in the world, but it was not until the 1870s that geographical interest really began to expand (Schneider, 1990). This can in part be explained through reference to the defeat of France in the war with Prussia (1870–71) which led people to look outside Europe for territorial expansion and colonial development (McKay, 1943; Freeman, 1961). However, once again partly through the influence of Pestalozzi, geography had been included in the primary school syllabus from 1857 as a means of developing children's powers of observation. It was therefore the need to provide qualified teachers to satisfy this demand that brought about many of the subsequent changes in geographical teaching at the universities.

Although during the third quarter of the 19th century geography was not well established as a university subject, the 1870s and 1880s saw a plethora of new French geographical societies being founded and journals published, inspired both by the need to provide a professional basis for geography, and also to make it more practical in its economic applications (Schneider, 1990). Most of the societies were concerned with colonial expansion, the development of commerce, and with the propagation of French civilization abroad (McKay, 1943). Among the most energetic advocates of the new geography in the period 1875–90 was Ludovic Drapeyron, whose efforts were central to the establishment of the Société de Topographie in 1876, and the launching of the *Revue de Géographie* in 1877 (Broc, 1974). However, it was not until the *Annales de Géographie* was founded in 1892 by Vidal de la Blache (1845–1918) and Dubois that an academic discipline of geography really emerged in France (Dickinson, 1969; Andrews, 1986). Vidal de la Blache began his academic career in the field of archaeology and ancient history, working for three years at the Ecole Française d'Athènes. He returned to France in 1870 and sought a teaching post in Paris without success. At the beginning of 1873, though, he began teaching history and geography at the University of Nancy, and was offered a chair in geography there two years later in 1875.

The expansion of geography in French secondary education at this time owed much to the reforms inaugurated by Jules Simon, Minister of Education from 1870 to 1873. These established a syllabus for 'teaching geography which proceeded from the concrete and familiar to the abstract and unfamiliar, rather than *vice versa* as was characteristic of the existing programme' (Andrews, 1986: 178), and they also outlined the basic texts and atlases to be used for teaching (Berdoulay, 1981). The resulting demand for more people qualified to teach geography at the secondary level necessitated an expansion of geography at university level, and Andrews (1986) has argued that it was this shift in institutional emphasis that was largely responsible for Vidal de la Blache's move away from ancient history to geography.

In 1877 Vidal de la Blache moved to the École Normale Supérieure in Paris, and then in 1898 to the Sorbonne. Two central characteristics of French geography at the end of the 19th century were well reflected in Vidal de la Blache's work. The first was that the subject was very closely linked with history. De Planhol (1972: 29) has thus argued that 'towards the end of the nineteenth century geography in the university became distinctively a branch of historical studies' and that since then 'geography has only slowly and partially broken away from its very close links with history'. Indeed, until 1942 *l'agrégation*, the main examination for anyone wishing to teach in universities and high schools, was a joint examination in geography and history (de Planhol, 1972; Bataillon, 1983), and elements of both subjects continue to be essential for anyone undertaking a degree in either geography or history. This close link with history resulted in part from the institutional development of the discipline, reflected in Vidal de la Blache's own origins as a historian, but it was also an outcome of the particular way in which French historians and geographers conceptualized the historical development of their country. At the end of the 19th century, many parts of rural France had been touched only indirectly by the dramatic industrial and urban changes that had occurred elsewhere, and had retained in their landscapes and societies a range of characteristics peculiar to themselves. It was the pursuit of this particular regional identity that provided the second central characteristic of French geography.

Buttimer (1971) stresses that the French geographical tradition owed much to the evolving conceptualization of society and milieu in 19th-century France, and in particular to the work of Frédéric Le Play (1806–62) and Émile Durkheim (1858–1917). Le Play's concerns with the relations between a society and its geographical milieu were in marked contrast to the positivists such as Comte and Saint-Simon, and had an important influence on geography through one of his disciples, Edmond Demolins (1852–1907). Although Demolins adhered strongly to the view that geographical conditions determined social systems, his ideas concerning the linkages between social groups and the environments in which they lived were important for the subsequent development of Vidal de la Blache's conceptualization of *genres de vie* (Dickinson, 1969). Moreover, Le Play's emphasis on an integrated social science, built around the three themes of place, work and family, was to have a lasting effect on the integration of French geography with other social sciences. The second major influence on the emerging discipline of geography in France, according to Buttimer (1971), was the work of Émile Durkheim, which she contrasts with the influence of Ratzel on German geography. Buttimer (1971: 30) thus envisages a fundamental distinction between their two approaches: 'the Ratzelian approach which studied world society in terms of spatial movements and ecological adaptation to nature; and the Durkheimian one which studied world society as an autonomous system possessing a "morphology" (formal patterns) and a "physiology" (life-styles, behavior) of its own'. While Ratzel was thus able to lay claim to world society as an object of study for his *Anthropo geographie*, Durkheim saw the society–milieu debate as lying firmly within social morphology.

### 4.3.3 Britain

The foundation of the Royal Geographical Society in London in 1830 came after those of Paris and Berlin, and likewise after a number of other British societies such as the Geological Society founded in 1807 and the Zoological Society created in 1826 (Mill, 1930; Cameron, 1980; Stoddart, 1986). The idea of establishing a geographical society was first proposed by members of the Raleigh Travellers' Club, which had been founded as a dining club in 1826 (Cameron, 1980). The chairman of the founding committee was Sir John Barrow (1764–1848), who had been Permanent Secretary to the Admiralty since 1803, and the other members were Robert Brown (1773–1858) Keeper of Botany to the British Museum, Lord Broughton (1786–1869) a politician, Sir Bartholome Frere (1778–1851) a diplomat, the Hon. Mountstuart Elphinstone (1779–1859) a diplomat who served much of his life in the East India Company, and Sir Roderick Impey Murchison (1792–1871) a soldier and later geologist (Gilbert and Goudie, 1971; Cameron, 1980). Admiral William Smyth, another member of the Raleigh Club, soon joined the committee, and under the patronage of King William IV, the society rapidly expanded. Most of the society's members were men of high social status, and this remained so throughout the 19th century during which time it also retained a considerable military emphasis. This was not necessarily to the academic benefit of the society, and Stoddart (1986: 61) has noted that 'Both of these components of the fellowship suggest a somewhat amateur, if not dilettante, approach to a subject not yet established in professional terms.'

The society nevertheless also included a number of noteworthy scientists, and it did much to sponsor scientific expeditions. Its initial aims were six in number: to collect and publish new facts and discoveries; to accumulate books on geography and maps; to procure instruments of use to travellers; to prepare instructions for travellers so that they might extend geographical knowledge; to correspond with similar geographical societies; and to communicate with societies of cognate disciplines (Cameron, 1980). It played an important role in sponsoring expeditions to the Canadian Arctic under the influence of Barrow, and to Africa, influenced largely by Murchison. Interpretations of the role of the society in African exploration vary greatly. Many of its founder members had also belonged to the African Association, established in 1788, and incorporated into the Royal Geographical Society in 1831. On the surface, there were three other reasons why Africa featured so prominently: in Cameron's (1980: 76) words, 'The continent was a paradise to the hunter, a challenge to the missionary and an affront to the opponents of slavery.' Underlying these motives, though, there was a deeper political reason, and one in which the activities of the soldiers and politicians among the fellowship of the society played a significant part.

From the 1850s, the opening up of Africa by explorers and missionaries paved the way for commercial exploitation and the subsequent partition of Africa between the colonial powers at the Congress of Berlin in 1884–85. The close links between geography and imperial policy were well exemplified in the work of Sir Bartle Frere, president of the Royal Geographical Society from 1873 to 1874, and an administrator and statesman serving in India and southern Africa. Thus Emery (1984: 345) has noted how 'Frere shared a widely-held contemporary view that the geographer's business was primarily that of widening knowledge of the earth by means of exploration.' Emery (1984: 346) goes on to note that 'Geographical exploration appealed to Frere not only because it opened the doors to fresh scientific investigation, but also because it reflected "the vital springs of active national life", extending British prestige throughout the world.' For Hudson (1977: 12) geography was promoted at this time mainly in order 'to serve the interests of imperialism in its various aspects including territorial acquisition, economic exploitation, militarism and the practice of class and race domination'. The reports of explorers were avidly read, giving rise to a particular view of Africa in the minds of the European public. These were not always, though, welcomed by the geographical establishment, as Driver (1991: 7) has so clearly traced in his analysis of the reactions to Stanley's expeditions in Africa.

The Royal Geographical Society also played an important part in the establishment of geographical education in British universities. Freeman (1980b: 4) has thus noted how 'In 1833 University College London, asked the Society to provide a small endowment for a chair in geography to which its secretary Captain Alexander Maconochie RN was appointed.' This venture proved to be remarkably unsuccessful. Few students were attracted, and when Maconochie resigned on his appointment as governor of Van Diemen's Land in 1836 no new appointment was made. At this time the only three universities in England were Oxford, Cambridge, and University College London, but these were soon followed by the establishment of King's College London in 1831, Durham in 1832, Bedford College for Women in 1849, and Owen College, Manchester, in 1851. Geography was not among the first disciplines to be taught at these universities and it was not until the 1880s that the discipline became firmly established as a university subject.

In the 1830s and 1840s there was a considerable overlap of interest between geographers and geologists, reflected in the organization of the British Association for the Advancement of Science, established in 1831. In 1834 geography and geology were grouped together as Section C of the association, and in 1841 the name was changed to 'Geology and Physical Geography' (Howarth, 1951; Beaver, 1982). In part this reflected the low esteem in which explorers and travellers were held by geologists, and as a result, in 1851 a new section, Section E was formed for 'Geography and Ethnology', with Murchison then president of the Royal Geographical Society also becoming president of Section E. In 1869, with the foundation of Section H for anthropology, Section E stood for geography alone. The growing separation and conflict between geologists and geographers in the 1850s and 1860s had important repercussions for the institutional development of the two disciplines. By the 1860s geology had become a professional discipline, well established in the universities, whereas geography remained somewhat vague and diffuse, still largely dominated by the social context of its founders (Stoddart, 1986). Nevertheless, during the 1870s geography became increasingly popular at the school level, and the Royal Geographical Society urged that professorships in geography should be introduced at Oxford and Cambridge. These pleas fell mainly on deaf ears, and the subject continued to suffer considerable criticism from geologists, who argued that the worthwhile parts of geography could largely be subsumed within geology. These differences were not only on academic grounds, though, and personal components also probably had a part to play in the confrontation. Stoddart (1986: 72) thus notes how 'The Survey geologists, especially in the 1870s, were generally fairly rugged outdoor types,' whereas 'The first academic geographers were, in general, the antithesis to this.'

In 1886 the Royal Geographical Society wrote to Oxford and Cambridge again, offering to provide financial support for the foundation of chairs or readerships in geography at the universities, and a year later they were rewarded. At the beginning of 1887 Halford Mackinder (1861–1947) had presented a major paper entitled 'On the scope and methods of geography' to the Royal Geographical Society, and by the end of the year he had been appointed reader at Oxford (Scargill, 1976). A year later, in 1888 Francis Guillemard was appointed to a lectureship at Cambridge, to be replaced six months later by John Young Buchanan, a fine scientist, who had established his reputation as a chemist and physicist on the *Challenger* voyage (1872–76), but of a somewhat retiring and cold disposition (Stoddart, 1975a). University geography in England thus got off the ground in two very different ways: at Oxford, Mackinder's emphasis on regional and political geography achieved considerable success, whereas at Cambridge Buchanan's lectures on the physical side of the discipline proved to be far from popular.

### 4.3.4 *Geography in the United States of America*

The establishment of geographical societies in other cities of the world did not go unnoticed by the merchants, publishers and philanthropists of New York, and in 1851 they established the American Geographical and Statistical Society of New York. In its early years the society was largely concerned with exploration and with the economic integration of the United States of America. Wright (1952) thus sees its four central concerns at this time as being the opening up of the western United States, the exploration and development of economic links with South America, the exploration of Africa inspired largely by reports of Livingstone's missionary activities, and exploration of the Arctic. In 1859, the year in which Humboldt and Ritter both died and also in which Darwin's *On the origin of species* first appeared, the first issue of the society's *Journal* was published. A particularly interesting feature of the society at this time was its emphasis on the gathering and publication of statistics on a wide range of different subjects from soils and agriculture to the postal services.

Two other prominent characteristics of the society were its concern with practical relevance, and the influence that German geographical scholarship had upon it. In his 1859 annual address, the vice president, the Rev. Dr Thompson, thus considered the value of geography to lie not only in the confirmation that it gave of the Bible, but also in the commercial gain that could be derived in the wake of geographical explorations. Arnold Guyot, since 1854 professor of geography and geology at Princeton, provided the main link with German geography, and in his commemoration of Ritter he emphasized the importance of geography's role in studying the organic unity of the human and physical worlds (Wright, 1952). Both Thompson's and Guyot's arguments were underlain by deeply teleological views, and reflected the considerable importance of Christian religion in shaping the thoughts of mid-19th century American geographers.

The American Geographical Society was not, though, the only influence on the development of geography in the United States during the 19th century. During the War of Independence, George Washington had recognized the need for a geographer-surveyor to provide maps and to describe the terrain of the areas of his campaigns, and in 1777 Congress resolved to empower him to appoint Robert Erskine to such a position. Subsequently the services of geographers were called upon widely in the military field. In 1818 a Department of Geography, History and Ethics was thus established at the United States Military Academy at West Point, and in the same year a Topographical Bureau was added to the Engineer Department of the army in Washington DC (Friis, 1981). Thereafter military surveyors provided numerous topographic accounts of the American West (Goetzmann, 1966), and eventually in 1879 the Congress authorized the establishment of the US Geological and Geographical Survey in the Department of the Interior. Subsequently the word 'Geographical' was deleted from its title, once again reflecting the substantial conflict between practitioners of the two disciplines, and indeed their relative strengths. However, under the leadership of John Wesley Powell the Survey continued to maintain a geographical emphasis, represented primarily by Henry Gannett, who was appointed as chief geographer of the US Geological Survey in 1879 and held that position until his death in 1914. Following the Civil War, during which he had an arm amputated, Powell was appointed to the professorship of geology at the Illinois Wesleyan University, Bloomington, and soon afterwards began his series of western expeditions, culminating in those to the Colorado River in 1869, and in 1871–72. In his report on the Colorado explorations Powell drew attention to three central theoretical issues: the principle of base level; the nature and potency of erosion; and the generic classification of landforms (Chorley, Dunn and Beckinsale, 1964). It was out of these ideas that much subsequent work by geologists and geomorphologists such as Gilbert and Davis was to develop. In summarizing Powell's achievements, Gilbert (1902: 638) has thus commented that

> With the novel ideas involved in the terms 'superimposed drainage' and 'antecedent drainage' were associated the broader idea that the physical history of a region might be read in part from a study of its drainage system in relation to its rock structure. Another broad idea, that since the degradation of the land is limited downward by the level of the standing water which received its drainage, the types of land sculpture throughout a drainage area are conditioned by this limit, was formulated by means of the word 'base level'. These two ideas, gradually developed by a younger generation of students, are the fundamental principles of a new subscience of geology sometimes called geomorphology, or physical geography.

Not only does Gilbert hereby assess Powell's achievements, but he also significantly emphasizes that by the turn of the 20th century physical geography was seen, at least in some people's minds, as being merely a subdiscipline of geology.

The institutional development of geography within universities in the United States was from its foundation closely allied with geology. While geography had been taught at Columbia by Gross from 1784 to 1795 and then by Kemp until 1812 (Dryer, 1924), the first chair of geography was that combined with geology held by Guyot at Princeton. Elsewhere during the 1870s and 1880s, as at Cornell (Dunbar, 1961), geography was usually included within departments of geology, with the emphasis being on the teaching of physical geography. William Morris Davis, the leading geographer of his day, was made professor of physical geography at Harvard in 1890, but this was still within the Department of Geology and Geography. It was not until 1898 that the first separate department of geography in a major university was created with the establishment of the Department of Geography in the College of Commerce at Berkeley within the University of California (Dunbar, 1961). Moreover, it was not until 1903 at the University of Chicago that the first Ph.D. in geography was offered in a United States university. Even then, under Rollin Salisbury, it was a department heavily influenced by physical geography, with its main emphasis being on the examination of environmental influences on life of all kinds (Pattison, 1961).

### 4.3.5 *Imperialism and the anarchist alternative*

Following the foundation of geographical societies in Paris, Berlin and London, others were rapidly established in different parts of the world: Mexico in 1833, Frankfurt in 1836, Brazil in 1838 and Russia in 1845. By 1869 there were twenty geographical societies world-wide, and by 1889 a further sixty-two had been established (Capel, 1981: 56). These represented the social formalization of the discipline, and tied it closely to the aims and enterprises of the capitalist powers in exploring and mapping parts of the world which, for them, were new. However, another geography also existed, an underground geography of critique, built upon the same earlier traditions of the discipline, but expressed in radically different practices. This was an anarchist geography, owing much to French and Russian social movements, epitomized by the efforts of the ill-fated Paris Commune of 1871. Its two best-known proponents were Elisée Reclus (1830–1905) (Giblin, 1979; Dunbar, 1981) and Pyotr Kropotkin (1842–1921) (Stoddart, 1975b; Breitbart, 1981; Alexandrovskaya, 1983; Potter, 1983).

Elisée Reclus was one of the most prolific geographical writers there has ever been, and is best known for his work on physical geography, *La terre* (1868–69) and for his nineteen volume masterpiece *Nouvelle géographie universelle* (1876–94). His father was a Protestant pastor in south-west France, and the young Reclus thus embarked on a theological education, including six months at the University of Berlin in 1851. Here he attended Ritter's lectures, and their influence was to have a marked impression on his later career. Although returning to France briefly, he opposed Napoleon III's *coup d'état* and for the next seven years attempted to find employment abroad in England, Ireland and North America. In 1857 he returned to France and eventually found a job writing travel guides for the publishers Hachette. During the next fourteen years he played an active part in the Paris Geographical Society, but he also joined the International Workingmen's Association and came under the influence of the anarchist Michael Bakunin. In 1868 Hachette had published the first volume of *La terre*, and this established his international reputation as a geographer. However, he was captured in the early days of the suppression of the Paris Commune of 1871, and then banished, choosing Switzerland as his place of refuge. Here he prepared the manuscript of his universal geography of the world, the *Nouvelle géographie universelle*, and also continued with his political activities, making a public statement of his anarchist beliefs in 1876. Although returning to Paris in 1890 he moved to Brussels in 1894, where he helped found the New University of Brussels, establishing a Geographical Institute there in 1898 (Giblin, 1979; Dunbar, 1981).

Prince Pyotr Alexeivich Kropotkin approached anarchism from a totally different background (Woodcock and Avakumović, 1950). His early career was spent as a military officer in Siberia, where he acutely observed both the physical environment and the social conditions of the population. After five years in Siberia he returned to Moscow, and during this period became increasingly conscious that it was not possible to study pure science devoid of any concern with social conditions. In 1872 he visited a small community of anarchist watchmakers in Switzerland, and thereafter became convinced of the need to adopt an anarchist version of socialism (Breitbart, 1981). He was arrested and imprisoned in Russia in 1874, but managed to escape to England in 1876. A year later he met Reclus in Switzerland, and thereafter the two men remained close friends and associates. Kropotkin, however, was again imprisoned, this time in France, between 1883 and 1886, and eventually thereafter settled in England, where he remained until returning to Russia after the revolution of 1917.

For both men their geographical work was part of their practice of anarchy. Reclus was less direct in his literary advocacy of anarchy, and his most anarchist work is his *L'Homme et la terre*, mainly published posthumously. In this he emphasized the importance of geography in determining the global distribution of resources and in enabling them to be used to the equal benefit of all people. His central argument was that communities of free labourers should determine decisions concerning production and consumption in order to ensure equitable distribution of the benefits (Dunbar, 1981). Kropotkin likewise emphasized the importance of communal action for the success of his anarchist vision. Influenced partly by the Social Darwinism of the late 19th century, he argued that the creation of a successful society depended on unity in diversity, what Breitbart (1981: 136) has described as 'a sense of mutual dependence on others for collective action, but also an opportunity to express individual difference'. These ideas were largely worked out in his *Mutual aid*, published in 1902, and designed to provide an overview of evolutionary history. Another central component of Kropotkin's geography was his view, influenced in large part by Humboldt, that the discipline could overcome the growing division between the human sciences and the natural sciences. Bowen (1981: 261) thus suggests that Kropotkin stood out as a notable exception to other practitioners of the discipline, by keeping alive 'Humboldt's model of the radical humanist geographer advocating synthesis, holistic views and social reform as part of a vigorous scientific empiricism' (Bowen, 1981: 261).

For the establishment geography of the day, both men were admired, but their anarchism was seen as being detrimental to their science. Reclus was honoured in 1894 by the receipt of a Gold Medal from the Royal Geographical Society in London for eminent services rendered to geography as the author of *Nouvelle géographie universelle* (Cameron, 1980), and Kropotkin, although honoured by the Royal Geographical Society at a banquet for his services to physical geography, declined election as a fellow of the society (Breitbart, 1981). It is indeed remarkable that an essentially highly conservative body of geographers should have held both men in such esteem. Kropotkin's scientific contributions to physical geography, in particular, were very highly thought of, and he maintained a close personal friendship with Keltie, Secretary of the Royal Geographical Society from 1892 to 1915. Nevertheless, even Keltie (1921: 319), in writing his obituary, expressed the view that 'This is not the place to deal in detail with Kropotkin's political actions, except to express regret that his absorption in these seriously diminished the services which otherwise he might have rendered to geography.'

Given the very close links that existed between the geographical establishment, represented by the societies and the newly founded university chairs, and the commercial and political representatives of the imperialist powers of Europe, it is not surprising that the views of Reclus and Kropotkin were subsequently neglected. However, Keltie's comment noted above, illustrates another factor which has had lasting effects on the practice of geography. Kropotkin's science was intimately connected with his social and political practice, but for Keltie and others, keen to establish a new basis for the subject, it became an essential element of their faith that science should be politically and socially neutral and value-free. Little did they realize that in adopting such a definition of science, albeit implicitly, the founders of much modern geography were eventually to divorce the discipline from one of its most fundamental concerns, a *critical* concern with the interactions between peoples and places. Although both men were outside the geographical establishments of their day, and had little immediate effect on the subsequent institutional development of the discipline, they retained alight a flame of critical social enquiry and practical action, which was elsewhere being extinguished by imperialist geography (Galois, 1976)

## 4.4 People, the environment and regional geography

The last third of the 19th century was a period when geographers were seeking to establish the formal basis upon which the newly institutionalized discipline would take its place in academia. They were therefore heavily influenced by the wider views within the scientific community about the nature of science, and also by the attitudes and opinions of practitioners of other disciplines with which they came into conflict. The publication of Charles Darwin's (1809–82) *On the origin of species* in 1859, was to have a profound influence on the long-established concerns of geographers with both the human and physical environments. In particular the wider acceptance of his ideas on evolution, led to the subsequent rejection of much of the geographical work derived from the teleological approach of scholars such as Ritter and Guyot. Armstrong (1985) thus sees one of the main effects of the Darwinian revolution on geography as being the breaking of its links with natural theology. Ironically, though, he also suggests that

> one of the consequences of Darwin's work was a lurch towards Neo-Lamarckism. Ideas on evolution – the notion of the gradual change of organisms, including man, through time – and the less radical concept of organisms being adapted to their environment, were apparently more easily assimilated than the central Darwinian idea of natural selection (Armstrong, 1985: 41; see also Campbell and Livingstone, 1983; Livingstone, 1984).

However, Darwin's ideas, allied with those of Spencer (Peet, 1985), also laid the basis for the development by geographers such as Ratzel of new ways of conceptualizing the relationship between people and the environment. Furthermore, when combined with Lyell's *Principles of Geology* (1830–33), Darwin's arguments provided a formidable challenge to those geographers eager to preserve the realm of the physical environment as their sole preserve. The result was that by the beginning of the 20th century geography had become firmly identified with two concepts: environmental determinism, and the region.

### *4.4.1 Darwin's influence on geography*

The influence of biological concepts on geographical thought in the 20th century has been of considerable importance. As Stoddart (1986: 159) has argued,

> Much of the geographical work of the past hundred years . . . has either explicitly or implicitly taken its inspiration from biology, and in particular from Darwin. Many of the original Darwinians, such as Hooker, Wallace, Huxley, Bates and Darwin himself, had been actively concerned with geographical exploration, and it was largely facts of geographical distribution in a spatial setting which provided Darwin with the germ of his theory.

However, geographers in the 19th century selected only a part of Darwin's corpus of work to incorporate into their newly emerging formal discipline. Stoddart (1986: 159) thus suggests that 'In geography . . . Darwinism was interpreted primarily as evolution,' in the sense of transformation over time.

In developing his arguments, though, Darwin (1888: 410) began with a discussion of the way in which the process of domestication was associated with 'much variability, caused, or at least excited, by changed conditions of life'. He noted that while 'Variability is not actually caused by man . . . man can and does select the variations given to him by nature' (Darwin, 1888: 410). Central to Darwin's ideas, evident in his sketch of 1842 and his essay of 1844, therefore, is the importance of initial random variability in nature (Darwin and Wallace, 1958). From here, he developed his argument on the 'Struggle for existence amongst all organic beings throughout the world' which he saw as being 'the doctrine of Malthus, applied to the whole animal and vegetable kingdoms' (Darwin, 1888: 3), and from which he eventually derived his concept of natural selection. Rather than being centrally concerned with evolution as transformation through time, Darwin's key considerations were thus to do with the origins of variability, and the process of selection. Interpretations of these themes can be seen to have had particular significance in three broad areas of subsequent geographical research and teaching: the relationships between people and the environment, the understanding of physical processes, and the use of the region as a subject for geographical enquiry.

### *4.4.2 'Man' and the environment*

Darwin's arguments had profound implications for the understanding of the relationship between people and the environments in which they lived. Like Humboldt and Ritter, Darwin included people 'in the living world of nature' (Stoddart, 1986: 167). In particular Darwin emphasized the interconnectivity of the elements of the living world, and provided the framework within which much subsequent ecological writing was to emerge (Stauffer, 1960; Vorzimmer, 1965). However, it was among biologists rather than geographers that these themes were explored to their fullest extent. Haeckel (1869), professor of zoology at Jena, was the first to coin the term ecology, and it was not until after the somewhat sterile debate concerning 'man and the environment' that some of the more relevant biological ideas pertaining to ecology were reintroduced into geographical enquiry (Barrows, 1923).

At the end of the 19th century and the beginning of the 20th century geographical concern with the relationship between people and the environment is widely seen as having been focused around two main themes: environmental determinism, and possibilism (Tatham, 1951). Johnston (1987: 36) thus sees these as two competing approaches which represented 'the first attempts at generalization by geographers of the modern period'. However, in the past the boundaries between these two positions have often been too starkly drawn. It was the German geographer Friedrich Ratzel (1844–1904) who has become most widely thought of as the founder of environmental determinism (Wanklyn, 1961; Buttmann, 1977), but although his *Anthropo-Geographie* (1882, 1891) did espouse the view that human activity on the earth was largely determined by the nature of the physical environment, it was primarily through the much more extreme arguments of American geographers such as Semple and Huntington that these ideas reached a wider audience in the English-speaking world.

The direct influence of Darwin on Ratzel has been much debated, but Bassin (1987a) has argued convincingly that Ratzel was primarily influenced by Moritz Wagner's elaborations of Darwin's ideas. In particular Wagner (1868) had argued that Darwin's schema lacked a spatial dimension, and it was largely as a result of his contacts with Wagner that Ratzel developed his ideas concerning migration, the territorial bases of species, and *Lebensraum*. These were expounded in the first volume of his *Anthropo-Geographie* (1882) in which he argued that the three main tasks of the new discipline were: '1) to describe the regions of the ecumene, and the distribution of humankind over it; 2) to study human migratory movements of all types in, as he put it "their dependency on the land"; and 3) to analyse the effects of the natural environment on the human body and spirit, both on individuals and on entire social groups' (Bassin, 1987a: 126). Ratzel's own arguments concerning the question of environmental determinism were, however, somewhat ambiguous and abstract, and it was therefore in the work of subsequent writers that the most direct causative linkages were claimed for the influence of the environment on people. These, in turn, need to be interpreted against the wave of intellectual support for the increasingly materialist and positivist views of the place of people within nature pertaining in the late 19th century. Here, the ideas of Herbert Spencer (1864, 1882) were also of importance in influencing the application of 'the scientific principles of organismic evolution conceived by Lamarck and Darwin to the development of the "social organism"' (Peet, 1985: 313).

The views that came to formulate the shape of geography over the first few decades of the 20th century were thus part of a much wider transformation in European and North American society, in which, in Livingstone's (1984: 22) words, 'the rationale of social legitimation was transposed from one theodicy, based on natural theology, to another, founded on the new laws of nature'. Such views also had important political significance, and this was a further factor behind the publication of geographical works claiming to indicate the superiority of white European and North American races over those people living in their African and Asian colonies, as well as over the indigenous inhabitants of the Americas.

In the United States of America, William Morris Davis (1906) was among the most prominent geographers to assert that the content of geography lay in the relationships between a controlling physical environment and human responses. He thus argued that 'any statement is of geographical quality if it contains a reasonable relation between some inorganic element of the earth on which we live, acting as a control, and some fact concerning the existence or growth or behaviour or distribution of the earth's organic inhabitants, serving as a response' (Davis, 1906: 71). However, although Davis promoted such ideas at a theoretical level, he did little to illustrate their connection in practice (Hartshorne, 1939). It was instead one of Ratzel's students, Ellen Churchill Semple (1911), who gave the most explicit expression to environmental determinism in her influential book *The influences of geographic environment*, subtitled *On the basis of Ratzel's system of anthropogeography*. It was largely through this work that the English speaking world was introduced to Ratzel's ideas (Wright, 1966; James, Bladen and Karan, 1983). Although Semple sought to rid Ratzel's work of what she saw as the detrimental influence of Spencer, particularly with respect to the organic theory of society and state which was subsequently to reappear as one of the claimed scientific underpinnings of Adolf Hitler's *Mein Kampf* (Peet, 1985), she was unable to escape from the overwhelming acceptance of Spencer's ideas, themselves heavily influenced by Lamarck, concerning the environmental determinants of human society (Campbell and Livingstone, 1983). The *influence of geographic environment* thus begins as follows:

> Man is a product of the earth's surface. This means not merely that he is a child of the earth, dust of her dust, but that the earth has mothered him, fed him, set him tasks, directed his thoughts, confronted him with difficulties that have strengthened his body and sharpened his wits, given him his problems of navigation or irrigation, and at the same time whispered hints for their solution (Semple, 1911: 1).

For Semple (1911: 2) the natural environment, which provides the physical basis of history, was essentially immutable, underlying all human activity, in contrast to 'shifting, plastic, progressive, retrogressive man'. Human temperament, culture, religion, economic practices and social life, could all be derived from environmental influences.

Thus Semple (1911: 620) argued that 'The northern peoples of Europe are energetic, provident, serious, thoughtful rather than emotional, cautious rather than impulsive. The southerners of the subtropical Mediterranean basin are easy-going, improvident except under pressing necessity, gay, emotional, imaginative, all qualities which among the negroes of the equatorial belt degenerate into grave racial faults.' Such deterministic views, often in an even more unfettered form, continued to be advocated well into the 1930s and 1940s by geographers such as Huntington (1925; 1945) and Taylor (1937). Paradoxically, both Semple and Huntington occasionally sought to temper their apparent espousal of environmental determinism (Lewthwaite, 1966), with Semple (1911: vii) thus having claimed that she 'shuns the word geographic determinant, and speaks with extreme caution of geographic control' and Huntington and Cushing (1934: 22) having noted that 'the geographical surroundings are only one of the great factors which determine the progress of a nation'. However, as Stoddart (1986: 171) notes, 'The questions which these determinists raised were posed in so gross a manner that they could only invite the grossest answers; most geographers realized this, and neither Taylor nor Huntington gained full academic acceptance.' The fundamental problem with environmental determinism was that its proponents derived their arguments from the existence of different human characteristics in different environments, but they generally failed to explore the processes whereby these distributions originated or were reinforced.

Although Semple's ideas were highly influential in North American geography during the early part of the 20th century (Hartshorne, 1939), they were not the only arguments to be voiced concerning the relationship between people and the environment. Another trend can be seen represented in the work of Marsh and Shaler (Olwig, 1980; Livingstone, 1980, 1987). George Perkins Marsh (1801–82) was 63 when he published his highly influential book *Man and nature* in 1864, just five years after Darwin's *On the origin of species*. In this, he emphasized the importance of people as active agents reacting to and changing the environments in which they live. This was a very different emphasis from that of the environmental determinists, and Olwig (1980: 45) argues that Marsh's work thus provided a core problematic for the subsequent conservation movement, 'in which man is not regarded as being to some degree a part of nature or as standing apart from it'. These ideas were taken up and put into practice by Nathaniel Southgate Shaler (1841–1906) (Livingstone, 1980, 1984, 1987). Shaler had been taught by the Swiss naturalist Louis Aggasiz, who like Guyot had left a chair in Neuchâtel to work in the United States of America (Lurie, 1960). Both Guyot and Aggasiz played an important part in introducing the ideas of Humboldt and Ritter to North America, but both also blended them with deep teleological convictions concerning a world ordered by a divine intelligence. Livingstone (1980: 370) thus suggests that Aggasiz's influence on Shaler 'prevented him from wholeheartedly embracing the evolutionary mechanism of Darwinian natural selection as a universal explanatory scheme'. As a result Shaler seems to have continuously struggled with the task of combining on the one hand a religious, aesthetic and preservationist view of the environment with, on the other, a scientific, technical and utilitarian conservationism.

While environmental determinism in its various forms dominated much geographical thinking during the first thirty years of the 20th century, the debate within Europe concerning the influence of the environment on human activity took on a different complexion, avoiding many of the excesses of Semple's interpretations of Ratzel. The most widely discussed, but in the end not particularly influential, alternative was that of possibilism proposed by the French historian Lucien Febvre (1922, 1925). The close contacts that existed between French historians and geographers enabled Febvre to build on the ideas of geographers such as Vidal de la Blache and Jean Brunhes, who had recognized that although natural features did influence human activity they could not be regarded as determining them. Brunhes (1925: 55–6) thus argued that 'without physical geography there could be no substantial human geography', and he suggested that the subject matter of research in human geography was formed by the 'bonds of interdependence and repercussion' by which 'The economic, social and political facts of population and material civilization are united with those of physical nature.' Febvre (1925: 236) channelled such ideas into his famous statement that 'There are no necessities but everywhere possibilities; and man as master of possibilities is judge of their use.'

### 4.4.3 *The divisions between physical and human geography*

During the latter years of the 19th century there was a clear overlap of interests between geographers and geologists in both Britain and the United States (James, 1967; Gregory, K. 1985). Indeed, in many instances this developed into outright antagonism. Keltie (1886: 31) referring to the situation in Britain, thus noted that 'the strongest opposition to geography as a distinct field of research comes from the side of geology: the pure geologist, especially, is unwilling to admit that geography has any existence apart from geology'. As Stoddart (1986: 69) has observed, 'The problem was compounded by a particular difficulty which beset the geographers. Geography itself appeared vague and diffuse, part belonging to history, part to commerce, part to geology.'

In the United States, Davis (1915) had argued powerfully that the unifying feature of geography lay in the causal links between inorganic and organic elements of the globe, and thus 'on the scientific dogma of evolution' (Herbst, 1961: 540–1). As Leighly (1955) has pointed out, Davis's work before the mid-1890s betrayed no evidence of this concern with the human implications of the physical environment, and even in 1900 he was able to exclaim that

> it is not altogether clear why geographers are so generally content to leave to geologists all treatment of matters so eminently physiographic as the weathering of rocks, the wasting of soils, the transportation of land waste by streams, the abrasion of land margins by

> waves. If these activities had occurred only in the remote past, geologists alone might lay claim to them; but, as a matter of fact, they are all part of the very living present (Davis, 1900: 6).

Nevertheless, Davis's subsequent insistence on the intellectual function of geography as resting 'on the presumption of an unbroken chain of causation linking the physical phenomena of the earth's surface, the organic realm, and human society' (Leighly, 1955: 312) was to have important repercussions for the future of geography in the United States. Even as Davis was propounding his Darwinian ideas, pragmatist social scientists at Harvard were arguing that the intellectual heritage of Herbert Spencer and the social Darwinists was inappropriate (Wiener, 1949). Davis's mechanical, monistic interpretation was thus 'rendered obsolete by the recognition that in human societies processes of an order different from the mechanical order were at work, processes that cannot be comprehended even by the categories applicable to organic phenomena' (Leighly, 1955: 313). The subsequent disintegration of Davis's unified view of the discipline in the United States, led the majority of geographers to seek to restrict the discipline purely to its cultural side. Those who did focus attention on its physical aspects, concentrated either purely on the inorganic world or drew attention to its role as a broader natural science including plants and animals (Herbst, 1961). None of these alternatives provided a sound basis for the successful institutionalization of the discipline. As Herbst (1961: 541) has noted, where the natural science geographers 'suffered from the dubious reputation of being interlopers and second-rate performers in the fields of geology, meteorology, geophysics, and plant and animal ecology the human geographers were soon derided as pseudo-sociologists, pseudo-political scientists, economists and historians'.

The stark division between the physical and human sides of geography was made even clearer in Harlan H. Barrows' presidential address to the Association of American Geographers in 1922, in which he advocated that geography was the science of human ecology with its aim being 'to make clear the relationships existing between natural environments and the distribution and activities of man' (Barrows, 1923: 3). As such he advocated a complete split from any vestiges of physical geography: 'In short, geography treated as human ecology will not cling to the peripheral specialisms to which reference has been made – to physiography, climatology, plant ecology, and animal ecology – but will relinquish them gladly to geology, meteorology, botany, and zoology, or to careers as independent sciences (Barrows, 1923: 4).

Such views caused considerable tension within the geographical community. While Davis had advocated the need for geography to study the causal links between the physical world and human activity, his own research was almost exclusively within the field of geomorphology. At the end of the 19th century most research in the United States on the surface features of the earth, in the wake of Powell and Gilbert, was undertaken within geology departments. However, Davis stood out within this field as being avowedly a physical geographer, and it was as a teacher of physical geography and meteorology, rather than geology, that he was first appointed at Harvard in 1878, before being promoted to an assistant professor of physical geography there in 1885 (Chorley, Beckinsale and Dunn, 1973). Davis dominated his field like few other people before or after him (Gregory K J, 1985). As Chorley, Dunn and Beckinsale (1964: 621) note, he 'was to become for many the embodiment of all past excellence in geomorphology' and his 'work formed the mainspring of half a century of research'; 'he to a great extent provided the backcloth against which the development of geomorphology between the 1880s and 1930s was played out' (Chorley, Beckinsale and Dunn, 1973: 6). According to Bowman (1934) Davis saw his own main contribution to geomorphology as being through the systematization of a sequence of forms through the concept of an ideal cycle and through the formalization of terminology. It was thus through his idea of the geographical cycle of erosion, comprising stages of youth, maturity and old age, with all landscapes being dependent upon the variables of structure, process and time (Davis, 1884a, b, 1889a, b, c) that Davis is best known. Above all, this was a model, which Davis (1905: 152) himself conceded, was 'not meant to include any actual examples at all', and it is largely because of its character as a model that it was so widely, and uncritically, accepted (Chorley, Beckinsale and Dunn, 1973). Despite criticisim, particularly from Germany by Hettner (1921), Albrecht Penck (1919) and his son Walther Penck (1924), Davis's methods provided the core focus of physical geography in the United States and Britain (Wooldridge and Morgan, 1937) until the 1940s (Beckinsale and Chorley, 1991), with staunch supporters such as Wooldridge (1955) continuing to advocate them well afterwards.

By the 1920s and 1930s the divisions within geography in the English-speaking world were giving rise to concern, and Herbst (1961) has seen this as being one of the main reasons for geography's low academic esteem at this time. The year after Barrows had apparently abandoned physical geography, however, Carl Sauer (1924) offered a very different alternative, in which physical geography was seen as playing an important role in providing the background to human activities. Sauer had studied for his Ph.D. under Salisbury at Chicago, and thus reflected a somewhat different tradition from that of Davis, of whom he was often critical (Chorley, Beckinsale and Dunn, 1973: 428). However, unlike those American geographers who had largely abandoned physical geography, Sauer 'never permitted his feet, or those of his students, to lose contact with the sustaining surface of the earth' (Leighly, 1963: 2). In 1925 Sauer published his famous paper on the morphology of landscape, which not only marked the death knell of environmental determinism, but also provided a schema for much subsequent geographical writing in the United State in the 1930s (Leighly, 1955). In this paper, Sauer (1925: 21) argued that 'area or landscape is the field of geography' and that 'Geography assumes the responsibility for the study of areas because there exists a common curiosity about that subject.' Moreover, he went on to argue for geography's status as a science in the following words:

> We assert the place for a science that finds its entire field in the landscape on the basis of the significant reality of chorologic relation. The phenomena that make up an area are not simply assorted but are associated, or interdependent. To discover this areal 'connection of the phenomena and their order' is a scientific task, according to our position the only one to which geography should devote its energies (Sauer, 1925: 22).

For Sauer, the geographical concept of landscape was equivalent to the historian's period; the geographer's facts are place facts, whereas those of historians are time facts. Moreover, he situated his study of geography firmly within a European tradition, quoting from Vidal de la Blache (1922) and Krebs (1923) to argue that 'geography is based on the reality of the union of physical and cultural elements in the landscape' (Sauer, 1925: 29). By providing a formal scheme by which geographical writing could be undertaken, Sauer's 1925 paper also, though, gave rise to a series of regional monographs, many of which have subsequently been decried. In Leighly's (1963: 6) words, 'The positive effect of this paper was, unfortunately, to stimulate a spate of detailed descriptions of small areas written in the ensuing twenty years, which had little value, either scholarly or practical.' Sauer himself recognized this, and in his later life repudiated the setting of such narrow limits to geographical enquiry (Sauer, 1941; 1956), arguing that

> The geographic bent rests on seeing and thinking about what is in the landscape, what has technically been called the content of the earth's surface. By this we do not limit ourselves to what is visually conspicuous, but we do try to register both detail and composition of the scene, finding in it questions, confirmations, items or elements that are new and such as are missing (Sauer, 1956: 289).

### *4.4.4 The region as an object of geographical synthesis and enquiry*

Sauer's attempt to bring physical and human geography together through studying the landscape, reintegrated geography in the United States with the wider tradition of regional geography established in Europe since the latter years of the 19th century. However, the interpretation of what was meant by regional geography varied markedly both between countries and over time. Herbst (1961: 543), in particular, notes an important difference between North American and European regional geography:

> Unfortunately . . . regional geography in America was seen primarily as a descriptive rather than a systematic study – a development which sprang in great part from the then fashionable rejection of systematic natural geography. The geographers of continental Europe, by con-

trast, did not share in this rejection of systematic natural geography, as also they had not shared the uncritical acceptance of social Darwinism. As a group they thus never accepted the antithesis of regional and systematic geography. This explains why in the European universities geography always had its acknowledged place.

The foundations for regional geography in Germany were laid by von Richthofen in the latter part of the 19th century, and closely built on the methodological framework established by Humboldt. Von Richthofen's essential argument was that 'geography studies the differences of phenomena causally related in different parts of the earth's surface' (Hartshorne, 1939: 92). Moreover, he also suggested that 'The actual purpose of systematic geography is to lead to an understanding of the causal relations of phenomena in areas . . . , an understanding which may be expressed in principles that can be applied in the interpretation of individual regions, *i.e.*, chorology' (Hartshorne, 1939: 92). Von Richthofen's view of geography was later extended and developed by Alfred Hettner (1895, 1903, 1927), (Schultz, 1980). In his early work Hettner had been closely influenced by the ideas of environmental determinism (Beck, 1982), but he gradually moved away from this position to argue later (Hettner, 1927) that the main role of geography was to bridge the gap that had increasingly developed between the natural and human sciences. For Hettner, geography's unity came from its chorological approach, central to which was 'the concept of the total causal relationship of an assemblage of phenomena at a certain place on the earth's surface, which causes each place to be considered as a whole and stamps it with individuality' (Elkins, 1989: 23). As founding editor of *Geographische Zeitschrift*, and through his many publications Hettner came to dominate German geography in the first thirty years of the 20th century. Although he vigorously encouraged the development of regional studies in Germany, Hettner also promoted systematic work, and saw geography as a combination of both. In Hartshorne's (1939: 94) words,

> Hettner introduced a somewhat unusual terminology designed to emphasize that there was no sharp separation between them. In a regional study of any extensive area it is necessary to study the notable variations in the individual geographic features systematically. On the other hand, the systematic study of a particular category of geographic features is not made with reference solely to that category, but rather in terms of its chorological relations to one or more other features.

In practice, von Richthofen's and Hettner's ideas found their expression in a number of regional studies, typical of which was Gradmann's (1931) work on southern Germany. These were based on the analysis of six basic elements (land, water, air, plants, animals and people), and the schema by which they were studied became known as the *länderkundliche Schema* (Hettner, 1932). Hettner's views did not, though, go unchallenged, and within Germany their main critic was Otto Schlüter, who felt that neither the regional concept nor the study of 'man–nature' relationships provided a sound basis for the delimitation of geography as a distinctive field of study (Elkins, 1989). Instead, he argued that geographical research should focus on the visible landscape, and that it should exclude all non-material aspects of human activity (Schlüter, 1906). For Schlüter the morphological analysis of the cultural landscape formed a central element in geographical enquiry, and this view was to have lasting influence on much German geography into the 1950s.

The French tradition of regional geography is intimately associated with the work of Vidal de la Blache. Unlike the geographers of the United States, the majority of whom were steeped in a geological background, Vidal de la Blache's academic training was in classics and history, and he thus approached the regional concept essentially from its human and cultural dimensions. Given the strength of French sociology, it was against sociologists rather than against geologists that he had to justify the existence of an independent geography, and he did this by arguing that it was the scientific study of places (Vidal de la Blache, 1913). Instead of treating nature as 'the passive stage for the drama of human life' (Buttimer, 1971: 45) as he accused sociologists and historians of doing, Vidal de la Blache (1922) incorporated nature as a dynamic element in his human geography. Central to his endeavour was the development of the three key concepts of *milieu*, *genre de vie* and *circulation*. The *milieu* was the basic differentiation of the earth's surface which tended to even out cultural variations in any one place; *genres de vie* were the life-styles of a particular region, reflecting the economic, social, ideological and psychological identities that were imprinted on landscapes; and *circulation* was the disruptive process by which human contact and progress took place between regions (Buttimer, 1971). For Vidal de la Blache the core focus of geography was thus the region, in which cultural and natural phenomena could be studied together, with each region being seen as a unique expression of the interaction between humanity and the physical environment. Such ideas, closely similar to those of Ritter (Dickinson, 1969), formed the basis for numerous detailed studies of different parts of France, such as Demangeon's (1905) monograph on Picardy, and Cion's (1908) study of eastern Normandy, and the regional monograph formed the basic model for French geography until well into the 1930s.

British regional geography reflects yet another series of influences, but at its heart was the same desire to identify a subject matter that could be seen as uniquely geographical, combining both human and physical elements. As Johnston (1984) and Freeman (1961) have pointed out, the British contribution to regional geography largely involved work at two scales: the one global (Unstead and Taylor, 1910), and the other local. At the end of the 19th century, Mackinder (1895) had argued that advances in physical geography had outstripped those in human geography, but that it was only possible to examine human geography in the context of geomorphology and biogeography. This he suggested could best be undertaken using a regional, rather than a systematic approach, arguing that 'the treatment by regions is a more thorough test of the logic of the geographical argument than is the treatment by types of phenomena' (Mackinder, 1895: 371). Added encouragement for the development of regional geography came in a circuitous route from France. One of Le Play's most energetic followers was the Scottish biologist, Patrik Geddes (1854–1932) (Beaver, 1962), who played a central role in the establishment of the Le Play Society, founded in 1930 to advance the cause of field studies. Geddes was a keen advocate of the method of regional survey, and had a strong influence on Herbertson (1865–1915), who began his career as an assistant to Geddes at Dundee. Having moved to Oxford in 1899, Herbertson published his paper on the major natural regions of the world in 1905. In part this appears to have been based on earlier work on climatic classification and on geological structure by Köppen and Suess, and he was acutely conscious of the need to avoid including political divisions in his essentially physical division of the world. Geddes also had an influence on H. J. Fleure (1919), but, in contrast to Herbertson, Fleure incorporated people into his system of regionalization. This consisted of global regions of increment, difficulty, privation and debilitation, each of which reflected a close interrelationship between human activity and physical constraint. At the other end of the scale were attempts to concentrate on much smaller regions. Fawcett (1917), for example, developed a scheme for the division of England into natural regions, again reflecting many of Geddes' ideas, and in 1933 Unstead proposed a formal classification of regions. This consisted of a nested hierarchy of different types of regions, beginning with stows consisting of features, and then moving up through tracts and subregions to major regions (Unstead, 1933).

In the United States the 1930s saw a plethora of studies of the region apart from the growing body of literature emanating from Sauer's work at Berkeley. In particular, James (1934) and Hall (1935) saw the use of regions as a way in which geography could claim recognition as a science. James (1934) thus envisaged the place of people within nature as being central to regional geography, with human occupance leading to the development of specific cultural landscapes. The science of geography then studied different levels of generalization, from the topographic to the chorographic and eventually world scale geographies. Hall (1935: 122) argued that the 'Major contribution of geography to the general field of science is the recognition, first of the ever varying aspect of the land, and secondly, that in spite of this variation, the land tends to be divided up into areas of more or less similarity' called regions. Moreover, he envisaged regions as serving four main functions: as a basis for taxonomy and classification; as a foundation for environmental and ecological studies; as an organizational unit for the advancement of human welfare; and as a device to replace partial correlation.

The methodological and theoretical debate over regional geography was, however, transformed with the publication in 1939 of Hartshorne's seminal monograph, *The nature of geography: a critical survey of current thought in the light of the past*. This work grew out of Hartshorne's frustration with what he saw as the lack of 'comprehension, accord or agreement among American geographers concerning the nature of their field' (Martin, 1989: 69), and it has become a classic of geographical writing. Its central concern was to examine geography's position among the sciences. In Entrikin's (1989: 1) words, 'Through his analysis of the German methodological literature and in his own arguments, Hartshorne sought to resolve a fundamental tension in the science of geography between its spatial perspective, that "sees together" the heterogeneous phenomena that constitute place and region, and the logical requirements of scientific concept formation.' In so doing, Hartshorne (1939: 368) argued that 'Since geography, in particular, must examine phenomena in the actual complexes in which they are found, it is impossible for it, in practice, to separate natural and human phenomena.' Moreover, he was strongly against the view that geography was a bridge between the natural and social sciences, and instead argued that it should be thought of 'as a continuous field intersecting all the systematic sciences concerned with the world' (Hartshorne, 1939: 369) (Fig. 4.1). In summary, Hartshorne (1939: 373) suggested, in words reminiscent of Kant's, that

> geography, like history, is to be distinguished from other branches of science not in terms of objects or phenomena studied, but rather in terms of fundamental functions. If the fundamental functions of the systematic sciences can be described as the analysis and synthesis of particular kinds of phenomena, that of the chorological and historical sciences might be described as the analysis and synthesis of the actual integration of phenomena in sections of space and time.

For Hartshorne, geography was thus the analysis and synthesis of phenomena in space: geography's unique role was to study 'the world, seeking to describe, and to interpret, the differences among its different parts, as seen at any one time, commonly the present time' (Hartshorne, 1939: 460).

Like Sauer before him, Hartshorne developed many of his ideas from his contacts with German geographers and from their writings, particularly those of Hettner. However, his conclusions concerning the place of geography were markedly different, and this led to the emergence of two very distinct traditions of American geography. Hartshorne's views were set firmly within the geographical context of the Midwestern and East Coast departments of Chicago, Wisconsin, Clark and Michigan, with its emphasis on empirical experience in contemporary regional analysis (Butzer, 1989). In contrast, Sauer's emphasis was much more on the historical role of people in changing the landscape. In his 1940 presidential address to the Association of American Geographers, Sauer (1941) thus criticized Hartshorne severely for relegating historical understanding to the outer fringes of the discipline. Both Hartshorne and Sauer were concerned with geography as the study of places or regions, but whereas Sauer (1941) included a concern with aesthetics and the subjective understanding of place within his interpretation of ge-

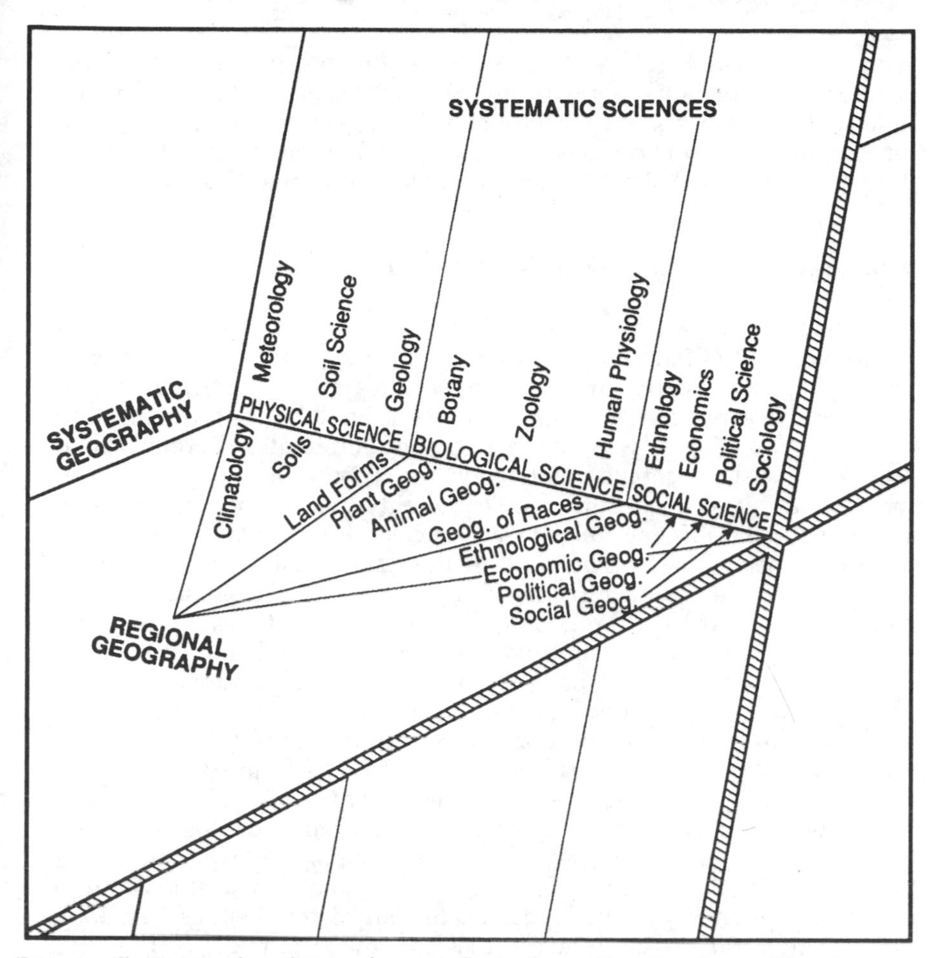


Diagram illustrating the relation of geography to the systematic sciences. The planes are not to be considered literally as plane surfaces, but as representing two opposing points of view in studying reality. The view of reality in terms of areal differentiation of the earth surface is intersected at every point by the view in which reality is considered in terms of phenomena classified by kind. The different systematic sciences that study different phenomena found within the earth surface are intersected by the corresponding branches of systematic geography. The *integration* of all the branches of systematic geography, focussed on a particular place in the earth surface, is regional geography.

**Fig. 4.1** Hartshorne's conceptualization of the relation of geography to the systematic sciences.
*Source*: Hartshorne (1939: 147)

ography, Hartshorne rejected it as being insufficiently objective and thus outside the bounds of 'science' (Entrikin, 1989).

Despite the criticisms of Sauer (1941), Hartshorne's monograph provided the standard, widely accepted view of the discipline until the 1950s. In part this was because it brought together much of what was already accepted as current geographical practice in the Midwestern universities, in part because of the intervention of the Second World

War, and in part because of its sheer length, which meant that few geographers were able to grasp the full complexities of his arguments. Indeed, *The nature of geography* has been used to symbolize the argument for geography as a chorological science rather more than it has been used as a work for critical re-examination and interpretation.

## 4.5 Geography in an institutional context

Ever since Varenius' formalization of geography as an intellectual discipline in the 17th century three central issues have beset geographers: the balance between geography as a regional (chorographic) and systematic discipline; its position as a science; and its conceptualization of the relationships between people and environments. There has been no one generally accepted solution to these issues, and in the light of the current vogue for postmodernist plurality this can perhaps be seen as being beneficial. However, the lack of clear direction and focus to the discipline meant that, particularly in the United States and Britain, it had difficulty in establishing itself as a 'true science' in the face of considerable criticism from geologists, biologists and sociologists. Geography departments were thus relatively late in being created in universities, and even then they were frequently initially merely adjuncts to geology departments. Geography's central role in the colonial and imperial expansion of European power, by providing knowledge about different parts of the world, was of great importance to the success of European expansionism. This was reflected in the establishment of numerous geographical societies during the 19th century. However, its image as being the discipline of exploration and descriptive accounts of travellers' tales, meant that it had little claim to scientific reputability in the great expansion of universities that took place at the end of the 19th century and the beginning of the 20th century.

In seeking to establish a firm basis for the existence of the discipline, geographers thus focused on two main issues: the interactions between peoples and environments, and the concept of the region. Although environmental determinism held sway as a dominant focus of opinion in the early 20th century, particularly in North America, it was by no means the only conceptualization of the relationships between people and places. Building on the work of Humboldt and Ritter, geographers in Germany and France, such as von Richthofen and Vidal de la Blache, were concerned with an understanding of the factors giving rise to regional variations of phenomena on the earth's surface, and thus with a return to chorography and the general geography of Varenius. This focus on the region served four important functions: it provided a framework within which the interaction between people and environments could be studied, without imputing any particular directionality of causal links; it served as an excellent teaching aid, by which knowledge of different parts of the world could be propagated; it provided geography with a sound classificatory tool, and thus with a hallmark of science; and it gave geographers a clear object of study, equivalent to the botanist's plant and the geologist's rock.

However, despite its apparent success, regional geography was not without its problems. Most notably, it had failed to resolve the position of systematic geographical studies, particularly the role of physical geography, and its emphasis on description set it apart from most of the other sciences, which were increasingly becoming concerned with the creation and testing of laws.

CHAPTER 5

# From region to process: the emergence of geography as an empirical–analytic science

> It is time for geography to mature as a predictive science and therefore it is so maturing. This period sees geography emerging as the science of locations, seeking to predict locations where before there was contentment with simply describing and classifying them. Geography has found it's [*sic*] central problem, has developed theory unique unto itself, and in the process has come to rely heavily on mathematics, especially geometry. It is fully endowed with its quota of martyrs, controversies, and eccentricities which seem to accompany periods of rapid intellectual growth. All the symptoms, side effects and glories of a scientific 'breakthrough' lay on the subject.
>
> Bunge (1966: xvi–xvii)

Regional geography, in all of its various guises, formed the basis for most geographical teaching and research in the period between the decline of environmental determinism and the late 1940s. However, by the 1950s growing disquiet within the discipline with geography's low reputation as a science prompted a series of substantial critiques of the regional concept. These focused primarily on the balance between regional and systematic studies, on whether or not geography was concerned with uniqueness or generality, and with the differences between description and explanation. Out of these debates there emerged a new and revitalized discipline, which found its unity in a particular type of methodology rather than with a specific kind of subject matter. The so-called 'quantitative revolution' thus sought to replace the traditional description of regional geography with an explanatory process-oriented science based on the testing of theories and the construction of laws. The adoption of many of the tenets of logical positivism in this process was, however, largely an unconscious process. It was the vision of creating geography as a real science that drove its practitioners forward; that this science was built upon the foundations of logical positivism was scarcely recognized.

## 5.1 *The demise of regional geography*

One of the difficulties of regional geography was that there was little agreement on the ways in which it should be written. French regional monographs were thus different from British ones, and in Germany the debate between the supporters of Hettner and Schlüter continued unabated. The years between 1939 and 1945, however, marked a turning point in the practice of geography. The war that spanned the globe from Europe to eastern Asia provided a rare opportunity for geographers once again to satisfy the task that Strabo had allocated them almost two millennia previously, that of providing military intelligence.

### 5.1.1 *Geographers at war*

In her survey of geography in the immediate post-war period, Taylor (1948: 137) described the activities of British geographers working for the Ministries of Information, of Economic Welfare, and of Supply, as well as for the Admiralty and the Meteorological Branch of the Royal Air Force, noting 'the sudden rise of geographical prestige which occurs in war time'. The requirements for maps, and detailed intelligence about countries where battles were being fought, meant that many geographers were employed in the Directorate of Military Survey and in Naval Intelligence (Wilson, 1946). The resultant Admiralty handbooks, written largely by geographers, were in effect comprehensive regional geographies of the countries in the war zone, comprising sections on their physical geography, history, peoples and economic geography (Freeman, 1980a). Within Britain, the Land Utilization Survey, undertaken during the 1930s under the directorship of L. D. Stamp (1947), also provided a firm basis upon which agricultural production during the war (Gould, 1985), and subsequent urban and rural planning could be undertaken. Moreover, the military requirements of warfare initiated a wide range of research in other fields of relevance to geography. King (1959: 37), for example, noted that 'The need for precise data concerning the character of sea and swell during the planning and carrying out of amphibious operations during the Second World War has led to a great increase in the knowledge and theory concerning all aspects of wave data.'

In contrast to this positive impression of the British experience, Ackerman (1945: 127) has argued that although American geographers were involved in the military effort, their 'literature provided adequate data for wartime geographic research on very few, if any parts of the world'. Moreover, he suggests that the failure of American geographers to meet wartime needs resulted from their 'inability to handle foreign language sources, and lack of competence in topical or systematic studies' (Ackerman, 1945: 122). In particular, the lack of substantive research on physical processes was seen as a significant weakness of geography. Tinkler (1985: 173) thus notes that, despite some conceptual continuity with the period before 1939, 'the War did lead to radical shifts in direction and marked changes of emphasis' in the practice of geomorphology. A similar view on the influence of the war on geography in the United States is reflected in Stone's (1979: 89) statement that 'World War II was the best thing that has happened to geography since the birth of Strabo'. In a more positive review than Ackerman's (1945), Stone suggests that the active involvement of many geographers in agencies such as the Office of Strategic Services (OSS), which later formed the core of the Central Intelligence Agency, and the Joint Army Navy Intelligence Studies led not only to greater inter-disciplinary collaboration, but also to a closer involvement of geographers in government service, both of which continued in the post-war era.

As in Britain and the United States, geographers in Germany were also involved in political and military action during the 1930s and early 1940s. The rise of the science of geopolitics, first under the Swede Kjellén and then under Haushofer, provided a link between the 19th-century work of Ratzel and the rise of the Nazi state. However, as Bassin (1987b) has stressed, the importance of geopolitics in influencing National Socialism has probably been overestimated in the past (Taylor, 1985). Haushofer seems to have had an ambivalent attitude to National Socialism, supporting the desire to expand Germany's *Lebensraum* (living space) and the creation of a pan-German state, but unwilling to accept the excesses of Nazi views on race. Despite his friendship with Rudolf Hess, Haushofer was never able to influence the mainstream of National Socialist thought, which instead found its underpinnings primarily in the ideological arguments of the 19th-century German *Volk* (people) movement (Bassin, 1987b). Nevertheless, several of Haushofer's ideas concerning Germany's strategic need for territorial expansion, derived in part from Raztel and Mackinder, appeared to provide sound scientific justification for German foreign policy in the 1930s. German geographers also played a key part in settlement planning through their work in the Reichsarbeitsgemeinschaft für Raumforschung (Reich Association for Area Research) established in 1936. This found a focus in the Arbeitskreis Zentrale Orte (Study Group on Central Places) founded in 1937 by Walter Christaller, whose dissertation on central places in southern Germany had been published in 1933 (Rössler, 1989). In 1940 Christaller was invited by Konrad Meyer, who was in charge of the Stabshauptamt für Planung und Boden (The Planning and Soil Office) to move to Berlin where he subsequently prepared numerous studies on the planned settlement of the east as part of Himmler's *Generalplan Ost* (General Plan of the East). At this time Christaller seems to have been an active member of the National Socialist Party, but after the war he became a member of the Communist Party, before joining the Social Democratic Party in 1959 (Rössler, 1989).

### *5.1.2 Problems of definition*

The weakness of geographers in undertaking systematic studies, noted by Ackerman (1945), was compounded by debates over whether or not regions with distinct personalities actually exist (Hall, 1935; Kimble, 1951; Gilbert, 1960; Freeman, 1961). On the one hand geographers such as Gilbert (1960: 158) continued to argue that 'the region is often so clearly distinguishable as a separate entity that it receives recognition in the shape of a distinct name'. In contrast, others such as Kimble (1951) considered the concept inadequate, arguing that regions do not exist in reality, cannot be perceived and do not have clear boundaries. At the heart of this debate lay two central concerns: whether there are processes or factors creating internal uniformity within particular areas, and whether it is possible to draw boundaries around such regions. Invariably the uniformity of regions was seen as being based upon environmental factors (Paterson, 1974). Renner (1935: 137), thus argued that 'It has been rather generally accepted among American geographers that a region is an area which is homogeneous enough in its physical character to possess either actual or potential unity in its cultural aspects'. In part this view was derived from Herbertson's (1905) concept of natural regions (Stamp, 1957), but it was also a result of the emphasis placed on rural societies, with the environment having a markedly more important role in determining agricultural rather than industrial practices (Wrigley, 1965). Those criticizing the regional concept tended to focus more specifically on industrial activities and urban societies, noting that few patterns of human activity had boundaries that coincided, and that the environmental conditions traditionally underlying regional studies were rapidly being changed by external factors. Paterson (1974) identified four other problems facing regional geographers:

1. The logical impossibility of providing a complete regional description in verbal form;
2. The problem of choosing an appropriate level of generalization;
3. The lack of detailed studies upon which regional synthesis could be attempted;
4. And the limited innovation possible in the format of regional geographies.

### *5.1.3 Uniqueness and generality; regions and systems*

Underlying the above problems with the definition of regional geography lay a much more fundamental tension, that between the search for uniqueness and the search for generalities. The specification of regions was centrally concerned with the identification of what it was that made each region unique. However, many regional geographers also saw their discipline as a science, or at least sought to gain wider recognition for it as such (Entrikin, 1981). Hall (1935: 122), for example, argued that 'The major contribution of geography to the general field of science is the recognition, first, of the ever-varying aspect of the land,

and secondly, that, in spite of this variation, the land tends to be divided into areas of more or less similarity.' The problem with attempts to claim that regional geography was a science was that the form of empirical–analytic science generally accepted within the wider scientific community was one that pursued general explanatory laws rather than one concerned with unique descriptions.

This distinction between two types of science, the one concerned with general laws and the other with uniqueness, owes its formal origins to the late 19th and early 20th century Baden school of neo-Kantianism, and particularly to the work of Wilhelm Windelband (1980) and Heinrich Rickert (1962) (Agnew (1989)); Entrikin (1989); Smith (1989). One of the central aims of this school was to reinsert critique into philosophical enquiry, and by so doing to challenge the increasingly dominant position of positivism, which had led to the pervasive view that philosophy was 'a worthless and contemptible enterprise' (Oakes, 1980: 166). In particular, Windelband and Rickert suggested that was infinite, and that, to make sense of it; individuals have to create rational concepts with their finite minds (Entrikin, 1989). In his 1894 Rectoral Address at Strasbourg, Windelband (1980: 175) argued that there were two kinds of empirical sciences, those seeking 'the general in the form of the law of nature', and those concerned with 'the particular in the form of historically defined structure. On the one hand they are concerned with the form which invariably remains constant. On the other hand, they are concerned with the unique, immanently defined content of the real event'. Windelband (1980: 175) introduced two new terms to describe such scientific thought: '*nomothetic* in the former case and *idiographic* in the latter case'. Furthermore, he suggested that this dichotomy reflected the division between the natural and the historical sciences. However, he also cautioned that the distinction was concerned only with modes of investigation and not with the contents of knowledge itself. Thus, 'it is possible – and it is in fact the case – that the same subjects can be the object of both a nomothetic and an idiographic investigation' (Windelband, 1980: 1975).

Within geography this distinction between a concern with the individual and the general was not new. It was reflected in Ptolemy's separation of geography from chorography, and in the 17th century in Varenius's distinction between general geography and special geography. From the 19th century, though, it found direct expression in the distinction between regional and systematic geography: regional geography tended to concentrate on the unique or what it was that gave regions their individuality, while systematic geography was concerned with the general. This distinction was particularly evident in the different approaches of Humboldt and Ritter, the former seeking to use a systematic approach to produce regional syntheses, and the latter concentrating on regional analyses in order to produce systematic generalizations. Both, however, saw the two as forming an integral part of the discipline of geography as a whole. At the beginning of the 20th century, this approach was expounded further by Hettner (1905), who took issue with Windelband and Rickert's distinction between nomothetic and idiographic sciences, suggesting that both types of approach were present in all sciences. This view formed an important element in Hartshorne's (1939) discussion of the relationship between the general and the unique in science, and between systematic and regional geography in particular. Following Hettner, Hartshorne (1939: 379) argued that 'these two aspects of scientific knowledge are present in all branches of science' (see also Hartshorne, 1955: 231). In referring specifically to two kinds of scientific knowledge, however, he seems to have misinterpreted Windelband's (1980) assertion that the nomothetic/idiographic dichotomy refers to *modes of investigation* rather than to the *contents of knowledge*. Hartshorne (1939: 379) compounded this error by suggesting that in some sciences there is a 'greater development of nomothetic knowledge' and in others of idiographic knowledge. By focusing on knowledge rather than on methods of investigation, Hartshorne thus rejected the nomothetic/idiographic distinction as inappropriate. However, his arguments are not in effect that far removed from those of Windelband. It is thus salient to note that Windelband (1980) specifically argued that some sciences, specifically the science of organic nature, could be both nomothetic and idiographic, and Hartshorne later modified his claim that both nomothetic and idiographic aspects 'are present in all sciences' (Hartshorne, 1939: 379) to the view that both generic and specific studies 'are of importance in nearly all fields of science' (Hartshorne, 1959: 149). Indeed, in his 1959 paper Hartshorne (1959: 164) argued that 'Since geography requires both generic studies and studies of individual cases – it is in part nomothetic, in part idiographic – there seems little point in attempting to measure the relative amount of the two types of study.'

With respect to the balance between considerations of the unique and the general, Hartshorne (1939: 383) argued that until the 19th century geography, like history, 'was largely limited to the study of the unique'. However he goes on to suggest that 'a geography which was content with studying only the individual characteristics of its phenomena and their relationships and did not utilize every opportunity to develop generic concepts and universal principles would be failing in one of the main standards of science' (Hartshorne, 1939: 383). For Hartshorne, both systematic and regional geography were essential parts of the discipline, forming the two different ways of organizing geographical knowledge. In his words,

> The ultimate purpose of geography, the study of areal differentiation of the world, is most clearly expressed in regional geography; only by constantly maintaining its relation to regional geography can systematic geography hold to the purpose of geography and not disappear into other sciences. On the other hand, regional geography in itself is sterile; without the continuous fertilization of generic concepts and principles from systematic geography, it could not advance to higher degrees of accuracy and certainty in interpretation of its findings (Hartshorne, 1939: 468).

Paradoxically, *The nature of geography* has subsequently been interpreted primarily as the standard-bearer of regional geography and the unique.

As Entrikin (1989: 10) argues, Hartshorne's detractors 'drew apart the idiographic and nomothetic and redefined them in terms of the unscientific study of the unique and the scientific search for lawful generalization. *The Nature* became the symbol of the study of the unique, and it still bears this emblem'. Although Hartshorne's (1939) work had been partially criticized by Sauer (1941), Whittlesey (1945) and Ackerman (1945), this formal separation was largely initiated by Schaefer in 1953 in his attack upon what he saw as a tradition of exceptionalism in geography.

Schaefer (1953) was roundly condemned by Hartshorne (1954, 1959) for a lack of scholarship and rigour. However, he caught the changing spirit of the age, and while his paper can best be interpreted as being symptomatic of the changes rather than initiating them, his ideas provide a useful starting point for an analysis of the introduction of logical positivist thought to geography. Schaefer's (1953: 227) central assertion was that geography should be 'conceived as the science concerned with the formulation of the laws governing the spatial distribution of certain features on the surface of the earth'. The whole purpose of his enterprise was thus markedly different from that of Hartshorne. *The nature of geography* was in essence a critical survey of previous geographical writing, whereas Schaefer was seeking to develop a new kind of geography. In Johnston's (1991a: 57) words, 'Hartshorne's was a positive view of geography – geography is what geographers have made it – whereas Schaefer's view, on the other hand, was a normative one, of what geography should be.'

Schaefer (1953: 227) argued that 'Description, even if followed by classification, does not explain the manner in which phenomena are distributed over the world. To explain the phenomena one has described means always to recognize them as instances of laws.' Two fundamental points need to be noted about this statement. The first is that Schaefer clearly distinguished between explanation, the task of science, and description. Secondly, he argued that explanation required the formulation of laws, and thus that a scientific geography should be concerned with the identification and application of such explanatory laws, rather than with the old descriptive practice of regional geography (Schaefer, 1953: 228). Once established, these laws then enable prediction to take place. The practice of geography as a science for Schaefer (1953: 229) begins with the recognition that

> Spatial relations among two or more selected classes of phenomena must be studied all over the earth's surface in order to obtain a generalization or law. Assume, for instance that two phenomena are found to occur frequently at the same place. A hypothesis may then be formed to the effect that whenever members of the one class are found in a place, members of the other class will be found there also, under conditions specified by the hypothesis. To test any such hypothesis the geographer will need a larger number of cases and of variables than he could find in any one region. But if it is confirmed in a sufficient number of cases then the hypothesis becomes a law that may be utilized to 'explain' situations not yet considered.

Schaefer's criticism of much previous geographical work was that an antiscientific spirit of historicism had entered the discipline through Hettner's reading of Kant. He thus asserted that 'Invoking the formidable authority of Kant, Hettner successfully impressed upon geography the exceptionalist claim in analogy to history' (Schaefer, 1953: 235). Schaefer saw geographers deriving their work from Hettner as having been concerned with the explanation of the arrangement of unique phenomena in space. The methodology of such geography was thus claimed to be unique, and it is to this position that he ascribed the term *exceptionalism*. Schaefer contrasts this Hegelian tradition of historicism with the growing strength of the positivist philosophy of science in the natural sciences in Germany in the 19th century. In opposition to the exceptionalist tradition Schaefer advocated that geography should become a positivist science, concentrating on the production of three kinds of law: the laws of physical geography, which he claimed were 'not strictly geographical' (Schaefer, 1953: 248); the morphological laws of economic geography, and particularly those of general location theory, which he saw as being truly geographic; and thirdly, process laws, the products of mature social science, but which are not morphological and thus not confined to geography.

Schaefer's essay was an unbridled attempt to launch geography firmly into the mainstream of positivist science. His early life had been spent in Berlin where he studied at the university from 1928 to 1932 (Martin, 1989), and he had thus been educated within the broad tradition of German political geography and political science. With the rise of Nazi power he was imprisoned, but eventually left Germany and settled in Iowa (Bunge, 1979a; Martin, 1989). Here he gained a position in the Department of Geography when it was formed in 1946 at the State University of Iowa under the chairmanship of Harold McCarty. At Iowa, Schaefer became friends with another refugee from Nazi Germany, Gustav Bergmann, who was in the Philosophy Department, and who had been a member of the Vienna Circle of logical positivists in the 1920s. Through Bergmann, Schaefer learnt much about logical positivism, and indeed it was Bergmann who read the proofs of his 1953 article, Schaefer having died of a heart attack at the age of 48 in the June of that year. Although this link provides a clear connection between geography in the United States and the work of the Vienna Circle, Schaefer himself published very little else, and was considered by his colleagues at Iowa, particularly McCarty, as being of relatively minor importance to the development of geography (Martin, 1989; but see King, 1979). Bergmann, however, appears to have been of more lasting personal influence. King (1979: 128), thus comments that

> At Iowa in the late fifties there was still Gustav Bergmann. He was the oracle outside the geography department, a logical positivist who had been one of the original Vienna Circle, a close friend of Schaefer, and the teacher in one way or another of most of us who went through the Iowa department. If Schaefer's paper was not, as I recall, required reading, Bergmann's book on the philosophy of science

certainly was! His influence on the work of the Iowa geographers was seen everywhere.

Hartshorne responded vigorously to Schaefer's paper, first in a short letter to the *Annals of the Association of American Geographers* (Hartshorne, 1954), and then in an article which provided a critique of Schaefer's (1953) paper section by section (Hartshorne, 1955). Hartshorne's main aim in these papers was to illustrate Schaefer's poor scholarship and 'trickery with words' (Hartshorne, 1955: 231), arguing that 'In total, almost every paragraph, indeed the great majority of individual sentences in the critique, represents falsification, whether by commission or omission' (Hartshorne, 1955: 243). Schaefer (1953) certainly failed to cite many of his sources, and, as Hartshorne (1955) illustrated so clearly, a number of his arguments were built on spurious assertions. Hartshorne's 1955 paper was the first part of his rebuttal of Schaefer's arguments, designed to demonstrate and correct the falsifications which he identified therein. The second part of his reaction was published in 1959 as a monograph entitled *Perspective on the nature of geography*. This argued that 'geography is that discipline that seeks *to describe and interpret the variable character from place to place of the earth as the world of man*' (Hartshorne, 1959: 47). In defining it thus, Hartshorne regretted the increasing division of the discipline into physical and human halves, which he saw as having emerged in the 19th century. In particular, he suggested that

> With the increasing prestige of the 'natural sciences', particularly the 'physical sciences', many geographers were stimulated to concentrate on the nonhuman aspects of their field and to construct courses and textbooks called 'physical geography'. Such collections of knowledge concerning particular categories of earth features, however scientific in quality, were lacking in coherence and divorced from the full context of reality; in consequence they had only limited appeal to the general student. At the same time the study of the human aspects of geography, in large part divorced from the physical earth features with which they are in reality interwoven, lost both scientific standing and student interest. The disastrous consequences for the status of geography in the secondary schools is well known (Hartshorne, 1959: 79–80).

In referring to the division between systematic and regional geography, Hartshorne continued to assert that there was no dichotomy or dualism between them, both being essential to geographical practice. However, in the *Perspective* he gave much more direct attention to the issue of whether geography should be concerned with the formulation of scientific laws. Indeed, he saw this as the most disturbing problem of concern to geographers at that time. In contrast to Schaefer, Hartshorne's view of science did not conform tightly to that of logical positivism. In particular, he argued vehemently that 'prediction is not the purpose of science' (Hartshorne, 1959: 165) and that the formulation of laws is likewise not its end-purpose (Hartshorne, 1959: 168). Instead he thought that science should be considered 'in the active sense of seeking to know' (Hartshorne, 1959: 168), and he thus argued that

> If then we understand the term 'scientific description' to include both what is known and what can be inferred, both of the phenomena and of the process relations and associations of phenomena, we may once again modify our statement of the purpose of geography to read: *the study that seeks to provide scientific description of the earth as the world of man* (Hartshorne, 1959: 172).

Hartshorne's view of science was not, however, generally accepted in the wider scientific community, and by arguing against the formulation of laws which had predictive capacity, Hartshorne was swimming against the tide. As Guelke (1977: 382–3) has argued, 'In the 1950s, geographers were given a choice between describing the unique or seeking scientific laws. The former alternative was, not surprisingly, unacceptable.'

### 5.1.4 *Geographical science and the art of geography*

Paradoxically, although Schaefer's own influence on the future direction of geography was slight, it was the ideas expressed in his 1953 paper that came to dominate the practice of geography in the 1950s and 1960s. Hartshorne's vision of science was rejected in favour of one which sought explanation rather than description, general laws rather than individual understanding, and prediction rather than interpretation. Regional geography, nevertheless, continued to find its supporters. Gilbert (1960), for example, sought to situate it within a much wider tradition of regional thought that emerged in Europe in the 19th century and included the regional novel and the political idea of the region as well as the geographical concept of a natural region. For Gilbert (1960: 159), regional geography was 'an art comparable with other arts', and he saw its lack of a rigorous scientific foundation as doing it no discredit whatsoever. A similar view was taken much more recently by Hart in his 1981 Presidential Address to the Association of American Geographers. Here he argued that

> Systematic geography generates theories to facilitate an understanding of regions, and regional geography is the proving ground where theories are tested empirically. The idea of the region provides the essential unifying theme that integrates the diverse subdisciplines of geography. The highest form of the geographer's art is the production of evocative descriptions that facilitate an understanding and an appreciation of regions (Hart, 1982: 1).

In criticizing the widespread adoption of scientism by the geographical community, he goes on to note that

> many geographers have been seduced by the aura that is perceived to emanate from Science, and they have assumed that geographers would be held in higher esteem if only they could somehow manage

> to convince the body politic that geography is a Science. These people have done some strange things in their attempts to look more 'scientific'. At one time, for example, it was high fashion among some geographers to run around in knee boots and red-checked flannel shirts, trying to look like geologists, who they assumed were more scientific than geographers. More recently the fashion among some members of the clan has been to hang around the computer center trying to look like econometricians (Hart, 1982: 3).

## *5.2 Models, systems and process: the implicit adoption of logical positivism*

### *5.2.1 Process and form in physical geography*

The debate between Hartshorne and Schaefer largely excluded the field of geomorphology along with much of physical geography in general. Hartshorne (1939: 423), for example, while accepting that in Germany geomorphology was a central part of geography, was somewhat ambivalent concerning the relationship between the two subjects in the United States. Although climatology, soils, landforms, plant geography and animal geography were included among his systematic geographies (Fig. 4.1) (Hartshorne, 1939: 147), he nevertheless recognized that geographers trained as geomorphologists faced particular problems in trying to incorporate their work into the practice of geography as a chorographic science (Hartshorne, 1939: 424). Indeed, physical geography in North America was encountering a serious crisis of identity in the 1930s and 1940s, with many geography departments abandoning it altogether, and most geomorphological research being undertaken in geology departments (Russell, 1949; Drake and Jordan, 1985; Tinkler, 1985; Vitek, 1989). According to Dury (1983: 91–2) the main reasons for this 'included a mix of the reaction against environmental determinism – itself reinforced by observation of the pioneering conquest of the wilderness – and the inability of geographers proper to compete with what was being done elsewhere, particularly in geology'. Similarly, Costa and Graf (1984) argue that 'By the late 1930s geography rejected the paradigm of environmental determinism, and geomorphology declined precipitously in importance.' In contrast to the decline of physical geography in the United States, it remained relatively healthy in Britain, where the field of geomorphology became dominated by geographers. Nevertheless, as Stoddart (1987b) points out the number of physical geographers in Britain in the 1930s was so small that they could only concentrate on a limited area of the subject.

The demise of the Davisian system of geomorphology was initiated primarily by engineers and geologists, and it was only slowly that new ideas infiltrated into physical geography from other disciplines. The central criticism of much previous work in physical geography was that it was descriptive, and failed sufficiently to explain physical processes (Tinkler, 1985). This was particularly evident in geomorphological research on hillslopes, aeolian landforms, rivers and coasts undertaken from the 1930s onwards. In a series of influential papers, for example, Horton (1932, 1933, 1935, 1945), an engineer by training, combined an interest in drainage basins and the infiltration capacity of soils to transform the study of hillslope processes. The particular significance of this work was its use of mathematics, and its formulation of laws designed to explain the physical shape of landforms. Despite criticisms (Pitty, 1971: 26–8), Horton's work influenced a number of subsequent studies on hillslope processes by geologists, and in particular those of Strahler (1950) and Schumm (1956a, b). Likewise, in his research on sand dunes, Bagnold (1941) stressed the need for a detailed understanding of the physics of sand movement, arguing cogently that it was essential for geomorphologists to explain the processes of dune formation rather than simply concentrating on the interpretation of form. Subsequently, he also emphasized the need for rigorous laboratory experiments to determine the processes involved in other kinds of sediment transport (Bagnold, 1954, 1966; see also Krumbein, 1955). The ideas of Horton and Bagnold, bringing together a concern with the statistical description of landforms and the explanation of process, found further expression in a range of studies on fluvial hydrology in the 1950s (Leopold, 1953; Leopold and Maddock, 1953; Leopold and Miller, 1956; Leopold, Wolman and Miller, 1964), which sought to explain channel form through a consideration of fluvial processes. Similar arguments concerning the relationships between form and process were also advocated in the coastal context by King (1959), whose book *Beaches and coasts* provided a clear example of the need to combine theoretical and experimental work.

Such views did not, though, pass without criticism. In particular, Wooldridge (1958: 31) argued that 'Geomorphology is primarily concerned with the interpretation of forms, not the study of processes,' and with respect to the rising tide of quantification he was also of the opinion that 'The direct attack by mathematical methods would seem to offer very limited chances of success' (Wooldridge, 1958: 32). He went on to caution against the adoption 'of a narrowly physico-mathematical approach to Geomorphology . . . for snobbish reasons – i.e. because it sounds impressive and it is fashionable to clothe our thought in the jargon of Mathematics' (Wooldridge, 1958: 33). The timing of this onslaught is interesting, because it illustrates that while these approaches were becoming increasingly accepted in the United States they had not yet gained wide acceptance in Britain, where geomorphological teaching and research still remained largely in geography departments. Within five years, though, the situation had changed dramatically, and by the early 1960s many British physical geographers were turning increasingly to the work of Strahler and Leopold for inspiration.

An increasing concern with physical processes and the development of quantitative models in geomorphology were not the only changes to take place in the conceptualization of physical geography in the 1950s (Gregory, 1985). Two other important alterations, again influenced largely by developments outside the discipline, also took place. These were the increasing attention being paid to the question of climatic change, and the growing acceptance of the concept of plate tectonics. Although Quaternary climatic change had been of interest to 19th-century geomorphologists such as Penck, de Geer and Geikie (Beckinsale and Chorley, 1991), increasing research in this field can be seen in part to have been related to the development of new techniques and a growing concern with accurate chronology. In particular it was the development of methods such as radio-carbon dating and the use of deep-sea cores in the 1940s that opened up entirely new possibilities for research into the precise dating of environmental change, which it was hoped would eventually lead to a greater understanding of the processes determining such changes (Antevs, 1928; Libbey Anderson and Arnold, 1949). Current concern with global warming at the end of the 20th century can in turn be seen as one of the outcomes of this avenue of research. New measurement techniques also provided the wherewithal for the development of research on ocean-floor spreading, and eventually to the acceptance of ideas relating to plate tectonics (Wegener, 1915). Although the term plate had been used by Sollas in 1904 and by Russell in 1936 (Beckinsale and Beckinsale, 1989), it was not until detailed research on the ocean floors during the 1950s and 1960s, influenced in part by developments in naval technology during the Second World War, that the concept of ocean-floor spreading became widely accepted.

These three developments in physical geography all represented an increasing concern with process, with precise quantifiable measurement, and with the development of laws and models, but at different space–time intersections. Geomorphological research increasingly began to focus on immediate processes at the small scale, Quaternary research on mid-term regional processes, and plate tectonics on very long-term global processes. In accounting for these changes it is important to consider the influence of the technological advances made during the Second World War, and also the increasingly applied nature of research in physical geography. The dearth of real knowledge in the 1930s concerning processes acting on slopes and coasts, for example, necessitated a substantial reappraisal of research programmes in these fields during the war. Tinkler (1985) thus comments on the importance of the war as a stimulus to new research on terrain evaluation and morphological mapping, and Morisawa (1985: 91) notes the importance of 'the impetus given to the study of waves and beaches as a result of the need for amphibious landings on Pacific Islands' in determining the progress of coastal research subsequent to the cessation of hostilities. More generally, funding bodies and in particular the US Geological Survey were increasingly encouraging research with practical applied applications, as in the fields of soil erosion, flood control, reservoir sedimentation and fluvial hydrology, in order to overcome some of the many environmental problems encountered in the inter-war period (Tinkler, 1985). Underlying such research was a growing desire to be able to predict the environmental repercussions of particular human actions through the development of models and laws of environmental processes.

### 5.2.2 *Theoretical approaches to a systematic human geography*

A similar concern with the development of geography as a theoretical and empirical science was also encountered in human geography in the United States during the 1950s. This focused mainly around teaching and research being undertaken in the departments of geography at the University of Washington at Seattle (Garrison, 1979), the University of Wisconsin at Madison, and the University of Iowa (McCarty, 1979). The excitement of the changes taking place in these institutions is well reflected in Morrill's (1984: 59) account of the atmosphere at the University of Washington in the middle and late 1950s:

> There was something electrifying about tilting with the dragons of the establishment. Perceiving ourselves as a subversive and feared minority gave us foolish strength and helped us to maintain a breathless pace of intellectual ferment, complaint and conspiracy. It was easy to personify Richard Hartshorne, whose work we studied in detail, as what we struggled against. We found heroes – notably Schaefer and Christaller – and many villains. Of course there was strong resistance, which made us all the more determined!

In a similar vein, Gould (1979: 140) recalls how this new generation of geographers 'was both sick and ashamed of the bumbling amateurism and antiquarianism that had spent nearly half a century of opportunity in the university piling up a tip-heap of unstructured factual accounts'. He goes on to recall how 'With the exception of one or two works of scholarship in historical geography, it was practically impossible to find a book in the field that one could put in the hands of a scholar in another discipline without feeling ashamed' (Gould, 1979: 140–1).

In reaction to this old establishment geography, the new research at the University of Washington was characterized by its quantitative nature and its search for theory. Under the leadership of Garrison, and to some extent Ullman, who had published a paper on a location theory for cities as early as 1941, a group of graduate students, including Dacey, Berry, Morrill and Bunge, sought to transform geography into an academic subject worthy of wider recognition. Morrill (1984: 67) records how their impression of geography was that it 'was intellectually and numerically weak, and held in low esteem by other disciplines', how it was descriptive and lacking in theory, and how its purpose was generally seen by outsiders as being simply for the training of teachers or the preparation of atlases. Their aim was to incorporate geography into mainstream science, by developing theories, testing them, and thus seeking to explain the organization and evolution of landscape. The context of such work was mainly within the fields of urban and economic geography, and it drew heavily on theoretical ideas developed in other disciplines, as well as on research by European geographers and economists.

This is well illustrated by Garrison's (1959a, b, 1960) survey of research on the spatial structure of the economy. This began with a review of books by Isard, Dunn, Greenhut, Ponsard, Lösch, and Boustedt and Ranz, in which he evaluated the relevance of economic theory developed in the 1950s to geography (Garrison, 1959a). In concluding this review, Garrison suggested that the establishment of the Regional Science Association in 1954 and its subsequent expansion indicated the growing concern of economists in problems of spatial arrangement which had long been of interest to geographers. In the second part of his review, Garrison (1959b: 482) argued that the use of algebraic notation and linear programming methods enable problems of location structure to be given an operational character, and that problems couched in such terms 'display the price interdependences associated with the location system in a manner which was not possible before'. Typical of the location problems for which he envisaged such a methodology as being useful were those of spatial price equilibrium and least cost locations for production given distributions of raw materials and markets. The final article in this trilogy considers a further set of problems associated with networks and input–output flows between regions, stressing the importance of equilibrium models. Garrison (1960: 372) concludes that it is important to consider 'the concept of the location system as a *combination of interrelated activities* at different places', and that such a system is a pattern of efficient combinations of activities, reflecting principles of maximization or minimization.

Similar research was also being pursued under the leadership of McCarty at Iowa and Robinson at Wisconsin. This stressed not only the need to incorporate theory (McCarty, 1953, 1954), but also the use of correlation and regression methods of analysing the spatial relationships between distributions (Robinson and Bryson, 1957). As with Garrison's work, that based at Iowa and Wisconsin reflected a direct concern with the establishment of laws and the development of models, approaches which were to receive widespread acceptance in the discipline as a whole in the 1960s. The general lack of previous geographical work in this vein, however, meant that geographers during the 1950s turned largely to the long tradition of theoretical models developed by German economists. Four works in particular came to dominate the geographical literature: von Thünen's (1826) *Der isolierte Staat in Beziehung auf Landwirtschaft und Nationalökonomie*, which established the theoretical basis of agricultural geography; Weber's (1909) *Uber den Standort der Industrien*, which laid the foundation of industrial location theory; and Christaller's (1933) *Die zentralen Orte in Süddeutschland* and Lösch's (1940) *Die räümliche Ordnung der Wirtschaft*, which provided the core of settlement geography. By the 1970s, these works had become so central to the practice of geography that they formed the core of standard secondary school texts such as that by Bradford and Kent (1977), with most readers probably failing to recognize that only Christaller had laid any claim to be a geographer.

A further influence on the development of this theoretical and mathematical tradition in the United States came from Sweden, through the work of Torsten Hägerstrand. Hägerstrand was introduced to the ideas of von Thünen and Christaller largely through the work of Edgar Kant, an Estonian geographer who took refuge at Lund in the aftermath of the Second World War, and for whom Hägerstrand worked as a research assistant (Hägerstrand, 1983). Influenced both by Kant and the ethnologist Sigfrid Svensson, Hägerstrand (1953) used mathematical models to examine the spread of innovations in central Sweden. In particular, he developed a probability model of the diffusion process which allowed the incorporation of stochastic, or random, variables. In 1959 Hägerstrand visited Seattle, and in commenting on the significance of this visit, Morrill (1984: 62) notes that 'The combination of theory and thorough field work expressed in the demonstration of the diffusion process, before they were available in English, had an electrifying impact.'

Two other important influences on the emergence of a quantitative, process oriented human geography, were the Chicago School of Human Ecology, and the social physics of Stewart and Zipf. During the 1920s and 1930s sociologists at Chicago under the leadership of Robert Park developed a range of models linking together social organization and spatial structure (Entrikin, 1980). As Jackson and Smith (1984: 65) have commented, 'Many of the traditions established at this time have left a lasting impression upon the way that urban geography is taught at school and university.' These traditions have usually been interpreted as lying centrally within a positivist conceptualization of social science, but Jackson and Smith (1984) note that much of Park's work reflects a deeply humanistic view, combining elements of neo-Kantianism with the pragmatism of James and Dewey. Park's attempts to combine these two divergent traditions created a series of dualisms in his work, which 'resolved around the pragmatic data requirements of a welfare-oriented discipline' (Jackson and Smith, 1984: 79). Mellor (1977) suggests that five central ideas underlay the work of the Chicago sociologists:

1. That cities could be seen as ecological communities;
2. That land values were a reflection of the natural order;
3. That cities were subject to a cycle of invasion and succession in space;
4. That natural areas reflected the fundamental spatial basis of all social organization;
5. That the distinction between the natural and moral orders of cities provided a structured organization to patterns of life-styles.

However, it was the linking together of spatial and social organization by the Chicago school (Burgess, 1964) that proved to be most attractive to geographers (Pooler, 1977). In particular, the models of urban land use proposed by Burgess (1925), Hoyt (1939) and Harris and Ullman (1945), seemed to provide geographers with a set of theoretical spatial statements which could then be tested and possibly elevated into laws.

The other main influence on the application of mathematical ideas to geography was through the work of social physicists. In 1947 Stewart, professor of astrophysics at Princeton, began to organize his 'Social Physics Project, with an original intention of considering thoroughly the transferral of methods and principles of physical science to the social field' (Warntz, 1984: 141). Once again, this involved the application of ideas external to geography, in this instance those of Newtonian physics, to questions concerning the distribution of phenomena in space. Its most influential expositions were the rank-size rule linking city size and rank, noted by Stewart (1947) and Zipf (1949), and the application of gravity models to human interaction (Carrothers, 1956; Olsson, 1965). Much of this work, however, was concerned primarily with noting empirical regularities, and with applying mathematical formulae to their description, rather than with any attempt at explanation of the regularities observed. Geographers contributed little to the early work of the Social Physics Project, but from the late 1950s many of these ideas were incorporated into geography through the collaboration between Stewart and Warntz (1963), the latter of whom managed to combine the positions of research associate at the American Geographical Society in Washington, research associate in astrophysical sciences at Princeton, and visiting professor of regional science at the University of Pennsylvania during the decade 1956–66 (Warntz, 1984). Eventually, this line of research was to lead to a consideration of geography as part of general spatial systems theory, in which the objective was to establish a single social science whose internal logic was isomorphic to that of the physical sciences.

A concern with modelling the spatial organization of society, and with the development of mathematical and geometrical descriptions of social relationships was thus well established in various different disciplines in the United States in the 1940s and 1950s, and their incorporation into geography should not be seen in isolation. However, it was in geography that this new approach reached a position of pre-eminence, since it appeared to offer geographers a fundamental justification for their subject. Conceptualizing geography as the science of space offered a dramatically different alternative to the increasingly discredited tradition of regional geography.

### *5.2.3 Systems and theoretical geography*

In almost all of the theoretical and conceptual works advocating a new approach to geography in the 1950s and 1960s little direct attention was paid to the philosophical underpinnings of the discipline. The perceived need was for geography to become more scientific; science was seen to be concerned with the explanation of process; and explanation in science required hypothesis testing and law building. Few authors, with the notable exception of Schaefer, paid even the scantest attention to the fact that this particular kind of science was grounded firmly in the tradition of logical positivism. As Guelke (1977: 381) has commented, 'The narrow philosophical choice offered to geographers in the realm of explanation made the wholesale adoption of the nomothetic approach all but inevitable. For many, geography was either science or mere description.' The overwhelming ideas of logical positivism, directly introduced at least to the Iowa geographers by Bergmann (1957), seemed so appropriate and valid that few geographers during the 1960s even considered the possibilty of developing a critique thereof.

In an influential review of the status of geography as what he termed a fundamental research discipline, Ackerman (1958) captured the essence of the discipline in the late 1950s. He drew seven conclusions concerning the conditions underlying the status of geography as it was then practised:

> that (1) the near universal characteristic of space-relations patterns is unceasing change; (2) several physical, biotic, and cultural processes are part of this change; (3) quantification is a major problem in describing the space-relations effect of these processes; (4) observational techniques need further development, to match the need for quantification; (5) a theory of abstract distributions is lacking; (6) the study of covariance in the significant processes is only beginning; and (7) the distributional effects which distinguish the several cultural processes significant in space relations are yet very imperfectly understood (Ackerman, 1958: 35)

These observations both illustrated the flavour of current research in the United States and also set an agenda for future research which was carried forward into the 1960s (see also Ackerman, 1963). Its central concerns were with space, quantification and theory building.

Two seminal works at the beginning of the 1960s captured the move towards a more theoretical geography. These were Scheidegger's (1961) *Theoretical geomorphology* and Bunge's (1962) *Theoretical geography*. Scheidegger's (1961) work can be seen as bringing together the conceptual and methodological advances made in geomorphology over the previous two decades, and presenting them in a mathematical and theoretical framework. It was surprisingly not widely acclaimed when first published, largely because of its inherent complexity. As Gregory (1985: 57–8) has commented,

> Such a theoretical approach was conceived from the viewpoint of geodynamics and did not receive the acknowledgement it deserved because it relied upon a mathematical theoretical foundation and did not cover the range of geomorphology completely. However, as an intriguing and stimulating approach it would probably have received more acclaim had it been launched up to a decade later to coincide with the movement towards a more theoretical foundation and particularly towards the properties of materials.

At a theoretical level, Scheidegger's work still stands as a foundation for contemporary geomorphology, and few works published since in the field have matched its rigour and eloquence.

In human geography, Bunge (1962) sought to establish the discipline as a predictive science of spatial locations. Building closely on the ideas of Schaefer, and influenced also by mathematicians at the universities of Wisconsin and Washington, Bunge reflected the advances made by the group of geographers working under the chairmanship of Hudson at Washington, and including Berry, Dacey, Garrison and Hägerstrand. Moreover, the importance of the Swedish connection in the development of theoretical geography is amply illustrated by the fact that Bunge's book was published at Lund rather than in the United States, where it had been 'Rejected savagely and repeatedly by the same reviewers' (Gould, 1979: 141). As this statement indicates, one cannot understand Bunge's role in the development of a quantitative and theoretical geography without some understanding of the emotions and personalities involved. Bunge (1968, 1979a, b) thus openly acknowledged his enmity with Hartshorne, and accused him not only of being responsible for his failure to pass his Ph.D. preliminary examinations at Wisconsin, but through his previous links with the OSS also of complicity in the FBI's harassment of Schaefer. As Martin (1989) has indicated, Bunge's (1968) essay contained a number of misrepresentations and unfounded insinuations, but these served to add a distinctive personal element to the intellectual disagreement between Hartshorne and Bunge. It is significant here that two of the geographers central to the emergence of logical positivism in geography, Schaefer and Bunge, both had strong socialist political sentiments, and felt themselves hounded by the establishment in which they sought to maintain their academic integrity. Even in the 1960s, the legacy of McCarthyism, with its profound hatred of any tinge of Communist or Marxist critical thought, had a considerable effect on intellectual practice in the United States.

Bunge's (1962) *Theoretical geography*, dedicated to Christaller, sought to establish geography as a strict science, deeply concerned with theory. He began by briefly establishing a general philosophy of science, based, but not explicitly so, upon the tenets of logical positivism. This was then followed by the development of a scientific methodology for geography, outlining the relationships between regional and descriptive, systematic and theoretical, and cartographic and mathematical geography, derived largely from the ideas of Schaefer. Thereafter, the book examined metacartography, combined maps and mathematics by demonstrating that shape can be described mathematically, and then applied descriptive mathematics to other areas of geography. The remaining chapters focused on increasingly abstract mathematics, considering theories of movement, central place and location, geometry, the meaning of spatial relations and patterns of location. Despite Bunge's attempt to combine both physical and human geography in his theoretical approach, most of his substantive examples were drawn from human geography, and his ideas consequently had little impact on the development of physical geography.

The subsequent development of quantitative and theoretical geography owed much to wider social and economic changes in North American and British society. In France the traditional strength of the Vidalian regional approach largely survived into the post-war period, and in Germany it was not until 1968, with the publication of Bartels's *Zur wissenschaftstheoretischen Grundlegung einer Geographie des Menschen* that any substantial attempt was made to incorporate quantitative and theoretical geography into the discipline (Lichtenberger, 1978). In Britain and the United States, the 1960s was a period of marked expansion in higher education (Stoddart, 1967), contemporaneous with considerable economic growth and apparent improvements in social welfare. Young staff and faculty, fired with enthusiasm for quantitative and theoretical geography, taught a rapidly increasing number of students this new approach to the discipline. Moreover, technological innovations during the 1960s, particularly in the development of computers, enabled large new data sets to be generated and analysed statistically with increasing ease and rapidity. This greatly enhanced the status and apparent scientific respectability of the newly emergent geographical methodology.

Paradoxically, though, the next main impetus to developments within this tradition came from Britain rather than the United States. In part this reflected the much stronger institutional position of geography within the British education system, but it was also a result of the rapid increase in the number of new geography graduates in Britain, eager to establish their careers through the production of new ideas. Until the late 1950s British geography had been little influenced by the movement towards a more theoretical geography. Thus, while Burton (1963) could argue from a Canadian perspective that a quantitative revolution had already taken place, few British geographers had yet accepted the methodological or theoretical implications of the new approach by the early 1960s. This situation, though, was soon to change rapidly, as a group of recently graduated students from Cambridge, many closely influenced by the economic geographer Caesar (Haggett, 1990), went to the United States for short periods as postgraduate students or visiting faculty. In particular the fortuitous reunion in California in 1962 of two Cambridge staff, Chorley and Haggett, the former of whom had been working with the US Geological Survey at Denver, and the latter of whom was teaching a summer school at the University of California at Berkeley, was to be of considerable significance. Sitting in the shade of the saloon bar at the abandoned mining town of Bodie, their minds turned to their plans for the following summer. As they recall,

> School geography in Britain was in something of the doldrums. Regional geography had become somewhat routinized and we were both concerned that some of what we saw then as exciting new developments in the universities on both sides of the Atlantic should be passed on to young geographers struggling through their sixth-form courses (Haggett and Chorley, 1989: xv).

Following discussions with staff at the Cambridge University Extra-Mural Board, it was agreed that they should organize a residential course for teachers at Madingley Hall in July 1963. The lectures given then and during the following summer were subsequently published as *Frontiers in geographical teaching* (Chorley and Haggett, 1965). This influential volume, divided into sections on concepts, techniques and teaching, provided a basic introduction to the new geographical ideas being developed on either side of the Atlantic, and drew particular attention to an increasing concern with quantification, to growing disquiet with the traditional idiographic approaches of British geography, and to the need for greater awareness of relevant changes in associated disciplines in the physical and social sciences.

Meanwhile, Haggett (1965) had also completed his major review of the theoretical changes in human geography (Gould, 1979), published under the title of *Locational analysis in human geography*. This was concerned with the need to identify order in geography, to examine the locational systems studied by geographers, the models used to describe them, and with the types of explanation derived from them. Significantly, Haggett's (1965) book was specifically concerned with human geography, but it also represented an attempt to incorporate this within a wider tradition of general systems theory (Bertalanffy, 1951), noting that attempts had already been made by Chorley (1962) to introduce such concepts into geomorphology and physical geography. Subsequently, systems and models were to provide an integrating framework for much geographical research during the 1960s.

The first major step towards this end was the publication of Chorley and Haggett's (1967) edited volume *Models in geography*. Unlike *Frontiers in geographical teaching*, in which not all of the contributors were as convinced by the new approaches to the discipline as were the editors, *Models in geography* was a deliberate attempt to summarize the achievements of quantitative and theoretical geography, and to present them as a central new focus to the discipline. Indeed, the editors consciously introduced Kuhn's concept of paradigms, albeit rather loosely, in order to support their argument that the new paradigm represented a fundamental break with the past represented by the older classificatory or regional paradigm. As they comment in retrospect, their 'most clearly definable aim was to make geography at all levels a more intellectually attractive and relevant subject' (Haggett and Chorley, 1989: xvii). Contributors to the volume provided accounts of the role of model building in chapters on both methodological issues and on systematic areas of the subject, such as industrial location. However, the term 'model' itself was not used entirely consistently by contributors, with it variously being considered as 'a theory, a law, a hypothesis, or any other form of structured idea' (Johnston, 1991a: 8).

For Chorley and Haggett (1967: 38), the new paradigm, built around the development of scientific generalizations, was designed to raise the profile of geography by modelling its form on mathematics and physics. Something of their enthusiasm for this venture is captured in the following retrospective account of their project:

> The early sixties was an optimistic period for geographical innovators. . . . For a brief period in the sixties geography PhDs were doubling in number every six years. Scientific methods which had shown such unprecedented success in solving technical problems in the physical and biological world showed some prospect of equal promise in the social world. Physical and human geography had

rarely been so united, the bridges between the different parts of the subject rarely so strong. It seemed, at least for a brief window of time, that all that was needed was to find the language which converted the man-made environment into the same terms as the natural environment, and that spatial structure would provide one such language, systems analysis another (Haggett and Chorley, 1989: xix).

In drawing parallels between the technical successes of the natural sciences, and the perceived potential of applying similar methods to the social sciences, this quotation neatly captures one of the key reasons why the new approach to geography achieved such popularity.

The quotation above also reflects another important aspect of the 'new geography', in that the focus on spatial laws, models and systems was seen by some as a way of reintegrating both human and physical geography. In the first half of the 20th century such integration had been made possible through the conceptual device of the region, but with the demise of regional geography, and the increasing attention being paid to process in physical geography, this central unity in the discipline had become fragmented. Chorley and Kates (1969: 1) thus note how 'In the early 1950s geomorphologists, especially in Britain, were able to look patronizingly at the social and economic branches of geography and dismiss them as non-scientific, poorly organized, slowly developing, starved of research facilities, dealing with subject matter not amenable to precise statement, and denied the powerful tool of experimentation'. By the end of the 1960s, however, the same authors comment that

> Little more than a decade has been sufficient to transform the leading edge of human geography into a 'scientific subject', equipped with all the quantitative and statistical tools the possession of which had previously given some physical geographers such feelings of superiority. Today human geography is not directed towards some unique areally-demarcated assemblage of information. . . . In contrast, most of the more attractive current work in human geography is aimed at more limited and intellectually viable syntheses of the pattern of human activity over space possessing physical inhomogeneities, leading to the disentangling of universal generalizations from local 'noise' (Chorley and Kates, 1969: 2).

During the 1960s there was therefore a fundamental tension between these two halves of the discipline. On the one hand there were physical geographers who, particularly in the United States, found themselves much more closely allied to geology and earth sciences, whereas on the other there were those seeking to reintegrate physical and human geography. In this latter vein, one solution advocated by Chorley and Kates (1969: 3) was 'to take a philosophical attitude implied by an integrated body of techniques or models (commonly called spatially oriented) and demonstrate their analogous applications to both human and physical phenomena'. The outstanding example of such an attempt to integrate physical and human phenomena in a single approach was the paper by Woldenburg and Berry (1967) in which they compared rivers and central places as analogous systems. However, it was in the emergence of a systems approach to geography that this tradition reached its fullest development.

An informal systems approach to geography existed long before attempts were made to use systems as a core uniting concept in the discipline (Bennett and Chorley, 1978). However, it was Chorley's (1962) paper on geomorphology and general systems theory, and then Haggett's (1965) *Locational analysis in human geography* that marked the formal introduction of a systems approach to the discipline. At its most basic level, such an approach concentrated on the identification of elements within each system, on the links between these elements, and then the links between systems. It was this connectivity, and the opportunity that systems provided for the measurement of elements and flows that was so appealing. Although a systems approach to geography seemed attractive, it is surprising how little research actually incorporated such an approach explicitly (Langton, 1972; Bennett and Chorley, 1978). The first substantial work to focus overtly on systems in geography was Chorley and Kennedy's (1971: vii) *Physical geography: a systems approach*, which had as one of its key aims the intention 'to show how the phenomena of physical geography can be rationalized and perhaps made to assume new significance and coherence when treated in terms of systems theory, statistical analysis, cybernetics, and other modern inter-disciplinary approaches to the features of the real world'. Chorley and Kennedy (1971) suggested that systems could be classified either according to functional criteria, into isolated, closed and open systems, or according to their internal complexity, into morphological, cascading, process–response and control systems. However, in addressing these different kinds of system, they also focused attention on processes such as energy flows, feedback, equilibria, entropy and self-regulation. These were key concepts in von Bertalanffy's (1956) general systems theory, which sought to provide an overarching theory accounting for the common features among systems in different areas of science. Although some, such as Greer-Wootten (1972) saw this as having considerable advantages for geography, others, most notably Chisholm (1967: 51) saw it as an 'irrelevant distraction'. The subsequent lack of interest paid to general systems theory would tend to reinforce Chisholm's conclusion.

In contrast to the lack of interest in general systems theory, a variety of different systems approaches to geographical research have continued to find their advocates, most notably Chapman (1977) and Bennett and Chorley (1978) (but see Kennedy, 1979). At the heart of such approaches lies the utility of systems as a framework for analysis in which it is possible to combine human and environmental phenomena. This integration between people and environments in a systems framework has found its clearest expression in the development of the ecosystem concept by biologists (Odum, 1963) and geographers (Stoddart, 1986). The term ecosystem was first used by the plant ecologist Tansley (1935) to refer to all of the plant and animal organisms living together in a habitat. Central to his definition was the idea that an ecosystem is maintained through interactions between organic and inorganic factors. This integration between the physical and the biological world closely paralleled the long tradition of concern with these issues in geography. In particular, Stoddart (1986) draws attention to the use of an organic analogy as a form of explanation in Ritter and Guyot's teleology, in Davis's concept of the geographical cycle, and in the Chicago sociologists' conceptualization of human ecology. Moreover, in the 1920s Barrows (1923) had advocated that human ecology formed the very basis of the science of geography. Stoddart (1965) has suggested that four aspects of the ecosystem are of particular relevance to geography:

1. The concept brings together human, biological and zoological elements within a single integrated framework;
2. Ecosystems are structured in an orderly way that enables them to be studied logically by geographers;
3. They function through a continuous flow of inputs and outputs, which can be measured; and
4. In general systems terms, they are examples of open systems tending towards a steady state.

Despite this apparent relevance, and the quite widespread advocacy of the concept (Chorley and Kennedy, 1971), it is surprising that ecosystems were not more extensively used as a framework for empirical research by geographers (but see Bayliss-Smith, 1982). At least three reasons can be identified for this failure: first, as Chorley and Kennedy (1971: 329) point out, 'The position and relevance of biogeography within physical geography has traditionally presented problems'; second, the concept originated within ecology, a branch of biology, and it is here that it has been most widely developed and defended; and third, biogeographers have generally been few in number with a marginal influence on geography as a whole (Stoddart, 1986: 230).

## 5.3 *Explanation, relevance and the social origins of concern*

By the end of the 1960s, the idea of geography as spatial analysis had become widely accepted within the discipline. Indeed, the decade finished with the publication of a work of similar magnitude and importance to Hartshorne's *The nature of geography* but advocating the new quantitative approach. This was Harvey's (1969) *Explanation in geography*. Nevertheless, by no means all geographers accepted this approach to the subject, and even as the new decade dawned, the social and intellectual seeds of discontent were coming increasingly to the fore.

### 5.3.1 *Explanation in geography*

Just as *The nature of geography* had formed a key text for a generation of geography students in the United States, so too did *Explanation in geography* for a subsequent generation of British students. However, Hart's (1979: 111) account of the significance of *The nature of geography* in the 1950s is probably also equally applicable to *Explanation in geography*: '*The Nature of Geography* had appeared just before the war, and all right-thinking graduate students slept with a copy under their pillows. A few had even read parts of it, and quoting it was one of our favourite indoor sports. Hartshorne was certainly our most quoted and least understood author'.

In *Explanation in geography*, Harvey (1969: viii) was concerned with 'the ways in which geographical understanding and knowledge can be acquired and the standards of rational argument and inference that are necessary to ensure that this process is reasonable'. Although Harvey (1969: 6) claimed that his book was 'concerned with methodology rather than with philosophy', in his preface he recognized that in order to understand the quantitative revolution he had had to adjust his own philosophical position. Indeed, he was one of the few people who was able to state quite categorically that 'The quantitative revolution implied a philosophical revolution' (Harvey, 1969: vi). For Harvey (1969: vii) 'it was the philosophy of the scientific method which was implicit in quantification', and he argued that for geographers 'the fantastic power of the scientific method' (Harvey, 1969: vi) had particular, but as yet unrealized, appeal. Herein, though, lies a fundamental problem with Harvey's (1969) project: it was built on the idea that there was but a single philosophy of the scientific method, and it failed to develop a critique of logical positivism upon which that philosophy was based. Indeed, as Cloke, Philo and Sadler (1991: 13) point out, *Explanation in geography*, 'the foremost scientific methodological text of the day, said virtually nothing about positivism as philosophy'. While Harvey (1969: 8) did reject the claims of 'some philosophers, logical positivists of the extreme variety, who have held that all knowledge and understanding can be developed independently of philosophical presuppositions', his own work was implicitly built upon the very foundations of that philosophy.

In *Explanation in geography*, Harvey (1969: 63–4) bemoaned the lack of attention paid in geography to the philosophy of science, and in particular to considerations of explanatory form. What was appealing in his view of the scientific method was that it provides control 'over the reasonableness and consistency of statements which we make about reality' (Harvey, 1969: 61). The vast bulk of *Explanation in geography* is thus about the methods geographers can use in achieving this ideal, through the development of hypotheses, laws and theories largely in the context of models and systems. In a nutshell, Harvey (1969: 482) establishes 'a simple structure for explanation in which initial conditions and covering laws are brought together to allow the deduction of the event to be explained'. Somewhat paradoxically, given his avowedly methodological aims, Harvey (1969: 482) concludes with a return to philosophical considerations, recognizing that 'it is not always possible to separate philosophy and methodology', and that while 'an adequate methodology provides a *necessary* condition for the solution of geographical problems; philosophy provides the *sufficient* condition'.

The tensions in Harvey's (1969) work closely reflect the arguments used by Habermas in his critique of empirical–analytic science. Thus Harvey's (1969) focus on methodology rather than philosophy is a fine example of what Habermas (1978) sees as the way in which positivist science protects itself against epistemological self-reflection. Likewise, Harvey's (1969) emphasis on the power and control offered by the scientific method is indicative of Habermas's (1978) concerns with the way in which science applies its methods without any consideration of knowledge-constitutive interests. Above all, Habermas (1978) reminds us that one of the central characteristics of logical positivist science is its aim of providing technically useful knowledge. It was just this aim that struck many geographers during the 1960s as being of such fundamental importance.

### *5.3.2 Boston 1971 and the relevance debate*

*Explanation in geography* concludes with a clarion call for the development of geographical theory, and Harvey (1969: 486) suggested that 'Perhaps the slogan we should pin upon our study walls for the 1970s ought to read: "By our theories you shall know us".' Despite the advances made in quantitative methodology, the incorporation of apparently rigorous statistical descriptions, and the attempts to develop spatial laws, the new geography had failed dismally to develop any major new theories to explain either human or physical features of the earth's surface.

Furthermore, a glance at the most important English language geographical journals, such as the *Annals of the Association of American Geographers*, or the *Transactions of the Institute of British Geographers*, let alone French or German journals, indicates that towards the end of the 1960s much research was continuing to be published in traditions very different from that advocated by proponents of the so-called quantitative revolution. There was scarcely the unanimity of agreement concerning the subject matter and method of enquiry appropriate to geography that would be required for the new approach to be termed a paradigm. While it was most readily adopted in process geomorphology, in economic geography and in urban geography, some systematic specialisms within the discipline, most notably historical geography remained relatively unscathed (see for example Baker, Hamshere and Langton, 1970; Darby, 1973; although note that Harvey's (1961) own Ph.D. thesis entitled 'Aspects of agricultural and rural change in Kent, 1800–1900', did adopt correlation and regression techniques in a historical context).

Moreover, as Taylor (1976: 138) has illustrated, the introduction of new methods and ideas in geography during the 1960s also had a generational dimension to it: 'the older geographers with little quantitative skills and the young geographers with some knowledge of statistical techniques'. As he points out elsewhere, 'For those near the bottom of the hierarchy, the best short-term strategy is clearly to challenge and possibly overturn the existing ideology' (Taylor, 1976: 132). In Britain, this generational conflict was reflected in a series of papers written by older geographers, countering what they saw as the excesses of the 'new' geographers, particularly their unbridled confidence (Stamp, 1966; Smailes, 1971; Steel and Watson, 1972; Farmer, 1973; Steel, 1974).

By 1971, disquiet within the geographical profession in the United States had, however, reached an altogether different level. The late 1960s had seen increasing evidence that the euphoria associated with 'the scientific method', and belief in its ability to solve either environmental or social problems had been ill-founded. In 1964 direct United States involvement began in the war in south-east Asia, to be followed in 1965 and 1966 by heavy bombing of North Vietnam; by 1969 over 550,000 US troops were involved in the theatre of war, despite the growing voice of the peace movement. The failure of the Civil Rights programme established in 1963 led to the first main race riots in Chicago in 1966, and these were enflamed still further by the assassination of Martin Luther King in April 1968. Both the Vietnam war and the Black struggle had a profound influence on many geographers, perhaps most notably on Bunge (1979b), who only a few years earlier had been at the forefront of the advocates of a theoretical geography based on logical positivism.

Whereas the late 1950s and the early 1960s had been a period of apparent affluence and economic expansion, the second half of the 1960s and the early 1970s witnessed increasing economic problems and social unrest in the capitalist states astride the northern Atlantic. This disquiet was in part fuelled by growing concern with the social distribution of the surplus that had been generated over the previous decade, and in Britain, for example, it found its expression in the series of prolonged and bitter strikes of 1972 and 1973. Moreover, the surplus itself enabled a generation of people to experiment with other forms of life-style, which had not been possible in the 1940s and early 1950s when European and North American societies were struggling to overcome the economic privations of warfare and its aftermath. Hence, the late 1960s was a period of social experimentation, reflected for example in the flower power movement, in the increasing use of hallucinatory narcotics, and in the rise of different clothing fashions and musical expressions. Interestingly all of these, despite their apparently anarchic sentiments, provided opportunities in a capitalist economy for the accumulation of considerable financial wealth to those disseminating the material products of the new fashions. Intellectual trends also paralleled these social movements. Above all, in Europe, never haunted by the spectre of McCarthyism, a radical tradition built in part upon the arguments of Marx, and reflected in works such as

Marcuse's (1964) *One dimensional man*, began to grow in strength. For a brief moment at the end of the 1960s, such intellectual ideals achieved real social and political expression, illustrated for example by the student protests of 1968 in France.

Within geography, these issues first came to the fore in the United States at the Boston meeting of the Association of American Geographers in 1971. Reporting on this meeting, Prince (1971) noted the increasingly difficult financial climate in which geographers were trying to undertake research, but more importantly he also reflected the growing concern with the irrelevance of much of that research. He thus commented that

> The power of technology to find remedies for the ills of the world cannot be taken for granted. . . . Methods currently used to forecast regional needs for transport, employment, housing, social services and recreation are known to be unreliable, cost benefit analysis and input-output studies are unsatisfactory guides for making locational decisions, and development studies in the Third World are of dubious value (Prince, 1971: 151).

As a result,

> At the annual general meeting members of the Association resolved to start putting their own house in order and to take notice of the sufferings of the outside world. Resolutions were passed inviting greater participation in the work of the Association by French and Spanish-speaking geographers, enlisting student representation on Council, setting up an inquiry into the status of women in the profession and calling for an end to American military involvement in South-East Asia. Whatever may or may not be done to implement these directives, geographers have been reminded that, collectively and individually, they have responsibilities extending beyond their classrooms and libraries (Prince, 1971: 152).

Similar views were echoed by Smith (1971: 153) who suggested that 'A new wind of change is beginning to blow, in the form of an emerging "radical" geography and an embryonic "revolution of social responsibility".'

This change was not, however, greeted everywhere with enthusiasm (Chisholm, 1971). Robson (1971: 137) thus asked whether concerns with black ghettos and American involvement in Vietnam were 'what geographers, *as geographers*, should be worrying about?', and argued that 'There is still a convincing case to be made for divorcing academic study from value judgements.' Likewise, Berry (1972: 77), noting the generational nature of such debates, asked 'How much of the noise is simply the current fad, new entrants to the field seeking their "turf", and how much goes deeper?' He suggested that neither the 'white liberals' nor the 'smaller group of hard-line Marxists' had '*any profound commitment to producing constructive change by democratic means*' (Berry, 1972: 77) and he argued strongly that 'an effective policy-relevant geography involves neither the blubbering of the bleeding hearts nor the machinations of the Marxists. It involves working with – and on – the sources of power and becoming part of society's decision making apparatus' (Berry, 1972: 78). Such arguments, though, did not receive whole-hearted support (Blowers, 1972; Dickenson and Clarke, 1972; Eyles, 1973), and thereafter the 1970s witnessed a plethora of attempts to provide a new foundation for geographical enquiry, that sought to make it of greater relevance to social justice. Most began with the development of a critique of the underlying philosophy upon which the 'new' geography of the 1960s had been based, that of logical positivism (Gregory, 1978).

### *5.3.3 The failures of logical positivism*

The above account of the emergence of geography as a spatial science has highlighted three important characteristics: first, it was initiated largely because of a desire by a new generation of geographers to establish the discipline upon rigorous grounds which were accepted by the wider scientific community; second, the majority of the concepts and methodologies adopted in this new scientific approach were borrowed from other disciplines, and few could be identified as being specifically geographical; and third, the goal of such enquiry was to establish laws and theories which, at one and the same time, could provide both explanation and prediction.

The views of Habermas outlined in Chapter 2 provide useful insights in accounting for the uncritical acceptance of the logical positivist model of science by many geographers in the 1950s and 1960s. Four aspects of his argument are particularly pertinent. First, the general identification of knowledge with science (Habermas, 1978: 4), among not only academics but also policy makers, meant that it was not easy for geographers who might have wished to develop other ways of knowing to have been able to argue for the inclusion of the discipline as a science. This was particularly important at a time when higher education was expanding, and in order for geographers to establish their discipline on firm footings, it was incumbent for them to reflect its 'scientific' merit. Second, at the heart of both Comte's original formulation of positivism and the later exposition of logical positivism, was the view that the phenomena of the human world could be analysed in precisely the same way as those of the natural world. For geographers, keen to integrate both the human and physical elements of the discipline, such an underlying concept, even if reflected only implicitly, was likely to have enormous appeal. If it was inappropriate, then there seemed to be every likelihood that the discipline would fragment into two entirely separate parts. Third, as exemplified by Harvey's (1969) discussion of the relationship between philosophy and methodology, a fundamental characteristic of logical positivism was the use of philosophy only in so far as it provided a means by which philosophy itself could be shown to be irrelevant and thus ignored (Habermas, 1978). The blind acceptance by many geographers of the logical positivist model of science, served to prevent any debate over the way in which knowledge itself was constructed. Fourth, the underlying technical interest of much geographical research in the 1960s further substantiates Habermas's (1978) account of the cognitive interest of empirical–analytic science. Such a technical interest in geography was not only reflected in the discipline's concerns with achieving technical solutions to social and environmental problems, but also in the explosion of techniques associated with the new quantitiative methodology and with the large sums of money spent on equipment to undertake such technical analyses. Methodology, frequently became an end in itself, and the mastery of statistical techniques spawned a plethora of books and courses on quantitative techniques for geographers (Garrison and Marble 1967a, b; Hammond and McCullagh, 1974).

The apparent inability of geographical methodology to offer solutions to many of the social, economic and environmental problems of the late 1960s forced a number of human geographers to look critically at the underlying philosophy upon which such methodologies had been developed. However, among physical geographers, and particularly geomorphologists, such movements were seen as being at best irrelevant, and at worst downright divisive. For physical geographers concerned with the low explanatory power of their models, the answer was to refine them still further, to develop new techniques, and to ally themselves still more closely with the successful 'hard' sciences. The apparent objective certainty of the natural world and the technical rigour of their methods convinced many physical geographers of the undoubted appropriateness of logical positivism to their field of study.

CHAPTER 6

# *Geography and historical–hermeneutic science: the quest for understanding*

> Most men will not swim before they are able to. Isn't it witty? Naturally they will not swim! They are born for the solid earth, not for the water. And naturally they won't think. They are made for life, not for thought. Yes, and he who thinks, what's more, he who makes thought his business, he may go far in it, but he has bartered the solid earth for the water all the same, and one day he will drown.
>
> Hermann Hesse, *Steppenwolf*
> (Harmondsworth: Penguin, 1966: 21)

Reactions to the perceived problems of the logical positivist tradition of geography have generally been identified as following one of two main alternatives: on the one hand a broadly humanist perspective, and on the other a radical neo-Marxist or historical materialist approach (Kobayashi and Mackenzie, 1989; Johnston, 1991a). While there is some overlap, and in the 1990s an increasing dialogue between these traditions, their practice in the 1970s and 1980s was sufficiently different in emphasis to justify their separation. This chapter therefore focuses on the former, whereas Chapter 7 addresses the radical tradition and the development of a critical geography.

One of the main criticisms of the logical positivist tradition in geography was that the laws and models developed during the 1960s failed sufficiently to address individuals and the human condition. This led many geographers (Entrikin, 1976; Tuan, 1976; Ley and Samuels, 1978a; Buttimer, 1979) to turn to the philosophical approaches of humanism, and in particular to its hermeneutic traditions. Central to these were a concern with reflection and *understanding*, as opposed to the logical positivist goal of *explanation*. In Dilthey's (1913–67) formulation, understanding provided the foundation of the human sciences (*Geisteswissenschaften*), whereas explanation was the goal of the sciences of nature (Bauman, 1978; Outhwaite, 1987). Consequently, many of those who advocated the incorporation of hermeneutic perspectives into geography effectively reinforced the human–physical duality of the discipline.

## 6.1 Geography: the magpie discipline

Humanism was one of the most dominant philosophical perspectives of antiquity. Indeed, as Ley and Samuels (1978b: 3) note, 'the idea of an alternative, nonhumanist perspective was anathema to the dominant modes of ancient thought'. However, modern 'western' traditions of humanism owe their origins to a combination of the self-consciousness born in the Italian Renaissance of the 15th century and the separation between the humanities and the social sciences that resulted from the 19th-century formulation of an objective science, largely devoid of human interpretation. The task of 20th-century humanists was thus 'to put man, in all his reflective capacities, back into the center of things as both a producer and a product of his world and also to augment the human experience by a more intensive, hence self-conscious, reflection upon the meaning of being human' (Ley and Samuels, 1978b: 7).

Prior to 1970 geography, particularly as it was practised in Britain and the United States, had been influenced by remarkably few of the broad developments that had taken place in the social sciences during the 20th century. In contrast, the work of French geographers was situated much more strongly within a humanist framework, one in which Marxist perspectives were also more readily acceptable. The next two sections thus briefly explore the reasons underlying the shift towards an interest in human behaviour, and also the French interface between geography and the social sciences.

### 6.1.1 Geography, human behaviour and space

From the end of the 1960s, two broad approaches were used to reincorporate people into human geography. On the one hand were attempts to overcome the assumptions of perfect knowledge and rational human behaviour (Wolpert, 1964) that underlay most of the spatial models developed during the previous two decades (Cox and Golledge, 1969, 1981; Bunting and Guelke, 1979). Such studies, generally referred to under the name of behavioural geography, remained essentially within a logical positivist framework, and still sought to develop models and theories that would explain group behaviour (Gold, 1980). On the other hand, were works which consciously sought to replace the epistemological and ontological foundations of the logical positivist tradition, through recourse to humanist philosophies, most notable phenomenology and idealism (Mercer and Powell, 1972). Before these two approaches are considered in more detail, it is important first to examine the relationships between geography and the social sciences in which such different discourses had been developed, because with few exceptions the emergence of behavioural geography and the adoption of humanist perspectives reflected the introduction of ideas developed in other disciplines rather than any substantial new contributions by geographers.

Early criticisms of the views of geography as spatial science concentrated on the inability of the models developed actually to predict spatial patterns of human behaviour in the very fields, such as industrial location and the organization of retailing, that they were attempting to explain. This led geographers such as Downs and Stea (1973, 1977) to turn to research on environmental images and perception by psychologists, sociologists and planners, such as Lynch (1960). In so doing, they recognized that such scholars had for a long time maintained and developed an interest in the spatial and environmental aspects of their disciplines, even if only tangentially. The recognition that all forms of human existence have both temporal and spatial expression thus forced geographers, such as Soja (1971), Olsson (1975) and Sack (1980), to abandon any attempt to use physical space as the exclusive organizing concept of their discipline. Paradoxically, the increasing discourse between geographers and other social scientists that this generated, has led more recently to the recognition that most social sciences have in fact paid insufficient attention to questions of space, and that one of the contributions of geographers to the contemporary development of social theory has indeed been their integration of space into social enquiry (Giddens, 1981; Gregory and Urry, 1985b; Harvey, 1985a, b).

Associated with this has been a shift in emphasis from conceptions of space as being absolute to the increasing acceptance of space as being relative (Sack, 1980; Sayer, 1985b). In classical Greek thought two views pervaded the discussion of objects or matter in space. On the one hand the Pythagoreans denied the existence of truly empty space, and argued that for an object to move it must replace an identical volume of something which they called space. On the other, the atomists, following the ideas of Leucippus and Democritus, distinguished between atoms and the void between them, and thus argued that space was merely a void in which matter existed (Smart, 1964b). The absolute, or substantialist, view of space, building on the ideas of the Pythagoreans, reached its clearest expression in the writings of Descartes and Newton, who in essence suggested that space and time have an existence independent of their contents (Newton-Smith, 1986). In contrast, in the early 18th century Leibniz proposed an alternative relative, or reductionist, view which considered space as 'merely a system of relations in which indivisible "monads" stand to one another' (Smart, 1964b: 6). Absolute space can thus be seen as existing in a real sense, independent of its observers, whereas relative space can only be comprehended through an understanding of relations between objects. The conceptualization of geography as a spatial science was largely based upon an absolute view of space, but the humanist critique that developed in the 1970s emphasized that 'space can only be understood in terms of the objects and processes that constitute it, with the implications that the study of space must be rooted in social theory' (Sayer, 1985b: 51).

### 6.1.2 *Geography and the social sciences in France*

Buttimer (1971: 1) has emphasized that 'Unlike other geography schools of the twentieth century, which tended to treat man individualistically or as the pawn of economic law, the French maintained an Aristotelian vision of collective man as *zoon politikon*, organized into spatially recognizable social groupings.' French geography during the first half of the 20th century thus maintained stronger links with broader traditions of social science than did the geographical community in Britain and North America, where the logical positivist emphasis had allied the discipline much more closely with the natural sciences (Ley, 1977). This was particularly evident in the links between Durkheim's social morphology and the human geography of Vidal de la Blache (Berdoulay, 1978). As Daudé (1937: 56) noted, 'Human geography and social morphology study the same phenomena. The former, however, studies them in terms of their connections with the geographical milieu . . . , while the latter studies them in terms of their connection with the social milieu.' Although, as Andrews (1984) has stressed, it is not easy to trace the differences and similarities between these two schools of thought, the central point to be grasped is that there were interactions between them, and that these changed over time. This meant that within France there was a dialogue between geographers and other social scientists; French geography in the tradition of Vidal de la Blache was always *la géographie humaine*, with strong social connotations, in marked contrast to the natural science emphasis which geography acquired in the Anglo-American realm (see Buttimer, 1978). Although some geographers, most notably Fleure (1947) in Britain, and Sauer (Leighly, 1963) in the United States, did indeed reflect a similar concern with the links between society and milieu, theirs was not the view that came to dominate the discipline.

In France, the success of *la géographie humaine* in maintaining a regional tradition of geographical research far longer than was the case in Britain, Germany or North America, meant that with the demise of regionalism in the post-1945 era there was little clear focus for the discipline. As Buttimer (1971: 137–8) has emphasized, French geography in this period 'lacked the scientific sophistication of the German *Allgemeine geographie,* nor could it be compared with the systematic human geography that developed in Anglo-American schools'. Nevertheless, geographers such as Cholley (1948), Le Lannou (1949) and Sorre (1961), continued to emphasize the social context of the discipline in their research, and George (1966) in particular maintained a close dialogue with contemporary sociologists such as Gurvitch (1958–60). Moreover, the long-established links between geography and history in France, ensured a continuing close relationship between the two disciplines in the broad area of historical geography, albeit one in which geographers played a largely secondary role.

### 6.1.3 *Geography and the social sciences in Britain and the USA.*

The previous chapter has illustrated that the form of geography that emerged in the United States and Britain during the 1950s and 1960s was heavily derivative from the work of a particular group of German philosophers, economists and geographers. However, in their original formulations in disciplines other than geography, many of the models and theoretical postulates associated with this geographical expression did indeed have close links with deeper currents of social enquiry. Thus Gregory and Urry (1985b: 1) note that von Thünen's model of agricultural land use was 'conspicuously informed by Hegel's political philosophy', and that Alfred Weber moved on from his abstract model of industrial location to develop links with 'a much broader cultural sociology which had much in common with his brother Max's remarkable research programme, in the course of which Alfred roundly rejected the possibility of an autonomous, purely geometric (or even singularly economic) location theory' (Gregory and Urry, 1985b: 2). Given these close links, it is salient to examine why it was that geographers prior to the 1970s failed sufficiently to maintain a dialogue with these other traditions of social theory.

At least three reasons can be adduced for this. First, although the traditions of environmental determinism and the regional approach to geography invited links with humanist perspectives (Glacken, 1967), the reactions to these approaches in the 1950s placed geography firmly within the dehumanizing context of logical positivism. In part, this was a result of the relative strength of the emerging process-oriented physical geography, but it also reflected the social and economic context in which the discipline was emerging. Second, geographers during the 1950s and 1960s were actively trying to create a niche for their discipline within the academic division of labour. This required them to focus on what they saw as being distinct about their discipline, rather than to explore the possible links with other subject areas, be they in the social or natural sciences. Thirdly, the political environment of the 1950s and early 1960s, particularly in the United States, was one in which radical traditions of social theory were actively discouraged. For social scientists to succeed professionally, it was incumbent on them to produce results that were deemed to be useful in furthering capitalist society.

The evident failings of the logical positivist approach to social understanding meant that from the late 1960s some geographers began to turn to theoretical debates in the other social sciences, and found there a wealth of ideas with which they were not as yet familiar. This led to a plethora of papers seeking to introduce new, potentially relevant, concepts to the discipline. The efforts of geographers to search out this new field of opportunity, however, were reminiscent of the actions of magpies, collecting nuggets of gold, but storing them up, largely unused.

## *6.2 Behavioural geography and the demise of rational economic man*

A central failing of the spatial models developed in geography during the 1950s and 1960s was that they did not sufficiently *explain* the phenomena that they were intended to. Thus, by their own criteria of scientific evaluation they were unsuccessful. As Johnston (1991a: 137) has summarized, 'The theory suggested how the world would look under certain circumstances of economic rationality in decision-making; that those circumstances did not prevail suggested that the world should be looked at in other ways in order to understand how people do behave and structure their spatial organization.' Consequently, geographers such as Brookfield (1969), Saarinen (1969) and Golledge, Brown and Williamson (1972) sought to improve their explanatory and predictive capacity through recourse to studies of human perception and behaviour. In so doing, they turned for their initial inspiration primarily to research in psychology, but also to the few earlier enquiries into the field that had been undertaken by geographers.

### *6.2.1 Geographers and the behavioural environment*

A concern with human environmental behaviour was not a new departure for geographers in the late 1960s. The form of cultural geography developed under Sauer at Berkeley was deeply imbued with the idea that it was through the interpretation of culture that human landscapes were forged from the environment (Leighly, 1963). Likewise, Wright (1947) had exhorted geographers to examine the role of imagination and the private worlds of individuals in their enquiries. However, it was a short paper by Kirk (1952) that provided the first real attempt to integrate geographical research on the behavioural environment with work being undertaken by psychologists.

Kirk's (1952) paper addressed the place of historical geography in the context of debates over possibilism and the relationships between people and the environment. His central concern was to find 'some working hypothesis in which nature and humanity are brought under one discipline' (Kirk, 1952: 158), and he suggested that this could be found in the *Gestalt* psychology developed by Köhler (1929) and Koffka (1929). Interestingly, he likened a *Gestalt* to 'a region in its dynamic aspect, a whole which is something more than the sum of the parts' (Kirk, 1952: 158). Using this concept, he then argued that the physical state of any human group or individual depends in part on the character of the physical environment, but that any action within the group will begin in the relief of stresses in an internal environment, which he termed the behavioural environment. Such stresses, he suggested, resulted from both the product of group culture and the act of observation of the physical environment. For Kirk (1952: 159) it was in the behavioural environment that 'the gap is closed between Mind and Nature'. Although Kirk introduced the *Gestalt* concept to geographers, he failed to make clear precisely how it explained the construction of his behavioural environment, and the obscure place of the paper's publication, in the Indian Geographical Society's Silver Jubilee volume, meant that it did not influence a wide audience during the 1950s.

Other early attempts to provide a conceptual framework for considering the behavioural environment included that of Lowenthal (1961), who took as his starting point Wright's (1947) concern with 'the relation between the world outside and the pictures in our heads' (Lowenthal, 1961: 241). In this, he specifically addressed the question of perception, drawing attention to the importance of personal geographies, arguing that 'Separate personal worlds of experience, learning, and imagination necessarily underlie any universe of discourse' (Lowenthal, 1961: 248). For Lowenthal (1961: 260), behaviour based on personal perceptions has its unique aspects, and must form a central focus for geographical enquiry, and he concludes that 'The geography of the world is unified only by human logic and optics, by the light and color of artifice, by decorative arrangement, and by ideas of the good, the true, and the beautiful'. Lowenthal's (1961) paper, however, is discursive rather than programmatic, and while bringing together discussion of perception and behaviour it did not intend to offer an agenda for future research. That agenda was provided by a series of studies in the later 1960s and 1970s, which concentrated on introducing aspects of human perception and behaviour into geographical enquiry (Cox and Golledge, 1969; see also Sonnenfeld, 1972), largely through a consideration of hazard perception, migration, landscape evaluation, mental maps and the development of time geography.

### 6.2.2 *Behaviour within the framework of logical positivism*

Among the earliest attempts by geographers to understand the relationship between perception and behaviour were a series of studies coordinated by White at the University of Chicago in the early 1960s on natural hazard perception (White, 1961; Burton and Kates, 1964; Saarinen, 1966). These were underlain by the argument that human behaviour was directly influenced by perception, and where it did not match the theoretical predictions of spatial science this was because of imperfect knowledge. The links between this research and the wider adoption of a systems approach to geographical enquiry were clearly illustrated in Kates's (1971) development of a systems model of human adjustment to natural hazards, in which cognition of a hazard is seen as creating a response mechanism which modifies the human use subsystem, the natural events subsystem or both (see also Brookfield, 1969).

One of the central theoretical bases underlying the ideas of White and his colleagues was Simon's (1957) concept that decision-making was bounded, as a result of imperfect knowledge. Accordingly, the standard assumptions of perfect knowledge and rational behaviour underlying many of the classical geographical location models could no longer be held to be true. Wolpert (1964) also adopted Simon's so-called satisficer model of human behaviour in his analysis of farmers' decision-making in Sweden. Here he illustrated that even if profit motives featured highly among farmers' goals, their finite ability to perceive and store information meant that they did not necessarily achieve optimal solutions.

A third strand in the emergent behavioural geography of the late 1960s and early 1970s was a concern with the measurement of spatial perception, typified in the research on mental maps (Downs and Stea,

1973; Gould and White, 1974). This generally sought to identify images held by individuals about particular areas, and then to relate these to social and economic characteristics of respondents. However, mental map research was based on the assumptions that individuals carry around with them already constructed map images, and that these then form the basis for their subsequent spatial actions. Neither of these assumptions has been rigorously tested, and both are open to substantial criticism. Moreover, given Lowenthal's (1961: 251) emphasis on the 'uniqueness of private milieus', the assumption that it is possible to make sensible statements about group generalizations derived from the mental images of a group of individuals is also open to debate.

In a substantial review of the achievements of behavioural and perception geography, Bunting and Guelke (1979: 448) suggest that the results of this 'research are of little value in the explanation of real-world geographical activity'. In particular they challenge two assumptions underlying the philosophy and methodology of most research in the field: 'that *identifiable environmental images exist that can be measured accurately*'; and 'that *there are strong relationships between revealed images and preferences and actual (real-world) behavior*' (Bunting and Guelke, 1979: 453). With respect to the first of these assumptions, they point out that the methods adopted by geographers to identify environmental images have often not been appropriate to the complexity of the subject. Thus, they argue that 'Many geographers would seem to have moved well-equipped from the quantitative revolution into behavioral frontiers. Yet no revolutionary breakthroughs have developed in response to the acknowledged need to measure individual behaviour' (Bunting and Guelke, 1979: 455). Moreover, they also express concern about the lack of generally accepted criteria against which mental phenomena can be evaluated (although see Potter, 1977). Turning to the second underlying assumption, Bunting and Guelke (1979: 460) note that 'there is effectively no empirical evidence to substantiate a clear and direct relationship' between environmental perception and behaviour, and they suggest that there therefore needs to be a 'refocusing towards overt or active behavior patterns' (Bunting and Guelke, 1979: 456) in geographical research.

In responding to this critique, Saarinen (1979; see also Downs, 1979) suggested that the achievements of geographical research on environmental perception were in practice much greater than Bunting and Guelke had given credit. In particular, he argued that the ideas generated by geographers and psychologists working in this field had 'helped to demolish the myth of economic man and led geographers to a more realistic search for factors influencing environmental decision-making' (Saarinen, 1979: 466). Furthermore, Saarinen (1979) stressed the interdisciplinary character of research on environmental perception, and criticized Bunting and Guelke for focusing too narrowly on a limited area of this much broader field.

An underlying problem with much behavioural geography, as Bunting and Guelke (1979) pointed out, was that it generally lacked any substantial grounding in theory. One solution was to turn to theoretical work in psychology, such as Kelly's (1955) personal construct theory, which Hudson (1981: 346) argued 'seemed to offer geographers a way forward. Not only did this theory place individuals' personal constructions – images – of environments in a pivotal role in the understanding of human behaviour, it prepared a flexible, valid and individually sensitive method of measuring personal constructions, the Repertory Grid.' One of the most appealing features of this methodology was, as Townsend (1977: 431) pointed out, that it appeared to provide a way of extracting people's 'own views with a minimum of interviewer interference or contamination'. However, once again, a central problem of personal construct theory is that, although offering a theory of cognitive structure, it fails to provide a sound basis for the explanation of behaviour. Moreover, it is also based on the premise that individuals have a free choice in making preferences, and it thus pays insufficient attention to the structural constraints which may restrict such choices.

Such research on human behaviour by geographers in the 1960s and 1970s sought to move away from the excessive dehumanization of geographical practice associated with many of the theoretical models being developed at approximately the same time. While challenging concepts of rational human behaviour and insisting on the inclusion of measures of perception, however, they generally failed satisfactorily to unravel the complex processes by which human behaviour is influenced by, and itself influences, social structure.

### *6.2.3 Time geography*

An alternative approach to understanding human behaviour was the time geography developed at Lund in Sweden by Hägerstrand (1975). Initially this grew out of his concern with diffusion processes, and with situating people in both space and time. It thus emerged within a logical positivist framework, and has frequently been classified as representing a behavioural view of geography (Carlstein, Parkes and Thrift, 1978; Johnston, 1991a). However, as Johnston (1991a) points out, more recent developments in time geography, particularly by Pred (1981, 1984), have sought to merge it with humanist and structuralist traditions.

Hägerstrand (1975) suggested that there are eight main conditions influencing human life and society:

1. The indivisibility of the human being;
2. The limited nature of human life;
3. Limited human ability;
4. The duration of each task;
5. The consumption of time by movement in space;
6. The limited packing capacity of space;
7. The limited size of terrestrial space;
8. The fact that all situations are rooted in past situations.

The starting point of his time geography was thus the development of a theory which would examine the interactions between these constraints. At its simplest this involved the notation of life-paths within time–space;

life-paths or trajectories could apply at all scales from individuals to communities, and time–space likewise could refer to all temporal and spatial scales, including for example day-paths and life-paths. However, Hägerstrand recognized that an examination of life-paths alone failed to explain the motives and contexts of individuals. Consequently, he invoked the concepts of *projects* and *dioramas* to account for the purposes behind events, and the complex interconnectivities produced by the flow of history.

In its original formulation time geography can be seen in part as a reaction to the dominance of spatial concerns within geography during the 1960s. One of its central aims was thus to reincorporate a sense of time into geographical enquiry, and to consider both space and time as constraints on human action. In his critique of Hägerstrand's time geography, however, Gregory (1985a) suggests that it still retains close connections with the physicalism and individualism of much of Hägerstrand's earlier work on diffusion, which was set firmly within a tradition of empiricism. He suggests that its failure to incorporate the knowledgeability of human subjects and the structures of social relations within which their experiences are sustained represent serious shortcomings of Hägerstrand's approach. To some extent these shortcomings have been addressed by the social theorist Giddens (1981: 4) in his theory of structuration, which seeks 'to connect the time–space constitution of social systems with structures of domination'. To do so, he introduces the concept of what he terms 'time–space distanciation'. Giddens (1981: 4) thus argues that 'The structuration of all social systems occurs in time–space, but also "brackets" time–space relations; every social system in some way "stretches" across time and space.' This process of stretching is what he refers to as time–space distanciation, and he uses the concept to examine the way in which different societies stretch over, or are embedded in time and space. Of particular importance here is the way in which different types of society are able to increase the extent to which they spread over space and time, through their ability to store resources and knowledge. Geographers such as Pred (1984) have, in turn, sought to incorporate some of the ideas of structuration theory into a reinvigorated time geography, which interprets place as a historically contingent process. Pred's (1984) concerns with language and structure, however, are far removed from the behavioural geography outlined in the earlier parts of this chapter, and in order to situate them more fully within the context of recent geographical enquiry it is essential to examine the developments associated with an explicitly humanist tradition within the discipline.

## 6.3 Humanist perspectives

Ley and Samuels (1978b) have noted that many different humanist perspectives were adopted by geographers during the 1970s. Most of these were underlain by the philosophies of existentialism and phenomenology (Buttimer, 1976; Entrikin, 1976; Relph, 1981), but idealism

(Guelke, 1974, 1976) and more recently pragmatism (S J Smith, 1984) and realism (Sayer, 1984, 1985a) have also found their proponents. What united them was a concern with the inability of logical positivism to provide a sound philosophical underpinning to the discipline. Unlike the behavioural approaches discussed in the previous section, many of which reflected an attempt to refine the model-building and law-seeking science of logical positivism, humanism offered a radically different alternative. In Tuan's (1976: 266) words, humanism provided geographers with the central task of reflecting 'upon geographical phenomena with the ultimate purpose of achieving a better understanding of man and his condition'. Tuan (1976: 267) drew the following contrast between scientific and humanistic approaches to the subject: 'Scientific approaches to the study of man tend to minimize the role of human awareness and knowledge. Humanistic geography, by contrast, specifically tries to understand how geographical activities and phenomena reveal the quality of human awareness.' For Entrikin (1976: 616) the humanist approach was a reaction by its proponents to 'an overly objective, narrow, mechanistic and deterministic view of man', with the appelation 'humanistic' reflecting their concern with 'the aspects of man which are most distinctively "human": meaning, value, goals, and purposes'.

Herein, though, lies a basic problem with the practice of humanistic geography: its aims and methods cannot be judged by the same criteria as those of logical positivist science. The introduction of humanist perspectives in geography occurred largely as a form of critique of the view of the discipline as a spatial science; it was not necessarily concerned with the production of new, and technically useful, knowledge. By the criteria for the success of logical positivist science, which include this very production of mechanistically useful technical knowledge (Habermas, 1978), the humanistic experience was thus seen as an irrelevant failure. Indeed, this contrast goes a long way to explaining the hostile reception given to historical–hermeneutic traditions of research by many physical geographers institutionalized within an empirical–analytic conceptualization of science. On a more practical level, while humanism offered human geographers an alternative philosophical view of the world, the lack of agreement over methodology proved to be one of its greatest drawbacks. This can be seen through an examination of the introduction of aspects of phenomenology, existentialism and idealism to geographical enquiry.

### *6.3.1 Phenomenology and the understanding of essence*

Among the major humanist traditions, it was the phenomenology of Husserl and Schutz that first attracted the attention of geographers in the early 1970s (Pickles, 1985). Thus Relph (1970), Tuan (1971) and Mercer and Powell (1972) all used the claims of phenomenology to develop a critique of the empiricist foundations of logical positivism. The publication of these papers in Canada and Australia interestingly reflects the emergence of this critical tradition, initially not in the logical positivist heartland of the United States or Britain, but in areas of the world which were actively developing their own newer geographical traditions, in which historical understanding played a central role.

The key attraction of phenomenology was its rejection of the assumption of objectivity, which enabled logical positivists to ignore the preconceptions and subjectivity upon which their laws and models were based. The concerns of phenomenology with intentionality and the constitution of knowledge thus appeared to offer entirely new avenues of research for geographers. However, Husserl's emphasis on pure reflection in the pursuit of essences meant that there were profound methodological difficulties in adopting his approach. As Gregory (1978: 125–6) has pointed out in referring to Husserl's philosophy,

> There is an essential difference between the contemplative intentions of his transcendental philosophy and the practical concerns of a social science, so that it is scarcely surprising that where geographers have aligned themselves with Husserl's project their efforts have been directed towards the destruction of positivism as a *philosophy* rather than the construction of a phenomenologically sound *geography*.

Geography according to Husserl's formulation of phenomenology would have been a highly personal exercise, involving the suspension of all preconceptions and the pursuit of the essences of the objects and concepts forming the empirical domain of the discipline. While such a transcendental exercise might have been rewarding in terms of individual understanding, its reflective rather than productive character meant that it was unlikely to be adopted by academics whose career paths were in part determined by the publications that they *produced*. This possibly goes some way to explain why so many geographers wrote about phenomenology, although there were very few actual examples of phenomenological research based upon Husserl's ideas.

A desire to escape this apparent impasse and to operationalize some of the more appealing aspects of phenomenology led geographers such as Tuan (1974), Relph (1976) and Buttimer (1976) to turn to the constitutive phenomenology of Schutz (1962) based on a search for meaning. As Ley (1977: 502–3) has argued, 'in all phenomenological traditions the question of meaning is a central concern, for meaning and perception speak of existence, of a subject in encounter with an object'. Schutz's particular contribution was to focus on the intentionality of human action, in order to gain an understanding of social meaning at the level of the life-world, rather than at the deeper transcendental level of Husserl (Gregory, 1978). For Schutz (1967: 52), 'meaning is merely an operation of intentionality' and it is only accessible through reflection. The clearest examples of the practical adoption of such an approach to geographical research are Tuan's (1974) *Topophilia: a study of environmental perception, attitudes and values* and Relph's (1976) *Place and placelessness*. Both of these works seek to reflect on the ties between individuals and the material environment expressed in the definition of place. In contrast to most previous analyses of place, both Relph and Tuan emphasize the social construction of places, taking into account such aspects as their emotional, aesthetic and symbolic appeal.

Such explorations of the human life-world were, however, few. As Gregory (1978: 137–8) has noted, their unfamiliar style and content

> met with resistance not only from positivist geographers who (understandably, if erroneously) complained that they had nothing to do with the proper conduct of science, but also from those who had never been persuaded by the tenets of positivism but still could not see that they had anything to do with the proper conduct of geography.

### *6.3.2 Existentialism: individuality and being*

Closely related to, and indeed partially derived from, Husserl's phenomenology is the philosophy of existentialism (Entrikin, 1976). As expressed in the works of Sartre and Merleau-Ponty, existentialism involves a rejection of well-defined academic philosophies, a return to the concrete world of being as the source of consciousness, and a rejection of idealism. According to Samuels (1978), existentialism provides a way of integrating geographical concerns with space and place:

> existential space involves the making of distance. Any spatial projection, including the projections of geometric analysis, is an example of existential space. But what the latter takes for granted (i.e., the fact of projection from someone), existential analysis elucidates. At root, existential space (meaning any spatial projection) is nothing more than the *assignment of place*.

For Samuels (1978) the appeal of the existentialist view of space is that it can combine the topophilia of Tuan with the geometric models of Isard and Berry. However, existentialism also involves a basic concern with the realities of the human condition, and as influenced by Marx, with the problem of alienation. Samuels (1978: 40) thus expresses the fundamental existentialist critique of modern science and social theory as follows:

> By revealing man's total dependency on his environment, the natural and social sciences successfully reveal man's estrangement from himself; i.e. they proclaim his total subjectivity, denying man's inherent alienation . . . , and, not incidentally, his 'freedom'. That most successful revelation and its employment in social management engineering, in turn, sparks the 'crisis' in modern philosophy, ethics and politics.

Existentialism's concerns with human beings in particular situations and with individuality also provided a potential focus for further geographical research, but despite pleas to this end by Entrikin (1976) and Samuels (1978) little subsequent attention has been paid to its agendas other than in general critiques of logical positivism. In part this is a result of existentialism's emphasis on the individual and freedom of choice, and thus its rejection of a social science concerned with the establishment of regularities governing human behaviour.

### 6.3.3 *Idealism and historical experience*

In contrast to existentialism's focus on reality *as being*, the idealist view of the world sees reality through its constitution by the human mind. As with phenomenology, geographical interest in idealism emerged, not at first in Europe or the United States, but among a group of historical geographers in Canada (Guelke, 1974, 1976; Harris, 1978). This connection with history is important, since it was out of studying the past that Harris (1978) was forced to confront the distinctive problems of historical data. For Harris (1978: 130) 'The historical mind assumes a reality that it seeks to understand,' and 'For the historical mind, the human landscape is the direct result of human action and, therefore, a product of thought, values, and feelings' (Harris, 1978: 127–8).

The most overt advocate of an idealist approach in geography during the 1970s was Guelke (1974, 1976), who defined the idealist project as follows: 'The idealist maintains that a rational action is explained when the thought behind it has been understood. In the idealist view human geography derived its autonomy as a field of geographical enquiry from the fact that it is largely concerned with the rational actions and products of human minds' (Guelke, 1974: 193). This view of idealism is taken largely from Collingwood's (1956) approach to history, which involved attempting to rethink the thoughts of people in the past. However, Guelke's comment above also indicates succinctly how he envisaged idealism as overcoming the problem of the distinction between logical positivist explanation and hermeneutic understanding. For Guelke, explanations of rational actions are achieved through an understanding of the thoughts behind them.

In focusing on thought and action, Guelke develops a concern with intention, which is somewhat similar to that proposed by Schutz, although without the emphasis that the latter gives to reflection. For Guelke (1974: 197), 'An intention provides the occasion on which an individual will apply the theory that he regards as appropriate.' This then requires that individuals have ready to hand a number of different theories which they can consider and apply to various lived situations. The task of the idealist geographer is thus to trace the links between such thoughts and subsequent action. However, Guelke is also concerned to draw similarities between his project and the verification of scientific theories, and it is here that his focus on explanation is most clear. He thus argues that 'scientific theories and interpretations of actions in terms of thought both attempt to explain the world of concrete appearances by postulating the existence of nonperceivable but real entities' (Guelke, 1974: 201).

Guelke's account of idealism is somewhat idiosyncratic and contradictory, directly following neither the metaphysical tradition in which reality is only mental, or the epistemological tradition in which understanding is restricted to the perception of objects. As Chappell (1976) and Harrison and Livingstone (1979) point out, Guelke seems to apply the term 'idealism' merely to describe an atheoretical approach involving the reconstruction of thought behind action. Moreover, his insistence on the explanatory power of his methodology, rather than offering a truly hermeneutic critique of logical positivism, instead confines his arguments to the very philosophy for which he is trying to provide an alternative. As Harrison and Livingstone (1979: 78) summarize:

> His acceptance of the verification principle, epistemological empiricism, and objectivity in science, led only to a reformulated positivism in which human thought was included in the raw material of science. Furthermore, his commitment to these principles has blinded him to the possibility of developing a radical, truly idealist geography, which would do full justice to the presupposition-based study of all human activity.

More recently Guelke (1981) has defended his position, seeking to integrate it within traditional idealist arguments. His central criticism of logical positivist spatial science is that it conceives of a *real* world; for the idealist reality can, in contrast, only be known in the mind. Despite Guelke's advocacy of idealism there have been few geographical studies which have sought to provide specifically idealistic interpretations. Instead, idealism has been used primarily as yet another way of illustrating the shortcomings of logical positivism. Nevertheless, in its emphasis on the links between thought and action, some of the tenets of idealism are beginning to be incorporated with those of phenomenology and existentialism in a very broadly defined humanist or cultural approach to geography.

## *6.1 The historical–hermeneutic alternative*

### *6.4.1 The context and practice of humanistic geography*

As the above account has illustrated, the central attraction of humanism to geographers during the 1970s was that it offered a series of philosophical positions from which to attack the logical positivist tradition that had emerged during the 1950s and 1960s. However, as critics of these early humanist reflections emphasized, they lacked substantial methodological and empirical expressions (Billinge, 1977). The spatial science of the 1960s can thus be characterized as having concentrated on its methodology and generally ignored its philosophical underpinnings, whereas the humanistic geography of the 1970s concentrated on its engagement with philosophy and tended to relegate its methodology to a secondary consideration. This contrast is closely paralleled by Habermas's description of empirical–analytic and historical–hermeneutic sciences, the former with its technical interest expressed through work, and the latter with its practical interest developed through the medium of language. It is thus no mere coincidence that many geographers turning to the humanist critique of logical positivism also began to address questions of geographical language and communication (Olsson, 1975, 1978; Billinge, 1983). However, as Porteous (1984: 372) cautioned, there remained a tendency for most humanistic geographers still to seek to impose their own particular theoretical system on the world: 'Humanistic geographers, from whom we might expect better things, are hung upon the hooks of Hegel, Habermas, Hempel, Hirschbergher, Husserl, Heidegger, and even Hartshorne (with forays into Kant, Cassirer and Collingwood). It is high time these bloodless philosophers dropped their articles and stepped out to see the world both for themselves, and for itself.' Few geographers have experimented with humanistic reflections or interpretations in poetry or painting (see Meinig, 1983), and most still seem to have ignored Porteous's (1984: 373) suggestion that 'The publication of geographical insights in non-traditional forms could be a first step toward the goal of silent place appreciation.'

While such criticisms can be applied to the early geographical engagements with humanism, they are less true of the 1980s, when several papers started to be published with a less overt call to a particular philosophical tradition, and a greater attempt at creating a truly humanistic geography (Cloke, Philo and Sadler, 1991). Early examples of this can be seen in some of the contributions to Ley and Samuels's (1978a) edited volume *Humanistic geography: prospects and problems*, but the development of concerns with literature, meaning and symbol can be found in works such as Pocock's (1981) edited *Humanistic geography and literature: essays on the experience of place*, Jackson's (1989) *Maps of meaning* and Cosgrove and Daniels' (1988) edited volume *The iconography of landscape*. Although these are all heavily situated in theory, their emphasis is primarily on the writing *of* humanistic geography rather than on writing *about* humanistic geography (see also Eyles and Smith, 1988).

These concerns have emerged primarily within the fields of historical and social geography, and they also reflect an increasing dialogue with the structuralist alternatives to logical positivism that are discussed in Chapter 7. During the 1950s and 1960s it was primarily those concerned with industrial location, urban settlement and transport who had turned to the methodological offerings of logical positivist science in order to provide knowledge in the form of explanations and predictions that could be used to advance the brave new world of capitalist society. The humanist tradition in geography thus emerged largely from a background that had been little influenced by the technical concerns of the 1960s. By the early 1970s, for example, there was widely seen to be an increasing rift between historical geography and other branches of the discipline (Baker, 1972), and rather than following the growing trend of quantification and spatial explanation, this tension led a group of young historical geographers, particularly in Britain, but also in Australia and Canada, to develop closer links with historians and other social scientists (see Baker and Billinge, 1982; Baker and Gregory, 1984). This engagement had profound effects for both the humanist and also the structuralist alternatives that were consequently introduced to geography.

The second main context in which humanist perspectives were introduced to the discipline was in the field of social geography (Cloke, Philo and Sadler, 1991). As with historical geography this was also associated with an increasing consideration of Marxist and structuralist alternatives, and is particularly well illustrated in the exploration of social geography offered by Jackson and Smith (1984; see also Jackson and Smith, 1981). Here they are primarily concerned 'with the possibility of mediating between the humanist's characteristic emphasis on subjective and intersubjective experience and the structuralist's typical stress on objective social constraints' (Jackson and Smith, 1984: 12). In part this argument builds on a reinterpretation of the work of the Chicago school of sociology, and an examination of Park's pragmatism. While pragmatism has been defined in many different ways, and can be seen as including elements of idealism, realism and materialism, Susan Smith (1984) suggests that it provides important methodological insights lacking in other humanist alternatives previously advocated by geographers. 'As a humanistic philosophy in the widest sense,' she argues that pragmatism allows 'society the freedom to shape an unfinished world' and to assign 'to human agency a responsibility for the future' (S J Smith, 1984: 366). In particular she argues that pragmatism adresses questions of morality, action, the links between intellect and common sense, and the structure–agency debate, all of which have recently become central to human geography. In practice, such concern with humanistic influence has led to a greater emphasis on individual experience, on meaning, and on the interpretation of place in social and cultural geography. This is well illustrated by Ley's (1983, 1988) interpretations of the way in which people make sense of urban life, and by Jackson's (1987) edited collection of essays on race and racism.

### 6.4.2 *Humanistic geography as a historical–hermeneutic science*

The above account of the development of humanistic geography has emphasized its diversity and the range of humanistic philosophies to which geographers have turned. However, almost all of these enquiries have been underlain by three central interests: a critique of geography based upon logical positivism; a concern with the reintroduction of the complex world of human subjectivity into geography; and a quest for understanding. These characteristics are closely similar to Habermas's conceptualization of historical–hermeneutic sciences, with their practical interest in human communication through the social medium of language, and their reflective critique of logical positivism. However, hermeneutics and humanism are far from synonymous, and there have been few direct attempts to formulate a tightly defined hermeneutic geography (see Mügerauer, 1981; Buttimer, 1983).

Habermas's (1978) concerns with hermeneutics mainly involved an engagement with Husserl's phenomenology, with Dilthey's quest for understanding and with Peirce's pragmatism. As far as human geographers have built on these traditions they can be seen as representing specific attempts at developing a historical–hermeneutic framework within the discipline. In more general terms, though, Habermas's broader conceptualization of historical–hermeneutic science does provide a useful overview of the wider humanistic reflections that have characterized human geography since the early 1970s. In social geography, for example, Jackson and Smith (1984: 17) see close parallels between their philosophical triad of positivism, humanism and structuralism, and Habermas's three types of knowledge, empirical–analytic, historical–hermeneutic, and critical. Much humanistic geography, particularly during the 1980s, has closely followed Habermas's (1978: 309) account of the methodological framework in which historical–hermeneutic sciences gain knowledge: the meaning of the validity of their propositions has not generally been constituted in the frame of reference of technical control; their levels of formalized language and objective experience have generally not been divorced; their theories have not been constructed deductively; experience has not been organized with regard to the success of their operation; and their access to 'facts' has been sought after through the understanding of meaning rather than observation.

At the end of the 1970s, Gregory (1978: 146) suggested that 'the interpretative movements in geography to date have in effect and with very few exceptions served only to conceal the tension which must exist between one frame of reference and another, the deep resonances and discordances which are struck by superficially similar clusters of meaning', and he concludes that 'the hermeneutic task must be to make these explicit and to clarify what makes such vital "immersion" possible'. To do so, he suggested, would mean that 'geography will have to dismantle the oppositions between subject and object, actor and observer, and emphasize the mediations between different frames of reference' (Gregory, 1978: 146). By the early 1990s humanistic geography, as expressed in works such as those by Ley (1983, 1988), Jackson (1989) and Cosgrove and Daniels (1988) has gone some way to achieving these ends.

### 6.4.3 *Physical science and human experience*

The humanist reaction was only one alternative to the conceptualization of geography as spatial science, and as Chapter 7 indicates, it is not easy to separate it from the more radical structuralist critique which developed alongside it. However, it is important to emphasize that many geographers remained largely uninfluenced by either of these developments. This is not only true of physical geographers, but also of the practice of much human geography. Spatial analysis and the scientific approach, by which is generally meant that derived from logical positivism, continue to have their advocates at both a theoretical level (Gattrell, 1985; Hay, 1985) and in terms of empirical research (Macmillan, 1989). A glance at most issues of major geographical journals such as the *Transactions of the Institute of British Geographers* and the *Annals of the Association of American Geographers* indicates the continuing strength and indeed expansion of interest in models, spatial analysis, quantitative methods and technical control.

The continued strength of logical positivism in human geography can be understood in terms of at least three main factors. First, there has been comparatively little recent expansion in higher education to match that which took place during the 1960s. Many of those academics who entered geography departments during the heyday of logical positivism are still in post, and see little reason to change their theoretical and practical approaches to the subject. Secondly, the recessions of the 1970s and the late 1980s, relieved only by the apparent economic success of 'Reagonomics' and 'Thatcherism' in the middle of the 1980s, have not been conducive to reflective or critical research. While the 1980s have seen a relative decrease in the level of central government funding of higher education, they nevertheless have also witnessed increasing direct political involvement in that funding. At a time of recession in the capitalist economies of the world, research funding has tended to concentrate on the explanation and technical solution to problems of inflation, industrial unrest and unemployment, rather than on their interpretation and understanding. This is not to deny that much reflective research has indeed taken place, but it is to suggest that logical positivist science, with its apparent capacity to explain, solve and predict, but most of all to serve those in power, has not surprisingly continued to find favour. A third factor involved in the propagation of logical positivist science has been the relative financial costs of different types of research. Much humanistic research is relatively cheap in terms of equipment and labour, and does not usually, for example, require expensive computers to undertake complex statistical analyses of large data sets. At a time when individuals, departments and institutions are increasingly being assessed in terms of the levels of grant that they attract, there is much pressure to obtain grants for high cost research projects, which frequently reflect the technical interest of logical positivism to maintain the social and political order.

While human geography in the 1970s and 1980s has been characterized by its multiplicity of philosophical approaches, the same can not be said of physical geography. Indeed, the emergence of a strongly humanistic geography during the 1970s has been one of the key factors enhancing the increased division between these two areas of the discipline. In their critique of logical positivism, many human geographers appear not only to have rejected the discipline's links with the natural sciences, but also to have broken their connections with physical geography. While the environment and landscape do find a central place in much humanistic geography, it is the human interpretations of that landscape that have become of central importance, rather than the physical processes that help to shape it. For most physical geographers concerned with the explanation of such processes, humanistic geography is seen not only as irrelevant but also as unscientific. As the languages of the two parts of the discipline have become increasingly differentiated, so has the level of communication between their practitioners diminished. While it is possible to conjecture that a phenomenological interpretation of a soil might involve its reflection in a thin section and the subsequent expression of its essence through the pages of a learned journal, this statement has little real meaning or utility.

For most physical geographers (K J Gregory, 1985; Clark, Gregory and Gurnell, 1987b), logical positivism still forms a sound philosophical foundation for their research and teaching; it produces useful results. Moreover, its dominance is so great that it usually simply passes under the guise of good scientific method. Clark, Gregory and Gurnell (1987b) thus comment that

> Physical geography has contrasted strongly with human geography throughout the last twenty years, and whilst there continue to be substantial overlaps of aim there is no indication that the two branches of the discipline will (or should) converge completely. Of all the roots of distinction, that which is deepest and thus most influential is the methodological contrast between the humanist and structuralist focus of much modern human geography and the scientific (positivist-type) approach to which physical geography has maintained a strong adherence. The undiminished acceptance of the scientific mode by physical geographers does not signify that the alternatives have been overlooked, but rather that they have been found to be less than ideal for many of the purposes of the physical geographer.

Significantly, in this quotation Clark, Gregory and Gurnell emphasize the methodological contrasts between human and physical geography, thus reflecting the classic logical positivist failure to recognize that methodologies are determined in part by their philosophical underpinnings. As Habermas (1978: 67) has reflected, 'by making a dogma of the science's belief in themselves, positivism assumes the prohibitive function of protecting scientific enquiry from epistemological self-reflection'. Moreover, all physical geographers are not in full agreement with this generally optimistic picture of the success of their subject. In particular, Haines-Young and Petch (1986: 199) comment that over the last few decades 'in physical geography there have been very few advances in our theories about, or our understanding of, the natural world'. They go on to suggest that 'The discipline, as it is taught and identified by professional geographers, can boast no major advances. In addition, the vast majority of journals and advanced texts still contain material which is either merely descriptive or an attempt to model some phenomenon by statistical or simple mathematical equations akin to those employed by engineers' (Haines-Young and Petch, 1986: 199).

Physical geography in the 1970s and 1980s has been characterized by an increased emphasis on techniques and on methodology. As Clark, Gregory and Gurnell (1987c: 384) comment, 'physical geography has become, and will remain, an effective natural science with a strong reliance on the development and application of accurate monitoring, analytical and modelling techniques'. Thus each aspect of the physical environment, from slopes and rivers to flora and fauna, has become subject to increasingly refined analysis; advances are made when physical phenomena are more accurately described by newer models. In following this line, physical geographers have found themselves increasingly allied to scientists in disciplines such as geology, engineering and biology. One corollary of this, at a time of increasingly restricted government funding for higher education, has been the amalgamation of various geography and geology departments into departments or schools of earth science, as has happened for example at the University College of Wales, Aberystwyth.

However, as Haines-Young and Petch (1986: 200–1) point out, 'In terms of theory and understanding, quantitative measures, statistical models and sophisticated apparatus have in themselves little to do with science.' They suggest that the key failing of physical geographers has been that they have been beguiled by the apparent success of logical positivism, in the guise of the scientific method. Rather than focusing on the sorts of questions that should be studied, many physical geographers have been happy merely to apply new techniques to old questions. Thus they conclude that

> We cannot condemn the lucky person who wants to study a problem of regional hydrology because he has just acquired a system for digital image processing on the grounds that he is not being scientific. But this situation is disappointing because there is no theoretical progression from any problem and no speculation is provided by equipment. And this is what has happened in physical geography (Haines-Young and Petch, 1986: 201).

### *6.4.4 Structure, constraint and the social context*

Most interpretations of humanistic geography, with their transcendental concern with understanding and reflection, and their emphasis on individual subjectivity, have been unable satisfactorily to address questions concerning the relationship between domination and constraint. These form the expression of Habermas's (1978) social medium of power through which his emancipatory cognitive interest is developed. More specifically, by concentrating on individual life-worlds, symbolic interpretations and realms of meaning, humanistic geography has tended to ignore the structural constraints within which individual lived worlds are expressed. Moreover, particularly in its phenomenological expression, humanistic geography has tended to be passive, lacking an active concern with mechanisms of social and political change. It is with these processes of transformation that critical science is concerned. As Habermas (1978: 310) has argued, critical science seeks to determine 'when theoretical statements grasp invariant regularities of social action as such and when they express ideologically frozen relations of dependence that can in principle be transformed'. Habermas's (1978: 310) critique of ideology is thus designed to set 'off a process of reflection in the consciousness of those whom the laws are about'. Although such self-reflection cannot make such laws inoperative, it can render them inapplicable and thus subject to transformation.

Paralleling the development of a humanist tradition in geography, the 1970s and 1980s also saw the emergence of a self-proclaimed radical tradition (Peet, 1977a, b). This initially took its direction from Marxist political economy, but rapidly diversified its focus to a consideration of other structuralist approaches. More recently, with the widespread acceptance that structuralism is too mechanistic and has failed sufficiently to consider the actions of human subjects, geographers have turned to realism and postmodernism in an attempt to combine elements of both the humanist and structuralist perspectives on society and space. It is with these attempts to form a critically relevant geography that the next chapter is concerned.

CHAPTER 7

# *Critical science and society: the geographer's interest*

> The career route in geography has nothing whatsoever to do with being oriented towards productive geography and everything toward 'playing the game' of personal career. It goes like this: At that point in their training where the student is supposed to do a significant piece of independent research, at last after starting in kindergarten as a total absorber of instruction, what does the typical geography graduate student do? He continues in his past pattern of trying to please his teachers. He cases the joint 'realistically' and rationalizes his sellout with the slogan 'after I get my union card'. Having conditioned himself into seeing his research as the symbol of his lack of integrity, to say nothing of his manhood, that is, having sold his thesis for his degree, he simply continues this pattern for the rest of his life. He publishes to keep from perishing. He sees tenure as the next 'union card'. And eventually he sees retirement as the goal of his existence. Along the way, he seeks out and finds a society of similar time servers, who rather than discussing what is wrong with themselves, the nature of geographers, they lash out endlessly, during marathon coffee hours, about the dismal nature of geography.
>
> Bunge (1977: 36–7)

To those concerned with effecting social and political change, the central failing of the humanist critique of previous geographical practice was its inability satisfactorily to produce knowledge with the capacity to enable people to transform the social conditions of their existence. With its focus on understanding and reflection, the hermeneutic tradition, while providing a cogent theoretical critique of logical positivism, failed to create a sound basis for geographical practice concerned with emancipation (Habermas, 1974). It was therefore to radical traditions of social and political theory that geographers seeking to challenge the very foundations of capitalist society turned in their quest for a critical examination of the power relations that upheld it. The changing fortunes of this critical tradition in geographical enquiry closely reflect the social and political context in which it has emerged, and it is therefore with a brief overview of this context that the present chapter begins.

## 7.1 The social context: geography in recession

Peet and Thrift (1989b: 6) note that radical geography began in the late 1960s

> as a critical reaction to two crises of capitalism at that time: the armed struggle in the Third World periphery, specifically United States involvement in the Vietnamese War, and the eruption of urban social movements in many cities, specifically the civil rights movement in the United States and the ghetto unrest of the middle and late 1960s in the United States, Great Britain and elsewhere.

Thus, particularly in the United States it was among young political and social geographers that the first interest in a radically relevant geography emerged. However, the recessions of the mid-1970s and the early 1980s served to curtail the expansion of higher education that had taken place in the 1960s, and provided difficult conditions for the development of a radical critique. Peet and Thrift (1989b), for example, see four reasons why the initial optimism of the radical geography movement was tempered during the 1980s: the strengthening critique of mainstream Marxist thought; the uncertainty of revolutionary politics; the replacement of the laid-back academic style of the 1970s by narrow professionalism in the 1980s; and the incorporation of some of the early radicals into the very establishment against which they had battled. The result, they suggest, was that radical geography became more sober and less combative. However, it can also be argued that, as with the changes that took place in the humanistic approach to geography, the developments in radical geography during the 1980s reflected a growing understanding of the intellectual linkages between geography and other social sciences. In particular, instead of the simple importation of ideas from Marxist political economy, geographers began to develop a fruitful dialogue with political economists and sociologists, which led to a substantial reappraisal of the interconnections between social relations and spatial structure (Gregory and Urry, 1985b). More recently, this dialogue has been extended as geographers have also begun to grapple with the critiques of science and society offered by realism and postmodernism.

### 7.1.1 Capitalist society in the 1970s and 1980s: power, recession and science

The apparent economic successes of the major capitalist states in the 1960s, and the superficial opening of access to the material benefits of that success in the burgeoning consumer society, opened the possibility by the end of the decade of a more reflective, less technically utilitarian, form of science. Society could readily accommodate not only those who wished to lead alternative life-styles, smoke cannabis and believe in flower power, but it could also afford an expansion in the liberal arts and the social sciences. As with the flowering of the Renaissance in 15th- and 16th-century Italy, there was the opportunity for an explosion of artistic and intellectual talent. As Chapter 6 has elucidated, one effect of this on geography was an expansion of interest in humanist philosophy. However, two aspects of the capitalist expansion of the 1950s and 1960s rapidly led to an increased awareness of the underlying contradictions which sustained it. First, its success was in part enabled by an increase in inequality, not only within the capitalist states, where although many of the poor became richer they did so in general at a less rapid rate than did those who were already rich, but more importantly between the capitalist states and the nations of what became known as the Third World. Second, moreover, the rapid expansion in the mass media and the communications industry meant that the population of the capitalist states could be made aware of these inequalities much more rapidly and extensively. Mass protest, both organized and disorganized, thus became not only much more feasible, but also more effective as a means of influencing political power.

The combined effect of these influences was that many geographers were brought face to face with the failures of capitalism and the empirical–analytic science which provided its technical support, at a time when there was an opportunity for them to express their disquiet in radical ways that had previously been impossible. Although there was still a profound fear of communism among the establishment and the political leaders of the capitalist states, particularly in the United States and Britain, the apparent success of capitalism meant that Marxism was no longer seen as such a threat, and that Marxist intellectuals, if not welcomed with open arms, could at least be tolerated. In France, where the broadly defined left had for many years retained a stronger influence over public opinion than had been the case in Britain and the United States, this found its expression in the acceptance of a range of Marxist ideas which were to form the basis for the widespread student riots at the end of the decade. At one extreme, there were those such as Marcuse (1964, 1972) who, in the light of the mass killings that had taken place under Stalin, turned to a more humanist interpretation of Marx's writings, focusing on concepts such as freedom, alienation and humanity. On the other, were those following Althusser (1969; see also Althusser and Balibar, 1970), who decried such revisionist tendencies, and advocated a return to the scientific historical–materialist core of Marx's later writings, and their central interest in understanding as a guide to action. In particular, it was Althusser's symptomatic reading of Marx, and his distinction between the writings of the young Marx, whose ideological problematic he saw as being inherited from Feuerbach, and those of the mature Marx whose problematic he claimed to be scientific, that were to set the scene for both theoretical debate and political action during the early 1970s (Lock, 1972; Macintyre and Tribe, 1975). Such arguments were to form a fruitful source of debate for a new generation of geographers entering the profession at this time (see Peet, 1977b; Castells, 1977; Gregory, 1978).

By the early 1970s, the capitalist global economy was in a period of crisis engendered by rising inflation and falling production. Following the devaluation of the US dollar in 1971 and 1973, and the 1973–74 hike in oil prices initiated by the Organization of the Petroleum Exporting Countries (OPEC), the crisis became a major recession. At first, this did not appear to have a great influence on higher education and scientific research other than serving to curtail the expansion of the previous decade, but with the second major rise in oil prices in 1978–79, which plunged the capitalist states further into recession, it became apparent that a substantial reorganization of higher education institutions was likely. Two trends were to coalesce in the early 1980s: on the one hand, the lack of graduate employment opportunities made many students focus attention much more directly on the career implications of their degrees than had previously been the case, while on the other governments became acutely aware of the need to fund research relevant to their needs and thus to the future success of capitalism. The subsequent economic revival under Reagan in the United States and Thatcher in Britain further encouraged the majority of students to forgo radical thoughts, to concentrate on learning useful knowledge, and to enter successful careers in banking, industry and finance. If in the 1970s Marx had appeared to students as exciting and even a little bit dangerous, by the 1980s he was seen by many as irrelevant. Indeed, by the end of the decade, with the collapse of the communist regimes in eastern Europe, mentions of Marxist theory in lectures were frequently greeted if not by derision, at least by total apathy.

### 7.1.2 *Geography and the production of knowledge*

Against this background, there were broadly four positions which could be adopted by geographers in the 1970s and 1980s. First, they could claim that they were pursuing pure value free science. This did not have to be seen as being of direct applied relevance, because all such science found its justification in the argument that it led to an advancement of knowledge that would eventually be of use to society. This empirical–analytic view of science, with its technical interest, was that frequently taken by physical geographers, who continued to pursue their explanatory research into the accurate description and modelling of physical processes (Clark, Gregory and Gurnell, 1987a). A second alternative was to seek to produce new knowledge, that would explain the recession and enable the social and economic problems associated with it to be resolved for the good of capitalist society. This was empirical–analytic research that the state and industry were eager to fund, and although not all of it was necessarily directed by the tenets of logical positivism (Bennett, 1985), much of it sought technical and empirical solutions to problems that more critical analysts argued lay at the very foundation of the capitalist enterprise (compare for example Bennett, 1980 with Massey and Meegan, 1982). For young geographers keen to rise in the academic hierarchy this was a sure way to progress (Beaumont, 1987), whereas for those who began their careers in the late 1960s it offered the hope of renewed relevance (Wilson, 1970, 1989; Wilson and Bennett, 1985; but see also Hay, 1985). Third, it was possible to retreat into the transcendental option offered by humanism, and discussed in Chapter 6. Again, this offered an illusion of pure scholarship, and while it helped to develop an understanding of the human meaning of recession, repression and inequality, it generally failed to offer practical solutions to their replication. Finally, there was the radical alternative, which self-consciously sought to produce a revolution in geographical theory and practice (Quaini, 1982). As Peet (1977b: 64) has summarized, two central assumptions underlay its practice: 'first, and most obviously, . . . there is no such thing as objective, value-free and politically neutral science, indeed all science, and especially social science, serves some political purpose; secondly, . . . it is the function of conventional, established science to serve the established, conventional social system and, in fact, to enable it to survive'.

### *7.1.3 The origins of radical geography*

The urban and racial unrest in the United States, together with the war in Vietnam, provided the context for young urban and political geographers, as well as those interested in development studies, to turn to Marxist political economy for a framework in which such expressions of capitalist structural contradictions could be interpreted. A new journal, *Antipode*, was launched at Clark University, in Worcester, Massachusetts, specifically as a forum for the publication of such self-proclaimed radical geography, and early issues capture both the fervour and commitment of its authors and editors (see Peet, 1977a). Its aim was 'to ask value questions within geography, question existing institutions concerning their rates and qualities of change, and question the individual concerning his own commitments' (Wisner, 1969: iii). Early issues of *Antipode* concentrated largely on the spatial manifestation of social welfare topics associated with such subjects as poverty, minority rights and access to social services, and as Peet (1977b) points out much of this adopted a methodology that had been developed within the existing framework of power relationships. It was not until the publication in 1972 of a paper by Harvey on revolutionary and counter revolutionary theory in geography in the context of ghetto formation that a new emphasis began to be revealed. This found its theoretical and practical grounding specifically in the writings of Marx, and from 1973 until the end of the decade radical geography became effectively synonymous with Marxist geography. As Peet (1977b) points out, this injection of Marxist theory came initially from Britain, where geographers had begun to explore Marx's writings in the late 1960s, and it was only during the 1970s that geographers in the United States really began to examine his corpus of work in detail.

Paradoxically, two of the strongest exponents of quantitative geography and logical positivism in the discipline during the 1960s, Harvey and Bunge, were also at the forefront of the introduction of a radical critique during the 1970s. Harvey (1973) recounts the way in which his theoretical and practical emphasis changed in the introduction to his book *Social justice and the city,* which rapidly became the flagship of the new radical geography. Having completed his examination of methodological problems in *Explanation in geography,* Harvey turned his attention to issues of social and moral philosophy, and to the ways in which they could be related to geographical enquiry. As he recounts, 'Since I had just moved to Baltimore, it seemed appropriate to use that city, together with other cities with which I was familiar, as a backdrop against which to explore questions that arose from projecting social and moral philosophical considerations into the traditional matrix of geographical enquiry' (Harvey, 1973: 9). Central to his developing ideas was the way in which social processes and spatial forms are related, and in *Social justice and the city* these are examined in the context of four particular themes: the nature of theory, the nature of space, the nature of social justice, and the nature of urbanism. For Harvey (1973: 17) the appeal of Marx's analysis was that it enabled him to achieve a reconciliation between such disparate topics, collapsing the 'dualisms without losing control over the analysis'. In conclusion, Harvey (1973: 286) suggested 'that the most important thing to be learned from the study of Marx's work is his conception of method. And it is out of this conception of method that theory naturally flows'. Subsequently, radical geographers began to examine a range of Marx's methodological and theoretical work, addressing in particular issues of underdevelopment–imperialism, rent theory, cultural evolution and spatial inequality (Peet, 1977b).

If Harvey's work can be seen as being seminal in the development of the theoretical and methodological implications of Marx's writing for the intellectual world of geography, Bunge's life from the mid-1960s, and in particular his efforts to bring geography to the poor, reflects the very different world of practical action. 'In 1967, Bunge was refused tenure at Wayne State University in Detroit on the grounds of obscenity (swearing during lectures)' (Peet, 1977a: 14), and in the following year he founded the Society for Human Exploration, designed to reinject into the discipline what he saw as the true meaning of exploration. He sought to encourage contributive explorations, rather than exploitative expeditions, and research that was community-people oriented rather than campus-career oriented. In practice, he was instrumental in arranging for courses directly relevant to central-city blacks to be offered at Wayne State University and in 1970 at Michigan State University (Horvath, 1971), but soon afterwards the principles of community control and free tuition which underlay this Expedition to Detroit proved to be unacceptable to the university authorities and the project was terminated (Peet, 1977b). Subsequently Bunge was forced to leave the United States, and moving to Canada he resurrected his agenda in the form of the Toronto Geographical Expedition which was established in 1972 (Stephenson, 1974). Bunge's tradition of radical practice, however, was by its very definition always going to be attacked by the establishment. For those willing to coexist within the present institutional structures of capitalist society, a new forum for debate was therefore necessary, and in 1974 a Union of Socialist Geographers was established to provide a more conventional forum for the organization of socialist practice within the discipline.

By the end of the 1970s, once the initial engagement with Marx's ideas was over, many geographers turned increasingly to other structuralist interpretations of the relationship between social and spatial structures, to realism, and eventually in the late 1980s to postmodernism. It is on these differing strands of a radical geography that attention now focuses.

## 7.2 Radical geography and a structuralist alternative

One of the key features of the development of a radical tradition of geography was that it brought geographers into close contact with other left wing social scientists. Radical geography thus formed but a part of a wider radical social movement. However, deeply embedded within it was a profound tension between its theoretical and practical interests.

### 7.2.1 Marxist geography

While Marx's work has been used by geographers in a substantive way to provide insights into the workings of capitalism, its greatest influence has been methodological. In establishing dialectical materialism as a form of scientific practice, Marx provided a framework in which much radical geography has subsequently been pursued. In its original form, the dialectic method as practised by Zeno, a disciple of Parmenides (Russell, 1961) in the 5th century BC, was simply the way of seeking knowledge through continuous questioning and answering, with one answer providing the basis for a subsequent question. It was used later in works such as Abélard's *Sic et non*, composed in 1121–22, but the dialectic achieved its fullest development in Hegel's philosophy, and particularly in his *Phenomenology of spirit*. In Hegel's almost mystic idealism, the aim of philosophical enquiry was the pursuit of the Absolute Idea, a form of self-consciousness in which subject and object are one. For Hegel nothing is ultimately real except the Whole, and knowledge of the Whole is reached through the dialectic method, involving the passing over of concepts or thoughts into their opposites, and eventually through continued reiteration of this method achieving a higher unity. Hegel's dialectic is usually characterized as being based on the triad of *thesis*, *antithesis*, and *synthesis*, but although much of his philosophical exposition did involve the use of triads, as in the instance of need–labour–enjoyment, he did not actually use this particular terminology. While recognizing that the concept of the dialectic had been used before, Hegel's innovative contribution was to insist that it included a conception of necessary movement. Other vital notions in his view of the dialectic included those of negation, scepticism and reason. Hegel also applied this dialectic procedure to the progress of society, arguing that 'the history of Civil Society was simply the progressive realization of the Idea' (Gregory, 1978: 109). Two key ideas that Marx developed from Hegel's dialectic were his insistence on the conception of necessary movement, and the idea of the dialectic as historical process. However, he rejected Hegel's suggestion that the driving force of the dialectic was Spirit. Marx's seminal contribution was to replace Hegel's idealist conception of the progress of society as determined by the human mind, with a dialectic materialism which reflected the materialist conditions of human life. For Hegel people thought; for Marx they laboured.

Six related theoretical abstractions underlay Marx's conception of history, and these can be found most clearly expressed in the preface to his *Critique of political economy* originally published in 1859. First, it was underlain by a concern with *relations of production*. He thus argued that things are always seen as objects, what he termed the fetishism of commodities, rather than as the social relationships which they embody. Second, his theory was *materialist*, focusing on the economic structure of society constituted from the totality of these relations of production. Third, this involved a particular conceptualization of *structures*, in which the economic structure, or infrastructure, is seen as providing the foundation for the particular expression of legal or political structures which form the superstructure and to which correspond particular types of social consciousness. In turn, fourth, he argued that the forces and relations of production constitute a *mode of production*, of which there have been four: the Asiatic, the ancient, the feudal and the modern bourgeois (Howard and King, 1985). It is at this junction that, fifth, the *dialectic* enters his argument, because Marx envisaged that each mode of production has its own internal dialectic of change. Each has its particular contradictions between the forces and class relations of production, and it is through the resolution of these that a new mode of production is forged. The aim of his research was thus, sixth, to identify the *historically determined laws* governing such processes, and it was this that he set out to do in *Capital*. Underlying these theoretical constructs, Marx had a clear practical interest: that of revealing to the proletariat the contradictions in the modern bourgeois, or capitalist, mode of production, in order to hasten the advance of global socialism through the forcible overthrow of existing social conditions.

Such ideas have found their clearest relevance in four main areas of geographical enquiry. First, in historical geography Marx's own writings had clear relevance to the study of the transition from feudalism to capitalism, in the rise of industrial capitalism, and in the colonial and imperialist extension of capitalist relations of production from their European hearth to the remainder of the world. Good examples of such work include Dunford and Perrons's (1983) account of the historical development of capitalism in Britain, which combines the work of a human geographer with that of a social economist, Gregory's (1982b,

1984) examinations of industrial change and class conflict particularly in the Yorkshire woollen industry, and Blaut's (1975) survey of Marxist theories of colonialism and imperialism. The second broad area in which geographers have built on Marx's ideas has been in an urban context. Two particular works stand out as being of early prominence in this field: Castells's (1977) *The urban question: a Marxist approach*, first published in French in 1972, and Harvey's (1973) *Social justice and the city*. Subsequently, Marxist approaches have been used in the planning context (Dear and Scott, 1981), in developing theories of rent (Harvey, 1974; Harvey and Chatterjee, 1974), and in housing (Boddy, 1976; Duncan, 1977). Third, a Marxist framework has been used in an attempt to understand regional inequalities associated with industrial restructuring (Massey and Meegan, 1979, 1982; Carney, Hudson and Lewis, 1980). Finally, although Marx himself wrote little about the Third World, his ideas on the development of capitalism, and its implications for social and regional inequality have had a considerable influence on research on the poorer countries of the world. This is typified by the work of Slater (1973), Santos (1974) and Buchanan (1972) on underdevelopment and empire.

What unites this work is its interest in class conflict, its focus on modes of production, and its pursuit of historically determined laws. More recently there has been a somewhat more critical appraisal of the links between Marx's arguments and the potential geographical contribution to radical science. In particular, Harvey (1982, 1985a, b) has drawn attention to Marx's lack of spatial awareness, and has sought to extend his analysis to include a comprehensive examination of the spatial implications of some of his theoretical statements, with particular reference to the urban context.

### 7.2.2 The place of radical geography

The different paths followed by Harvey and Bunge reflect a central tension within radical geography: that between theory and practice. As Bunge's experiences illustrated, a practice designed overtly to overthrow the institutions of capitalism will be challenged systematically by the establishment against which it is arrayed. In contrast, a purely intellectual critique, without a corresponding practical commitment can lay claim only to self-indulgence. For those seeking to pursue a truly radical geography, designed to overthrow the social and economic repression of capitalism, three alternative paths are open. First, a radical geography might be created outside the higher education institutions of the capitalist states. Such a path, however, assumes that geographical teaching and research have something to contribute to revolutionary practice, and there have been few, if any, clear attempts to substantiate such a claim. Nevertheless, taken to its extremes, and once again returning to Strabo's connection between geography and the requirements of military commanders, such practice would involve trained geographers leaving the safe confines of universities and entering the uncertain and dangerous world of armed revolutionary struggle. A second alternative is for geographers to remain within their institutional contexts, satisfying the requirements of their capitalist paymasters through teaching and research, but at the same time becoming involved in local or national political action. Although this represents a much safer strategy, and while a number of geographers are indeed active, particularly in local politics, it is remarkable how few geographers have actually entered national political arenas. The third alternative is to continue within the institutional structure of higher education, but to use that structure to reveal the contradictions of capitalism through teaching and the practice of research.

At an undergraduate and postgraduate level, teaching designed to challenge the basis of capitalism continues to take place, but during the 1980s it faced increasing pressure from four directions: the apparent economic success of capitalism encouraging students to participate in the material benefits associated with the Thatcher and Reagan era, typified by the 'yuppie' syndrome; increasing government intervention in higher education, through the tighter monitoring of courses and systems of repressive staff appraisal; a secondary education system designed increasingly to propagate useful, largely technical, knowledge, rather than critical enquiry; and the collapse of communism in Eastern Europe and the former Soviet Union. The full influence of the last of these factors has yet to be realized, but the abortive attempt by hard-line communists to stage a coup in the Soviet Union in August 1991, and the subsequent swing to capitalist relations of production, under the guise of democracy, has presented a fundamental challenge to radical social scientists and philosophers. The central problem here is that Marx, while providing a substantial critique of capitalism, largely failed to outline the economic, social and political framework within which he envisaged socialism as being practised; the shape of socialism was to be determined by the revolutionary practice of the proletariat. Consequently, in the minds of most people in capitalist society, the political, economic and social experiences of the Soviet Union between 1917 and 1991, under the name of communism, have erroneously become equated with Marxism. However, the collapse of communism does not necessarily mean that capitalism is in any way the only, or best, form of economic system, nor does it mean that Marx's writings are irrelevant as an analysis of contemporary capitalism. The challenge facing radical geographers in the 1990s is to lay bare the contradictions still present in capitalism, and to reveal them to their students and those among whom they conduct research.

The central failing of much Marxist geography has been its focus on theoretical and philosophical critique rather than on practical action. This is typified by the arguments of many radical geographers in the 1970s decrying direct involvement in the implementation of social and economic change. In reviewing the field of development studies at the end of the 1970s, Harriss and Harriss (1979: 576), for example, suggested that there were only two approaches then current:

> on the one hand, varieties of the liberal position entailing at best a multidisciplinary approach in analysis and committed to intervention

through 'planning'; and on the other, radical positions acknowledging various connections with Marxian theory and generally critical of interventionism, even when this involves practical programmes with the apparent objective of ameliorating conditions of poverty.

For most radical geographers, practical intervention or 'the practice of development by national and international agencies' (Harriss and Harriss, 1979: 582) was shunned because of its links with capitalism. A second factor influencing the lack of a practical content in radical development studies was that Marxist and other radical critiques of logical positivism were in part built upon a critique of its predictive capacity. If radical geographers were criticizing logical positivism for its failed efforts to explain and predict, it was difficult for them therefore to justify any predictions of their own.

Nevertheless, one area of substantial theoretical and also practical importance where a radical stance has begun to reach fruition has been in the development of a feminist approach to geographical enquiry, focusing on issues of gender inequality and the oppression of women (Bowlby *et al.*, 1989; Peake, 1989). This has been concerned both with substantive research themes and also with the institutional reconstruction of the discipline. Thus, although professional geography is still dominated by men, there has been increasing recognition of gender bias in the profession, and a number of bodies, such as the Association of American Geographers and the Institute of British Geographers, now have equal opportunities statements as part of their constitutions. Moreover, the sexist language that has dominated modern geographical writing and that is evident in many of the quotations cited in this book, is increasingly being replaced by a more gender aware vocabulary. In terms of research agendas, feminist geographers have also opened up a wide spectrum of issues that have previously been ignored, and the 1980s have seen the emergence of a strong tradition of such research exemplified by McDowell's (1983) analysis of the gender division of urban space, Mackenzie's (1986) examination of women's responses to economic restructuring, and Momsen and Townsend's (1987) edited volume entitled the *Geography of gender in the Third World*. While the rise of feminist geography has had substantial repercussions for the discipline, the same can not be said of concern with other forms of oppression and inequality. In particular, the dearth of black geographers in the profession, and, with a few notable exceptions (see for example Jackson, 1987), the lack of substantial research on racism are matters for continued concern.

Overall, though, and as a result both of the countering effects of the establishment, and of its inability to achieve practical results, radical Marxist geography has had a relatively small effect on the discipline. As Johnston (1986b: 386) has commented, 'Although very active, those committed to radical geography have not made great inroads into the discipline's establishment, in part because of their revolutionary aims, their overt political goals and their threat to the status quo.' By the early 1980s the initial somewhat naïve engagement of geographers with

Marxist theory had given way to a much broader examination of the wider context of the social theory of which it was a part. In particular this has led geographers such as Gregory (1978) and Sayer (1984) to examine Marxist theory within the broader context of structuralism and realism.

### *7.2.3 Structuralist alternatives*

The exploration of Marxist political economy during the 1970s introduced geographers to a range of different structuralist philosophies. In particular Peet and Thrift (1989b) note the contrasts between Castells's (1977) adoption of Althusser's structural Marxism, and Harvey's (1973) more eclectic combination of Piaget's (1971) structuralism with Marx's political economy. Moreover, Gregory (1978) draws attention to the assumptions and procedures which Piaget's epistemology shares with that of Lévi-Strauss, and thus reintegrates his discussion of structuralism with the debate between Vidal de la Blache and Durkheim over the relationships between geography and sociology. What the wide range of structuralist philosophies hold in common is that the empirical world of observable phenomena is determined by underlying structures. In contrast to systems, which are concerned with empirical reality, structures cannot be touched and measured, but are nevertheless assumed to be real. Structuralism therefore offered geographers another source of ideas with which to counter the empiricism of the logical positivist version of geography that dominated the 1960s. Moreover, this property of structures also provided geographers with a neat solution to the problem of description and explanation, because if such structures exist then explanations of surface phenomena can be achieved through a description of the underlying structures. In particular changes in the spatial distribution of surface features could be explained through recourse to a description of their underlying structural transformations. However, there remains considerable debate over the way in which knowledge about such underlying structures can be achieved. Glucksmann (1974) thus contrasts the approach of Lévi-Strauss, which seeks to make theoretical abstractions from empirical reality, with that of Althusser, which begins with theory and thus works from the underlying structure to surface reality.

Lévi-Strauss (1953, 1963) was an anthropologist whose concern with social structure emerged from his development of a methodology designed to enable him to understand aspects of kinship, myths and symbols (Leach, 1974). This methodology was based on the initial assumption that everything in life is made up of pairs of opposites, or binary oppositions, such as light/darkness or naked/clothed. Lévi-Strauss's method was then to define a phenomenon under study as a relationship between two or more terms, to construct a table of the various possible permutations of those terms, and then to use this table as the basic object for analysis. The empirical phenomenon chosen at the beginning thus becomes but one possible combination of the complete system. His goal was eventually to identify the common features underlying all systems of myth and kinship from these basic tables. In so doing he combined ideas from linguistics with mathematics to suggest that kinship systems can be interpreted as examples of algebraic structures (Piaget, 1971). In particular his schema closely parallels that of de Saussure's (1916) synchronic view of language, which was concerned with the relations that bind terms together to form a system in the collective minds of speakers. Just as de Saussure's systematic linguistics was concerned with determining the underlying rules governing communication, so Lévi-Strauss's anthropology was intended to reveal the structure underlying human society. However, such a conceptualization is based on the premise that any differences in social phenomena are but variants of the same underlying structure. Consequently, there is no dynamic of change and no possibility of progress. Lévi-Strauss's conceptualization of historical change has therefore been termed 'categorical' in contrast to Althusser's 'dialectical' conception (Gregory, 1978).

While Lévi-Strauss was essentially concerned with anthropology, and with understanding universal truths of the human mind, Althusser's focus was on philosophy, and in particular with a reinterpretation of Marx's historical materialism. This involved a discussion of the concept of structure at three related but distinct levels. First, in identifying an epistemological break in Marx's writings, Althusser (1969) focused on the structures underlying the written text to suggest that prior to 1845 his problematic was ideological whereas subsequently it was scientific. This is not to suggest that the former was in some way false, and the latter correct, but rather that scientific and ideological knowledge serve different purposes. For Althusser, science is

> a form of knowledge which works with concepts as a means of production to produce its own object and order of proof, and thus produces new knowledge. In contrast to this, ideological knowledge can produce only variations of the original, since its problematic does not break away from the context of practical–social problems, and merely re-translates these practical–social problems into different forms (Macintyre and Tribe, 1975: 18).

Marx's achievement was thus to move from a system of thought which was ideological to provide the basis of a new science which would enable social formations to be analysed and thus changed.

However, secondly and in more general terms, according to Althusser's formulation, ideology is a structure enabling people to think and act. In contrast to some interpretations of Marx which conceived of ideology as a device used by the ruling class to deceive the mass of the population and thus maintain it in a state of subservience, Althusser envisaged ideology as combining both conscious and unconscious thoughts. Ideology, for Althusser, thus serves an important role 'representing our relations to us and enabling us to regulate our behaviour' (Macintyre and Tribe, 1975: 20).

This then, thirdly, required Althusser (1969) to provide an alternative to the crude economic reductionist interpretation of Marx's works, which envisaged the economic base or infrastructure as always determining the form of political and ideological expression in the superstructure. Althusser did this by drawing a distinction between dominant and determinant instances (Althusser and Balibar, 1970). He thus suggested that social conflict could be formulated in any of three instances of social activity: economic, political or ideological practice. The instance in which this happened he termed the dominant instance, and he suggested that the economy then represented the field of possibilities in which the dominant instance could operate. It was in this sense that the economy was determinant. As Gregory (1978: 113) has summarized, 'In the capitalist mode of production, therefore, the economic level is both dominant and determinant, whereas in other modes of production other levels occupy the dominant position, but still as an effect of the conditions of existence of the economic level.'

Although variants of structuralism had important influences on biology, linguistics, mathematics and psychology as well as anthropology and philosophy, it was Althusser's structural Marxism that was to be of most significance to geography. This largely reflected the importance of Castells's (1977) work in introducing Althusser's interpretation of Marx to geographers, as well as Gregory's (1978) analysis of the potential contribution of structuralism to geographical enquiry. However, it also reflects the relatively limited links that had been established with other social sciences, particularly psychology and linguistics, prior to the late 1970s.

### *7.2.4 Space, time and structuration*

Despite the power of the structuralist critique of empiricism and logical positivism, there was surprisingly little geographical research undertaken during the 1970s within an *overtly* structuralist framework. Although much radical geography can be interpreted as being broadly structuralist in approach through Althusser's interpretation of Marx's work, the main emphasis of most Marxist geography focused on its Marxist rather than its structuralist content. For those concerned with empirical research a central problem of Althusser's framework was its profoundly theoretical basis, which necessitated the interpretation of surface features through the prior construction of theory. In practice, neither Lévi-Strauss's or Althusser's extreme position is tenable; our knowledge of deep or underlying structures is in part determined by our experience of surface reality, but that experience itself is closely influenced by the economic, social, political and ideological structures which underlie it.

More formally, this problem reflects growing concern during the late 1970s and early 1980s with the inability of structuralism to deal with individual human actions. Thompson (1978) thus argued that Althusser's version of structuralism reduced men and women merely to passive carriers of structural determinants. This led to a widespread debate within Marxist social science (Benton, 1984), between those advocating a more humanist interpretation and those continuing to support Althusser's version of structural Marxism. In geography Duncan and Ley (1982: 30) thus suggested that structural Marxists had created a holistic mode of explanation in which 'reified entities such as capital are treated as the formal cause while people are regarded as the efficient cause, the mere carriers of structural logic'. Critical of structuralism, they suggest that macroscale social structures 'do not have autonomy or an existence that is not ultimately reducible to cumulative human actions and interactions' (Duncan and Ley, 1982: 32), and they conclude that

> the intersection of human geography with structural Marxism has led to a passive model of man that is conservative and results in an obfuscation of the processes by which human beings can and do change the world. Furthermore, philosophical holism is extremely difficult to apply in empirical research, the result being that in some cases the explanations are totally inadequate with causal power attributed to abstract mental constructions, while in other cases theoretical structures are almost completely divorced from the empirical analysis (Duncan and Ley, 1982: 54).

However one regards the philosophical critique of logical positivism provided by structuralism, this quotation once again emphasizes that it has proved very difficult satisfactorily to undertake empirical research based upon Althusser's structural interpretation of Marx.

One solution to the problem of how to combine human agency within a structural perspective was to seek to integrate elements of hermeneutics with structural Marxism. This had been attempted in sociology by Giddens (1979; see also 1981), whose structuration theory was also advocated from a geographical perspective by Gregory (1981) and Pred (1984). In essence, structuration theory sees structures as being both the outcome and the medium of human agency, and its aim is to analyse both the production and the reproduction of such structures. Giddens's (1981) structuration theory can be summarized in ten propositions. First, it distinguishes between structures and systems; structures have only a virtual existence in time-space, whereas social systems are constituted of situated practice. Second, 'Structures can be analysed as rules and resources' (Giddens, 1981: 26), and power is therefore as integral a part of social life as are meanings and norms. Third, there is the notion of duality of structure, by which Giddens means that structures are both the medium and the outcome of social practices. As he argues, 'The concept of the duality of structure connects the *production* of social interaction, as always and everywhere a contingent accomplishment of knowledgeable social actors, to the *reproduction* of social systems across time-space' (Giddens, 1981: 27). Fourth, 'the structural properties of social systems are embedded in *practical consciousness*' (Giddens, 1981: 27). Fifth, the study of structuration implies an analysis of the conditions of the continuity, change and dissolution of social systems in a non-functionalist style. Sixth, all reproduction is contingent and historical; 'the knowledgeability of actors is always *bounded*, by *unacknowledged conditions* and *unintended consequences* of action' (Giddens, 1981: 28). Seventh, Giddens identifies three layers of temporality in the analysis of social systems: the immediate, the contingency of life in the face of death, and the long-term reproduction of institutions. Eighth, the theory of structuration is specifically concerned with power and domination. Ninth, 'the integration of social systems can be analysed in terms of the existence of "systemness" as *social integration* and as *system integration*' (Giddens, 1981: 29). Tenth, he differentiates between contradiction, the opposition between structural principles of a social system, and conflict, the struggle between actors.

Although the above summary reflects the dense style of Giddens's writing, there are two central points to be grasped from his overall approach. The first is that it seeks to understand the interactions between human agency and structure, and second, this is undertaken through an introduction to social theory of relative views of time and space. For Giddens (1981: 30) 'Time–space relations are portrayed as constitutive features of social systems, implicated as deeply in the most stable forms of social life as in those subject to the most extreme or radical modes of change.' Following Leibniz, Giddens (1981: 30–1) suggests that 'We can only grasp time and space in terms of the relations of things and events: they *are* the modes in which relations between objects and events are expressed.' For geographers, one of the most interesting features of Giddens's structuration theory, particularly through his reference to Hägerstrand's time geography, is thus that it represents an attempt to bring geographers' traditional concern with space into social theory. In so doing, Giddens (1985) seeks to develop the theoretical foundations of time geography through a consideration of the ideas of locales, the settings of interactions, and of regionalization. Although attempts have been made to incorporate these ideas into geographical research, such as Pred's (1984) examination of place, and Duncan's (1985) brief analysis of political legitimation in Sri Lanka, it has as yet mainly been discussed within a theoretical context.

A somewhat different means of incorporating space into structuralist and Marxist theory, but still focusing on a relativistic interpretation thereof, has been advocated at a theoretical level by Harvey (1985a, b, 1989a). In the continuing development of his understanding of capitalist urban society and consciousness, he suggests 'that the very existence of money as a mediator of commodity exchange radically transforms and fixes the meaning of space and time in social life and defines limits and imposes necessities upon the shape and form of urbanization' (Harvey, 1989a: 165). Harvey's overall argument is that 'Command over space . . . is of the utmost strategic significance in any power struggle' (Harvey, 1989a: 186). Following Lefebvre (1974, 1991), he suggests that the created space of society is the space of social reproduction, and thus that 'control over the creation of that space also confers a certain power over the processes of social reproduction' (Harvey, 1989a: 186). Harvey (1989a: 196) thus interprets the urban process under capitalism as being fraught with political confusions, which can be understood through an examination 'of how urbanization is framed by the intersecting concrete abstractions of money, space, and time and shaped directly by the circulation of money capital in time and space'. Moreover, in order to examine changing urban spatial practices, and again building his arguments from those of Lefebvre (1974, 1991), Harvey (1989a) establishes a grid, based on three dimensions, namely those of (a) material spatial practices (experience), (b) representations of space (perception) and (c) spaces of representation (imagination), which he sees as intersecting with three aspects of spatial practice, namely (a) accessibility and distanciation, (b) appropriation and use of space, and (c) domination and control of space. Harvey's use of the term 'space', though, is somewhat ambivalent. As is evident from the above, in general he adopts a relativistic view of space, arguing that it is something that can be commanded (Harvey, 1989a: 165) and conquered. However, elsewhere, for example, he suggests that money 'permits the separation of buying and selling in both space and time' (Harvey, 1989a: 175), and that social power can be concentrated in space (Harvey, 1989a: 176), both of which would appear to represent absolute views of space.

In an attempt to operationalize the theoretical work on human agency and spatial structure in an empirical context, Massey (1984) has focused on an analysis of the way in which economic and social change vary in different places. This has led her and others (Massey and Meegan, 1982; Cooke, 1989) to examine spatial variation in the economic restructuring of Britain. In so doing, Massey has developed a line of argument somewhat similar to that of Giddens, but devoid of much of the latter's obfuscation. Moreover, in criticizing Giddens's concept of locale as being too vague, passive and lacking social meaning (Johnston, 1991a), this research has led to the coining of the term 'locality' to refer to the space where people's working and consuming lives are lived (Cooke, 1989). Localities are thus seen as the totality of social structure and human agency in space, as centres of collective consciousness, and as the expressions of social and political interest. However, as Duncan (1989) stresses, the term 'locality' has been used in many different, and sometimes contradictory ways. In particular, he argues that 'The idea of locale should not be equated with locality. The terms are asymetrically related (localities may be locales, few locales will be localities), locale is not a dimension of social organization in the way that locality should be – rather it is a mediation of social relations, and where locale is typical locality is unique' (Duncan, 1989: 247). Furthermore, such an attempt to integrate structural Marxism with empirical research has suffered from criticisms on the grounds both that it is a return to empiricist studies of place and also that it is an attempt to give structural Marxism a human face (Cochrane, 1987; Smith, 1987). Nevertheless, as Duncan (1989) points out, the term 'locality' became one of the most popular geographical organizing concepts of the 1980s, and formed the focus for substantial research grant funding. As such, it represents one of the few examples of substantial empirical research resulting from the interface of structural Marxism and a concern with human agency.

## 7.3 Realism and postmodernism

By the mid-1980s the humanist and structuralist critiques of logical positivism had provided powerful arguments for the rejection of a conceptualization of geography as spatial science. However, they had concentrated largely on epistemological issues, concerned with the claims to knowledge advanced on behalf of different theoretical positions. Moreover, the numerous different theoretical alternatives advocated provided a highly fragmented framework for the discipline. For those seeking to understand this fragmentation, there was a need to find an overall way of interpreting this diversity of approaches to understanding in the social sciences. This has been achieved through two main approaches, namely those of realism and postmodernism. Whereas realism seeks to provide an overarching meta-theory within which the philosophical diversity of the last twenty years can be understood, postmodernism rejects such a possibility, and instead encourages an attitude of mind through which to interpret these changes.

### 7.3.1 Reality and realism

The contrasts between idealism, which conceives of the world as only being known and constituted by the human mind, and realism, which admits of a real world independent of human perception, were briefly alluded to in the earlier discussion of humanist approaches to geographical enquiry. During the 1980s, though, a new form of transcendental realism emerged in the social sciences largely through the influence of Bhaskar (1978, 1986; see also Keat and Urry, 1981), and this has begun to have a significant influence on geographical enquiry (Gregory, 1982a; Sayer, 1984, 1985a). Its particular attraction, as Cloke, Philo and Sadler (1991: 135) note, is that 'proponents of realism claim that all of the post-positivist tendencies in human geography can be interpreted as roads towards realism, be they labelled Marxist, humanist or even other "mainstream" geographies'. While realism can therefore be seen as a way of uniting such diverse critiques of positivism, this very inclusivity means that it is not easy to characterize it as a single philosophy; there is a real danger that realism can mean all things to all people. Indeed, Gregory (1986) has suggested that Habermas's critique of empirical–analytic sciences can be applied not only to logical positivism, but also to realism.

Outhwaite (1987) has argued that one of the key features of realism is that rather than focusing on epistemological issues it has addressed questions of ontology, seeking to examine the features of the world that make it possible for knowledge to exist. Bhaskar (1978) has thus suggested that there have been three main ontological traditions within science: classical empiricism in which the source of knowledge ultimately derives from what he terms 'atomistic events'; transcendental idealism in which knowledge derives from mental constructions imposed on the world; and transcendental realism, which sees the basic objects of knowledge as the structures and mechanisms that generate phenomena. For Bhaskar (1978), in transcendental realism, these basic objects of knowledge are neither phenomena, as characterized by empiricism, nor human constructs imposed upon them, as would be the case with idealism, but rather real structures which endure outside our knowledge and experience.

Outhwaite (1987: 45) has summarized three basic realist ontological principles. First, there is a distinction between transitive and intransitive objects of science, the former being such things as models and concepts used by scientists, and the latter being the real objects making up the social world. This has the epistemological consequence that both empiricism and conventionalism, the latter of which implies that knowledge is merely the conventions adopted by scientists, are rejected, and this leads to the adoption of the concept of real definitions which are 'statements about the basic nature of some entity or structure' (Outhwaite, 1987: 45). Second, transcendental realism divides reality into three realms: the real, the actual and the empirical. 'The last of these is in a contingent relation to the other two' (Outhwaite, 1987: 45). Third, there is the idea that causal relations are tendencies determined by the interactions of generative mechanisms. Such interactions need not necessarily produce events, and if they do, these events need not be observed. In turn this has the epistemological implication that 'the realist conception of explanation involves the postulation of explanatory mechanisms and the attempt to demonstrate their existence' (Outhwaite, 1987: 46). In practice this leads to two methodological procedures for realists: the need to identify how something happens, and the need to establish how extensive a phenomenon is. More formally, this involves the identification of both causal mechanisms and empirical regularities. From what has been said so far, it is evident that this schema has close relationships with structuralism, hermeneutics and critical theory, and there has been much debate as to the relationships between these various philosophical positions (for a summary see Outhwaite, 1987). Bhaskar (1980), for example, makes clear that in his version of realism all science should be critical and emancipatory, both of which are key features of Habermas's critical theory. Moreover, realism has closely engaged the debate over human agency and structure (Sayer, 1984), although it has avoided making this the foundation of its approach, rather seeing structuration theory as one of a number of possible interpretations of the relationship between individuals and structures.

One of the key advantages offered by realism in Bhaskar's formulation is its potential to link together both the natural and the social sciences. Bhaskar (1978, 1986) thus suggests that similar ontological and epistemological foundations underlie both the natural sciences and the social sciences. However, not all agree, with Harré (1986) and Benton (1981) in particular seeking to reformulate Bhaskar's project, stressing that by their very definition social structures, unlike structures in the natural world, are not independent of individual human agents.

Within geography, the work of Sayer (1984) has been of seminal importance in developing the practical implications of realism. He thus identifies four main types of research which geographers can undertake: abstract theoretical research concerned with structures and mechanisms; concrete practical research focusing on events and objects produced by structures and mechanisms; empirical generalizations concerned with the establishment of the regularity of events; and synthesis research, which combines all of these types of research in order to explain entire sub-systems (Sayer, 1984). Furthermore, Sayer (1984) suggests that there are two fundamental varieties of concrete research, which he terms intensive and extensive. The former focuses on producing causal explanations pertaining to a small number of individual cases, whereas the latter seeks descriptive generalizations based on surveys of large populations. These, he emphasizes, address different types of question, use different methods, and define their objects in different ways.

Such ideas have provided the basis for a small but increasing amount of research specifically seeking to put into practice realist perspectives. Among the earliest such attempts were Allen's (1983) approach to property relations and landlordism, and Lovering's (1985) analysis of defence industries and the structuration of space in South Wales and the Bristol area of England. As this last example indicates, though, there is a close overlap of interest between realist approaches and those building on the structuration theory of Giddens. This similarity of interest has been emphasized by Soja (1985: 121) who suggests that 'The realist philosophy of social science seems almost ready-made to sustain and rationalize the theoretical directions taken by the contemporary materialist interpretation of spatiality.' Such attention to the significance of space has also been addressed by Sayer (1985b), who argues that the failure of much concrete research to develop satisfactory explanations results in large part from its failure to consider spatial form. Realist perspectives on the relationships between spatial and social structures have also led Gregory (1985b) to combine an interest with places, a new regional geography and spatial structure in order to develop an understanding of the role of space in society that also owes much to structuration theory.

### *7.3.2 Concrete buildings and postmodern alternatives*

At the close of the 1980s, the increased links between geography and the other social sciences led some geographers, notably Soja (1989; but see also Dear, 1988; Gregory, 1989; Harvey, 1989b), to turn to the postmodern critique of previous social theory. It is extremely difficult succinctly to summarize the broad spectrum of ideas encapsulated within postmodernism. At one level it is a body of social theory that has been derived from the critique of the modernist style of architecture developed following the First World War by architects such as Le

Corbusier, Bauhaus and Gropius. This style, full of optimism and functional simplicity, gave rise to the concrete tower blocks so characteristic of much housing and office construction in Europe and North America during the 1950s and 1960s. It was a Utopian, technical vision built on the grand theory that efficient, uniform buildings and town planning would lead to a well-organized society with an ability to maintain a highly efficient capitalist economy. However, in practice this monotonous concrete jungle environment, with its stark alleyways, and streets dominated by metal monster cars, devouring and polluting the atmosphere, provided the context for widespread alienation and dehumanization. When urban riots ensued and graffiti artists flourished, planners blamed the people living in the concrete and asphalt New Towns for deviant behaviour. They were unable to recognize that grand theory had failed; that humanity was crying for its freedom.

At a different and deeper level, though, postmodernism reflects a much wider artistic and cultural experience. For Olsson (1991) it is fundamentally a question of language and representation, of the relations between collective and individual unconsciousness, of revealing the hidden in the taken-for granted. As he goes on,

> It is with such questions that I now invite you to join me on a dangerous excursion into the Land of Thought-and-Action. This is unknown territory, for therein lies the geography of the future. To find our way we must rely on maps of the invisible, themselves invisible, and on a compass whose needle points not to the magnetic North pole, but to the socially taken-for-granted. Pushing into the unknown, we will gradually discover that it is when a discipline is turned against itself that it reveals its extraordinary power (Olsson, 1991: 85).

This involves an extremely radical programme of deconstruction, one in which the codes of the human mind, the very essence of culture, are destroyed in order that they might be rebuilt anew. For Olsson (1991), the key to this process is through an understanding of language and signs; postmodern human geography involves destroying the power of the concrete example; it requires challenging the intellectual heritage that finds its truth in concrete exhibition.

Postmodernism is concerned both with a critique of modernism, and more importantly with a search for difference. As such, it has at its basis a rejection of grand theory, and a desire to create a new intellectual and physical environment for human existence. Gregory (1989: 69–70) thus sees postmodernism as combining three main features: it is suspicious of systems of thought that claim to be complete and comprehensive; it is 'hostile to the "totalizing" ambitions of the conventional social sciences (and, for that matter, those of the humanities)'; and one of its key characteristics is its accent on heterogeneity and difference. Postmodernism can therefore be seen as encompassing a range of arguments critical of the all-embracing theoretical certainty associated with logical positivism and capitalism.

For Soja (1989) a central feature of modernist social theory has been its acceptance of the theoretical primacy of history and time. Although he is pessimistic about its success, seeing 'the present period primarily as another deep and broad restructuring of modernity rather than as a complete break and replacement of all progressive, post-Enlightenment thought' (Soja, 1989: 5), Soja argues that postmodernism needs to be informed by a reassertion of the importance of space. In developing a political programme he thus asserts that 'We must be insistently aware of how space can be made to hide consequences from us, how relations of power and discipline are inscribed into the apparently innocent spatiality of social life, how human geographies become filled with politics and ideology' (Soja, 1989: 6). In so doing, he seeks to construct a spatialized ontology built on Giddens's structuration theory, in which

> one can see more clearly an existentially structured spatial topology and *topos* attached to being-in-the-world, a primordial contextualization of social being in a multi-layered geography of socially created and differentiated nodal regions nesting at many different scales around the mobile personal spaces of the human body and the more fixed communal locales of human settlements. This ontological spatiality situates the human subject in a formative geography once and for all, and provokes the need for a radical reconceptualization of epistemology, theory construction, and empirical analysis (Soja, 1989: 8).

This concern with a reassertion of space is also at the heart of Harvey's (1989b) interpretation of postmodernism. In seeking to explain the changes that have taken place in capitalism since the 1960s, Harvey thus argues that an era of mass production and mass consumption has now been replaced by a much more flexible system of production and consumption, which can make use of an increasingly variegated space. He goes on to assert that this economic transformation has given rise to the cultural condition of postmodernity. In this formulation, Harvey (1989b) can thus be seen to be unwilling to reject the basic theoretical arguments of Marxism concerning the relationships between base, expressed as economic transformation, and superstructure, the cultural condition of postmodernity. By focusing on the explanatory power of Marxist theory, he is unable to accept the essentially anarchic practice of postmodernism.

Despite Olsson's (1991: 85) call for geographers to embark 'on a dangerous excursion into the Land of Thought-and-Action', there have, as yet, been few attempts to construct a truly postmodern geography. Among those who have sought to put these ideas into practice, the most explicit has been Soja's (1989: 223) attempt to take Los Angeles apart, by providing 'a succession of fragmentary glimpses, a freed association of reflective and interpretive field notes'. His target is thus

> to appreciate the specificity and uniqueness of a particularly restless geographical landscape while simultaneously seeking to extract

> insights at higher levels of abstracion, to explore through Los Angeles glimmers of the fundamental spatiality of social life, the adhesive relations between society and space, history and geography, the splendidly idiographic and the enticingly generalizable features of a postmodern urban geography (Soja, 1989: 223).

While this is, and can only be, a single interpretation, it does capture the essential element of a postmodern approach: an attempt to understand the variety and diversity of human society. As Gregory (1989: 91) has suggested, 'that there is more *disorder* in the world than appears at first sight is not discovered until that disorder is looked for'. Postmodernism seeks to reveal the illusory coherence that common sense imposes on the world's complexity.

## 7.4 Geography as critical science: the environmental conscience of society

Radical traditions in geography, in all of their very different guises, have sought not only to provide a critique of logical positivism, but also to effect social and political change. Given the apparently rampant success of capitalism during the 1980s, and the collapse of the communist regimes of eastern Europe during the early 1990s, radical geography has manifestly failed to achieve all of its practical intentions. In examining the reasons for this, it is salient to return to the five key features of Habermas's (1974, 1978) critical theory: his concern with the relationships between theory and practice; his theory of cognitive interests; his theory of communicative competence; his interest in empancipation; and his practice of self-reflection. Although critical theory is widely discussed in the geographical literature (Gregory, 1978; Johnston, 1991b) there have been remarkably few attempts to integrate its practical implications into the subject (although see Ley, 1980). Parallels have certainly been drawn between Habermas's critical theory and both realism and postmodernism (Cloke, Philo and Sadler, 1991), but geographers have tended to follow the latter two directions rather than address the central aims of critical theory. In so doing, they have concentrated on theory, to the detriment of practice. Habermas (1987a) has a central concern with the way in which reason can become practical, and it is this that he suggests is lacking in postmodernism. As McCarthy (1978: xvii) has concluded,

> Habermas agrees with the radical critics of enlightenment that the paradigm of consciousness is exhausted. Like them he views reason as inescapably situated, as concretized in history, society, body and language. Unlike them, however, he holds that the defects of the Enlightenment can only be made good by further enlightenment. The totalized critique of reason undercuts the capacity of reason to be critical. It refuses to acknowledge that modernization bears developments as well as distortions of reason.

### 7.4.1 *Institutions, career profiles and the funding of science*

Higher education institutions serve two main roles in society: research and teaching. Neither of these activities is socially or politically neutral. As Habermas (1978) has illustrated, the dominant type of science in capitalist society, namely empirical–analytic science, has a profoundly technical interest. It is supported only so long as its research and teaching produce and propagate knowledge that is useful to the enhanced sustainability of existing social and economic relations.

This is achieved not only through the direct funding of particular research projects, but also through the institutional structures in which teaching and research are undertaken. In recent years academic careers and promotion prospects have been influenced by two main criteria: the number of publications and the research income obtained by an individual or research group. Increasingly a third criterion, the number of students opting for particular courses, is now also being taken into consideration. Significantly, all of these indices are measures of quantity, and quality only rarely enters the assessment procedure. Under such conditions, there is much pressure for academics to obtain large research grants, to publish numerous papers, and to teach popular courses. None of these actions is conducive to the practice of critical enquiry. In most instances large research grants are provided either by government research councils or industrial funding, neither of which are likely to support research overtly designed to transform the status quo. The plethora of papers and books published on theoretical, rather than practical issues, particularly in human geography, can therefore largely be accounted for by the second of these factors, the pressure to publish. If radical geographers cannot obtain large research grants, then one way in which their careers can be advanced is through the maintenance of a successful publication record. However, if they are not undertaking *practical* research, which is extremely difficult in the absence of funding, then their publications must focus on *theoretical* and philosophical issues. Moreover, undertaking empirical field research is usually a lengthy task, and the administrative and teaching commitments of established staff further serve to minimize the periods available for such research. In contrast, it is possible to participate in seminars, to read books and papers, and to write critical articles within the usual institutional structures of most higher education institutions, even in term time. This goes some way to accounting for the theoretical rather than practical focus of much radical geography.

As far as teaching is concerned, there have been few studies of the types of courses for which students opt. However, it would appear that many students choose courses that are either perceived to be relatively easy, or that provide knowledge which is of direct relevance to their future employment. The whole primary and secondary education system is largely designed to provide students with a desire to learn and regurgitate accepted knowledge. Courses in higher education that require them to challenge such knowledge are therefore very likely to meet with resistance.

Such a scenario is based on the assumption that those in power in capitalist societies are intent on maintaining the social and economic structures upon which those societies are established. In turn, this implies that they consider capitalism to be sustainable, along with the numerous inequalities and oppressions upon which it depends. Although the demise of communism in what was the Soviet Union and in eastern Europe would appear to reinforce the vitality of capitalism, there is sufficient evidence of increasing inequality, both between groups of people within the major capitalist nations, and also between their economies and those of the poorer countries of the world, to suggest that this vitality is but illusory. The task of critical geography is to reveal these inequalities, to persuade those in power of their likely repercussions, and to be actively involved in the creation of new forms of social and economic organization. In short, it is to recognize the disturbed being of our society, to adopt a self-reflective stance towards it, and to serve as the psychoanalysts of the condition of which we are part.

### *7.4.2 Towards a critical geography*

In concluding this chapter it is possible to suggest four ways in which geographers could begin to adopt a more critical theoretical approach to the discipline. The final chapter will then explore some of the practical implications of such an enterprise.

A central feature of Habermas's (1974) critical theory is his conceptualization of a particular kind of relationship between theory and practice. This applies not only to the way in which 'facts' are related to 'theories', but also to the institutional practice of science. For Habermas, the objectivism of logical positivism deludes scientists with the idea that there is a self-subsisting world of facts structured in a lawlike manner. In contrast, a critical approach to geography would be concerned with revealing the way in which facts are constituted. This applies equally to the physical and human parts of the discipline. Moreover, Habermas argues that theoretical knowledge (1974, 1976, 1978), freed from human interests, is the only kind of knowledge that can orient action. It is thus essential that a critical geography seeks such action-oriented knowledge. Our research needs continually to be re-created out of a dialectical relationship between theory and practice. The implication of Habermas's work is that it is only critical science that can change the social formation of which it is a part. Critical theory places the emphasis of change back on those who undertake research. They have the choice, either to work for the poor, the oppressed and the exploited of the world, or to continue to support those whose very existence is dependent on such poverty and oppression.

Secondly, the relevance of Habermas's theory of cognitive interests needs further examination in the context of geographical enquiry (Gregory, 1978, 1986). While his critique of empirical–analytic and historical–hermeneutic sciences has provided a basis for considerable debate (see Thompson and Held, 1982a), the key feature of his overall approach is that it recognizes that different types of science have different aims and intentions. The implications of this statement for the various approaches that have recently been adopted in geography require a much more detailed examination than has been possible in the confines of this book, and it would be instructive to develop a formal framework for analysing each approach according to such characteristics as its cognitive interest, its form of knowledge, its conceptual domain, its validity criterion and even its mode of social organization.

Thirdly, Habermas has drawn attention to the importance of communication and language, not only in the practice of science, but also in the relationships between science and society. In developing his theory of communicative competence, Habermas (1984) was centrally concerned with establishing normative–theoretical foundations for the practice of social enquiry. To do this, he was forced to address issues concerning the motivation of communication, questions of truth and meaning, and the distorted patterns of communication resulting from the contradictions of capitalism. The full implications of his theory of communicative competence have not yet been rigorously appraised by geographers, but his concern with language provides a useful indicator of an avenue which geographers might explore further. The increasing divide between 'physical' and 'human' geographers, for example, is in part related to the different language systems that they adopt; many human geographers have difficulties in comprehending the mathematical notation of physical formulae, whereas physical geographers frequently decry what they see as the obscure waffle of much writing in human geography. Perhaps of more importance, is the observation that both sides generally fail to understand the purpose or intention behind the use of such language. This suggests that if any progress is to be made in bringing the two sides of the discipline back together, it needs to begin with an attempt to understand their differing modes of communication.

Although geographers have paid some attention to the structural linguistics of de Saussure and Chomsky (Gregory, 1978), it is indeed remarkable how little attention has been paid to linguistic philosophy as a source of inspiration, given its importance in the wider debates in 20th-century philosophy (but see Olsson, 1975, 1979, 1980, 1982, 1991). In particular, the two different philosophical positions outlined by Wittgenstein (1961, 1967) in *Tractatus logico-philosophicus* first published in 1921, and *Philosophical investigations* first published in 1953 warrant serious reconsideration by geographers because of their direct concern with the relationship between thought and language. Wittgenstein's fundamental aim in both works 'was to understand the structure and limits of thought, and his method was to study the structure and limits of language' (Pears, 1971: 12). However, to do this, it was essential for him to examine the relationships between reality and language. The key difference between his two philosophies was thus that in his earlier work he asserted that the structure of reality determines the structure of language, whereas in *Philosophical investigations* he argued that our language determines our view of reality. A second important difference between his two philosophies is closely related to debates that have subsequently taken place concerning structuralism in social theory. Pears (1971: 13) summarizes this as follows: 'In the *Tractatus* he had argued that all languages have a uniform logical structure, which does not necessarily show on the surface, but which can be disclosed by philosophical analysis', whereas in his later work he held the view that the diversification of linguistic forms 'actually reveals the deep structure of language, which is not at all what he had taken it to be. Language has no common essence, or at least, if it has one, it is a minimal one, which does not explain the connections between its various forms. They are connected with one another in a more elusive way, like games, or like the faces of people belonging to the same family' (Pears, 1971: 13–14). It is interesting to note here that such a changed viewpoint can also be seen as presaging aspects of a postmodern critique: from a position in which he held that language had a uniform theoretical structure, he later argued that there was no common essence to language. In describing the creative process of *Philosophical investigations,* Wittgenstein (1967: vii) thus comments that

> After several unsuccessful attempts to weld my results together into such a whole, I realized that I should never succeed. The best that I could write would never be more than philosophical remarks; my thoughts were crippled if I tried to force them on in any single direction against their natural inclination. . . . And this was, of course, connected with the very nature of the investigation. For this compels us to travel over a wide field of thought criss-cross in every direction. . . . The philosophical remarks in this book are, as it were, a number of sketches of landscapes which were made in the course of these long and involved journeys.

This process closely parallels Olsson's (1975: 29) geographical examination of thought, action and language in the first chapter of his book *Birds in egg*: 'What I have done in this chapter is to provide a set of impressionistic vignettes of what will follow. In each vignette I have conveyed essentially the same message. I have done so by viewing the same topic of thought and praxis from several different angles. The idea is of course that the only false perspective is the one which pretends to be unique.' Olsson (1975, 1979, 1980, 1982), moreover, provides a direct linkage between Habermas's concerns with justice, truth, theory and practice on the one hand, and Wittgenstein's investigations of language, thought and reality.

Fourthly, Habermas's concern with emancipation through self-reflection can be applied at the levels of individual experience, teaching and research. Geographers in their everyday practice can thus seek to pursue their own emancipation, by striving to achieve a state where their labour is no longer a commodity, to be used by those in power merely to accrue surplus profit. In their teaching, they can seek to enable students to achieve a state of emancipatory knowledge, in which they are free from self-imposed coercion. Rather than teaching being considered as the propagation of objective facts, this would involve providing students with the grounds from which they can make decisions enabling them to understand the conditions of their 'enslavement'. As Geuss (1981: 58) has argued, 'a critical theory has as its inherent aim to be the self-consciousness of a successful process of enlightenment and emancipation'. Finally, in their research, geographers could seek to reveal to individuals and society their conditions of false consciousness and unfree existence. This is broadly the task associated with the growing body of development theory concerned with empowerment (Friedman, 1992), by which is meant research and action designed to empower the poor in their own communities, and to mobilize them for widespread political participation.

CHAPTER 8

# *The place of geography*

> Since antiquity geographers have explored and analysed the earth's surface from two related perspectives: that of the spatial differentiation and association of phenomena with an emphasis on the meaning of space, spatial relations and place; and that of the relationship between man and his physical environment. The two are closely related because the meanings of space and place depend on the interrelationships among physical and human activities located in space, and man's relationships to the environment occur in the context of space and place.
>
> (Sack, 1980: 3)

One of the most salient characteristics of geographical practice in the last twenty years is that geographers have increasingly accepted the inherent diversity of the discipline and have in general ceased trying to identify a single core to the discipline. Rarely are arguments today promulgated to suggest that the central aim of geographical enquiry is, for example, to create a spatial science, or that systems analysis forms a unifying methodology for the discipline. Some, moreover, have argued that 'there is no *need* for geography and the other presently constituted fragments of social science, since they *must* be rejected' (Eliot Hurst, 1985: 60). In particular, this period has been characterized by an increasing division between the human and physical sides of the discipline. Stoddart (1987a: 330) has eloquently described this situation as follows:

> The result is clear enough. Across geography we speak separate languages, do very different things. Many have abandoned the possibility of communicating with colleagues working not only in the same titular discipline but also in the same department. The human geographers think their physical colleagues philosophically naive; the physical geographers think the human geographers lacking in rigour. Geography – Forster's, Humboldt's, Mackinder's – is abandoned and forgotten. And inevitably we teach our students likewise. Small wonder that the world at large wonders what we are about.

Stoddart (1987a: 330) goes on to suggest that two dangers arise in such a situation: first, physical geography loses its coherence outside the more general framework provided by its linkages with human geography; and second, 'Human geography as an exclusively social science loses its distinctive identity – it competes with sociology, economics, anthropology – but on their ground, not on ours.'

Although there is widespread acceptance and indeed reinforcement of this fragmentation of the discipline, its recognition has precipitated at least some geographers to reinterpret past views of the discipline and to move towards a new conceptualization of the links between its constituent parts (Johnston, 1989). Johnston (1991b: 132), recognizing the need to promote geography within the context of university planning and politics, has thus sought to identify the core of the discipline in 'the nature of regions – or of places' in his preferred terminology. While, in one sense, as the above quotation indicates, this involves a resuscitation of the traditional terminology of regional geography (Gregory, 1978), it is a resuscitation suffused by the more recent philosophical concepts of realism and structuration theory. For Johnston, who has in the past advocated a clear separation of physical and human geography (Johnston, 1991a), this would appear to represent a substantial move towards a *rapprochement* between the two main divisions in the discipline. However, as his examples (Johnston, 1991b, c) indicate, his concern with an interpretation of place is as yet largely uninfluenced by contemporary practice in physical geography.

A somewhat different approach, but one that still focuses on the identity of place, is that offered by Entrikin (1991). In this he consciously seeks to examine the full dimensionality of the concept of place, implicitly recognizing the very different standpoints of empirical–analytic and historical–hermeneutic science. Entrikin (1991: 5) thus argues that

> To understand place requires that we have access to both an objective and a subjective reality. From the decentered vantage point of the theoretical scientist, place becomes either location or a set of generic relations and thereby loses much of its significance for human action. From the centered viewpoint of the subject, place has meaning only in relation to an individual's or a group's goals and concerns. Place is best viewed from points in between.

The significance of place for Entrikin (1991) is that people as actors are always situated in particular places at particular periods, with the context of their actions contributing to their sense of identity. His focus, however, is fundamentally concerned with the lived experiences of people, and his conceptualization of place therefore has little to say about the contemporary process-oriented concerns of physical geographers.

An alternative attempt to provide a firm basis for geography which would enable it to claim the high ground of intellectual endeavour is provided by Stoddart (1987a). While suggesting that the building blocks of geographical research are 'Location, position, distance, area', and that they can be assembled to 'build regional geography' and 'to show the distinctiveness of place' (Stoddart, 1987a: 331), he argues that this is just the beginning. For Stoddart (1987a: 331) 'The task is to identify geographical problems, issues of man and environment within regions – problems not of geomorphology or history of economics of sociology, but geographical problems: and to use our skills to work to alleviate them, perhaps to solve them.' Stoddart's vision is of a geography that addresses central questions concerning the human use of land and resources:

> It is of a *real* geography – a reasserted *unified geography*, built on Forster and Humboldt, and at the same time a *committed geography*, seeking to honour Kropotkin's resolve. . . . It is a geography which will teach our neighbours and students and our children how to understand and respect our diverse terrestrial inheritance (Stoddart, 1987a: 333).

In conclusion this chapter seeks to explore the ways in which such ideas might be further developed in geographical research and teaching. In so doing, it suggests that some of the most fruitful lines of enquiry involve a return to issues which have been central to geographical practice since classical antiquity, most notably a concern with the place of people in nature, and with the meanings of space and time. Having reflected on past geographical practice, though, it is necessary to forge a new identity for the discipline, firmly situating it within the contemporary society of which it is a part.

## *8.1 Geographers and the environment*

During the 1980s environmental issues and green politics have risen to considerable prominence in the capitalist nations of the world (Sandbach, 1980; Porritt, 1984; Owens, 1986; Kuzmiak, 1991). In referring to environmental and resource issues, Emel and Peet (1989: 50) have thus commented that 'It would be difficult to find a set of issues which symbolizes more vividly the torment of a way of life gone astray, which captures more exactly the transformative urge propelling political–economic work.' However, geographers have had remarkably little involvement in this trend, either through their research or their political actions. There are notable exceptions to this generalization (O'Riordan, 1976, 1989a, b; Sayer, 1979; Pepper, 1984; N Smith, 1984; Redclift, 1987; Rees, 1989), but despite these it remains the case that professionals from other disciplines, most notably ecologists (Gorz, 1979) and even economists (Turner, 1988), have made substantial contributions to a field which in the first third of the 20th century was seen as being central to geographical research and teaching. In referring to the 1960s, Rees (1989) thus comments that 'Despite the fact that geographers once defined their subject as the study of the relationship between human society and the physical environment, this reawakening of concern over natural resources largely by-passed human geography', and she goes on to suggest that 'it is still true to say that the contribution of human geographers to the analysis of resource problems has been relatively slight'.

One reason for the diversity of disciplinary concern with environmental issues is that the analysis of resource use and management strategies requires a comprehension of many different types of phenomena and social practices. As Rees (1989: 365) has noted, 'an understanding of the problems involved in resource exploitation and the development of management policies . . . must involve inquiry into physical systems, economic processes, social organizations, legal and administrative structures and political institutions'. However, this does not by itself explain why such issues have only been addressed by a limited number of geographers. Another factor that needs to be taken into consideration is the institutional restructuring of higher education that took place, particularly in Britain, during the 1970s and 1980s. One aspect of this was the combination of departments in some institutions, and the creation of interdisciplinary research and teaching units in others. In particular, the rise of subjects such as environmental science and conservation as disciplines in their own right can be seen as having focused on much of the subject area that might once have been considered as part of geography. This is, for example, well reflected in the career of O'Riordan, one of the most prominent geographers concerned with environmental issues, but whose title is now that of a professor of environmental sciences, rather than that of a geographer. A further recent example of this trend has been the growing number of degrees being offered in such fields as environmental geology, environmental chemistry and environmental biology, all of which have sought to ride on the crest of the popular environmental wave but which represent a substantial overlap with much ground that has traditionally been covered in geography courses. Underlying these observations, though, it has been the increasing division of the discipline into a physical geography, still profoundly influenced by the tenets of logical positivism, and a human geography, searching for its credibility through association with the social sciences, that has been instrumental in preventing more geographers from tackling the central issues of environmental concern in contemporary society.

### *8.1.1 Applied physical geography*

Recent surveys of physical geography (K J Gregory 1985; Clark, Gregory and Gurnell, 1987a) amply illustrate that its main focus remains the explanation of both past and present physical processes in the environment. For the majority of its practitioners, physical geography is concerned with explaining changes in state of an objective world of facts. This is achieved through rigorous experimental design, model building or hypothesis testing, and its success is judged by the closeness of fit between its predictions, formulated through such expressions as mathematical formulae, and observations made in an empirical world of reality. Since the 1970s, though, increasing attention has been paid by some physical geographers to questions of an applied nature. Ken Gregory (1985: 187) notes, for example, that this has been reflected 'in the appearance of a final chapter devoted to applications in many books', and also in the growing number of books specifically devoted to applied issues (Goudie, 1981; Verstappen, 1983). Cooke (1987) has argued that there are four interrelated explanations for such a trend. First, he suggests that a shift in emphasis by physical geographers towards research on contemporary processes has made their work of direct relevance to planners and engineers. Second, the improvement in techniques associated with this has meant that they have been able to provide advice 'in a form that was both intelligible and acceptable to those outside the subject' (Cooke, 1987: 273). Third, he suggests that these changes coincided with increasing international concern with environmental issues, and fourth, he comments that an increasing number of geomorphologists have consciously sought 'to try to serve the needs of environmental managers' (Cooke, 1987: 274).

Most such applied physical geography retains a close adherence to the underlying philosophical principles of logical positivism. To return to Habermas's (1978) description of empirical–analytic science, it is concerned with the production of useful knowledge, and its technical interest is with prediction and control. This is expressed particularly clearly in the following description by Cooke (1987: 275) of the work of applied geomorphologists:

> Most geomorphological work for environmental management serves the requirements of clients who are almost certainly not geomorphologists. The research problems are posed by the clients, and the answers must be provided in a form that the clients can understand. As a result, most such applied geomorphological research – while it may be innovative in terms of methods and may ultimately feed back into geomorphological theory – must be cast in a form that is imposed from outside.

Application of the theories and methods of physical geography is essential for the management of the environment, but, as the above quotation illustrates, only very rarely are physical geographers involved in the decision-making processes behind the perceived need for environmental management. More importantly, they have little to say about who is managing the environment and for whom. Such applied physical geography is thus far removed from the concerns of environmentalism as they are posed, for example, by O'Riordan (1976) and Sandbach (1980). To the four explanations for the rising involvement of physical geographers in applied work noted by Cooke (1987) above, it is therefore necessary to add a fifth: the recent political intervention in higher education, which has forced academics to obtain increasing amounts of external funding for their research and teaching activities. During the 1980s and 1990s it has become evident that departments and institutions are being assessed increasingly in terms of their income-generating capacity, with much less emphasis than before being placed on the quality of their research. Furthermore, this cannot just be explained by the greater ease with which it is possible to quantify such things as contract income compared with intangible concepts such as research quality. Instead, it reflects a fundamental transformation in the way in which society, through the actions of its elected representatives, considers the nature of academic research. The shift in emphasis towards externally funded and applied research implies that increasing amounts of such research will be oriented to the maintenance and support of the social system that pays for it, rather than to research that seeks to reveal the contradictions within that society.

### *8.1.2 Landscapes and the domination of nature*

Those human geographers who have developed a research interest in the environment have generally approached the subject from a very different stance. At least two important traditions of such research can be identified. First, some geographers (see Cosgrove, 1984; Cosgrove and Daniels, 1988), focusing on the historical–hermeneutic quest for understanding and interpretation, have sought to examine the human meaning of landscape. This has involved studies not only of the symbolism of surviving landscapes, as represented in parks and gardens, but also of the symbolic representation of landscapes in paintings and literature. Nevertheless, as Cosgrove (1984) has amply illustrated, the complexity of symbolic representation in and through the landscape has proved very difficult to account for in theoretical terms. Among the most interesting avenues for such enquiry has been the consideration of elements in a landscape as analogous to words in a text. However, such an approach fails satisfactorily to conceptualize the landscape also as an image. This has led Cosgrove to turn to the methodology of iconography formulated by Panofsky (1939), and to identify three different levels at which landscapes can be recognized: that of the formal recognition of images and composition; that of the symbolic recognition of references made by elements and images; and that of iconology, or the situation of landscape in its social, historical, geographical and cultural context (Cosgrove and Daniels, 1988).

A second approach adopted by human geographers has been to turn to Marxist conceptualizations of the relationships between society and nature. According to these, 'Human activity reshapes nature but, at the same time, this necessary activity shapes the human character and the social relations between people – there is a constant interaction of human subject and natural object in the historical process' (Peet, 1989: 44). Two important implications underlie this approach: the first is that the relationships between people and nature are mediated through labour, and the second is that they are socially and historically structured (Sayer, 1979). Marx (1976: 283) thus argued that 'Labour is, first of all, a process between man and nature, a process by which man, through his own actions, mediates, regulates and controls the metabolism between himself and nature.' Moreover, for Marx (1976) human labour, by changing nature, also changes the nature of humanity. Furthermore, it is through the increasing human domination of nature, that some people are increasingly able to dominate others. In essence, what differentiates humanity from nature is its social context. However, two contrasting positions have been adopted in explaining the relationships between nature and society under capitalism. On the one hand, Neil Smith (1984) suggests that nature becomes merely an appendage of the production process; in effect nature is socially produced. In contrast, Peet (1989: 46) argues that 'Nature as origin and never-transcended inevitability (food, death etc.) makes human action better characterized as reproduction, i.e., we *reproduce* ourselves and our environment rather than *producing* nature.'

While analyses of landscapes and the human domination of nature have tended to provide the main focuses of attention for human geographers concerned with the links between nature and society, they have generally emphasized the primacy of human society over nature. In complete contrast are views which suggest that human society itself is essentially a product of environmental forces; that our mental images are merely the products of biochemical reactions. Such views, which are becoming increasingly widely held by sociobiologists and geneticists, popularized in books such as Dawkins's *The selfish gene* (1976) and *The blind watchmaker* (1986), provide a substantial challenge to the arguments of many human geographers. However, it is a challenge to which there has as yet been little response for two main reasons. On the one hand, there is still a heritage of concern over the excessive assertions of the proponents of environmental determinism earlier in this century. The exaggerated claims of Semple (1911) and Huntington (1925), have, for example, tended to discourage geographers from re-evaluating the influences of the environment upon human society and culture. On the other hand, the recent emphasis of much social theory, particularly that with a Marxist perspective, which has tended to treat such a view merely as another form of ideology, has also served to limit the attention paid to it by human geographers.

Habermas himself has also paid little attention to this debate, and has chosen to concentrate his attention primarily on the social world. Nevertheless, underlying his critical theory is the view that there is a two-way flow of interaction between people and nature, which operates at a range of scales from the individual to that of societies. For Habermas (1978) knowledge of nature comes from human interaction with nature, and yet at the same time such knowledge also stimulates the development of social labour. Habermas's critical theory, however, by focusing almost exclusively on the social realm, has failed sufficiently to examine the precise interactions between people and nature that are implied in the above argument. There is therefore a substantial opportunity for the development of a critical theory explicitly addressing these relationships in greater depth.

A further way in which geographers have made tentative steps towards an interpretation of people–environment relationships in the context of societies and space has been through a consideration of the concept of territoriality. Much of the initial emphasis on the idea of territoriality was prompted by analogies between animal and human behaviour. As Sack (1986: 1) has commented, 'Perhaps the most well-publicized statements on human territoriality have come from biologists and social critics who conceive of it as an offshoot of animal behaviour. These writers argue that territoriality in humans is part of an aggressive instinct that is shared with other territorial animals.' In contrast to such views, Sack (1986: 2) sees territoriality 'as socially and geographically rooted'. He goes on to argue that 'territoriality is intimately related to how people use the land, how they organize themselves in space, and how they give meaning to place' (Sack, 1986: 2). Likewise, Dodgshon (1987) in his analysis of the European past draws specific attention to territoriality in his argument that different stages of societal development found their expression in different systems of spatial order. Taking a long temporal perspective, Dodgshon (1987) examines the way in which the emergence of different types of society involved the creation of different patterns of spatial order. Thus, referring specifically to the development of a heightened sense of territoriality, he suggests that 'With the emergence of farming, we can say that territoriality acquired a sharper edge to it as territories slowly became segments of space over which groups establish exclusive rights of access and use and in which they had invested labour' (Dodgshon, 1987: 67). Territoriality can thus be seen as one expression of the way in which society and space are interrelated, and by examining changing concepts of territoriality it is therefore possible to explore the various ways in which the domination of nature has been used to influence power relationships within human societies.

Such concerns with landscape and nature are far removed from the process-oriented approaches of most physical geographers, reflecting the very different philosophical and methodological traditions in which they have emerged. Yet the last decade has seen increasing numbers of both physical and human geographers moving towards a reinterpretation of one of the discipline's traditional core themes. For those who wish to identify a focus for a revitalized geography, Stoddart's (1987a) vision of a discipline addressing key issues of human use of the environment, is thus one that is not only highly relevant to contemporary humanity, but also one with a considerable diversity of interests and approaches. In his description of the apparent impasse between physical and human geography quoted earlier, Stoddart (1987a: 330) emphasized the crucial point that 'we speak separate languages'. Any effort to bring both sides together again must therefore address ways in which geographers communicate, not only among themselves but also in their wider interactions with society. It is here that some of the ideas of Habermas's (1984, 1987a) theory of communicative competence may be of relevance, with their emphasis on the importance of language in reinforcing social life, and in particular on an explanation of social disorder through an understanding of the structure of communication. The use of separate languages by physical and human geographers, well illustrated for example in the contrasting pages of the journals *Progress in Physical Geography* and *Progress in Human Geography*, serves not only to provide a common corpus of understanding among practitioners of each part of the discipline, but as each language becomes more refined and distinct it also serves to make communication between the two groups more difficult. If this condition is seen as being indicative of disorder within the discipline, then one remedy might be for geographers to begin to learn each other's languages, or indeed to begin to create a new language satisfying Habermas's ideal speech situation. However, this will not be easy, largely because of the different truths that physical and human geographers have come to pursue: truths of explanation and truths of understanding.

## 8.2 *Space(–)time and geography*

One of the clearest differences in the use of language between physical and human geographers is to be found in their reference to space. While the idea of geography as a spatial science was highly influential as a unifying concept during the 1960s, the subsequent two decades have seen considerable divergence of opinion concerning the conceptualization of space. Most physical geographers continue to accept a view of space 'defined as three-dimensional Euclidian in which action occurs by contact' (Sack, 1980: 56), whereas many human geographers now interpret space as being socially constructed.

Geographers have had a surprisingly insignificant role in contemporary philosophical debates concerning the nature of space. In the social sciences and humanities, it has been philosophers such as Lefebvre (1991) and social theorists such as Giddens, (1981, 1985) who have in a sense rediscovered the importance of space in human society. Lefebvre (1991: 412) thus notes that 'It is impossible, in fact, to avoid the conclusion that space is assuming an increasingly important role in supposedly "modern" societies, and that if this role is not already preponderant it very soon will be.' In the physical sciences, it has likewise not been geographers who have been at the forefront of debates concerning the nature of space. Instead, it has largely been astronomers and physicists who have made substantial advances in understanding the physical significance of space and time (Smart, 1964b; Davies, 1974; Flood and Lockwood, 1986a; Hawking, 1988).

### *8.2.1 The social production of space*

One reason for the dearth of geographical considerations of space during the 1970s was that an important element of the critique of logical positivism within human geography focused on what was seen as the spatial fetishism of geographical practice during the 1960s. Moreover, the increasing engagement of human geographers with wider traditions of social theory, and their castigation of much physical geography as being imbued by logical positivism, made them reluctant to examine the implications of advances in theoretical physics in the wake of Einstein's general and special theories of relativity. For physical geographers, such advances were also not seen as being particularly relevant, because they dealt with scales both much smaller and much larger than that of processes operating at the surface of the earth. This is not to deny that some geographers were actively interested in questions concerning the nature of space and time (Carlstein, Parkes and Thrift, 1978; Bird, 1981; Morrill, 1985), but it is to suggest that most of the original intellectual contributions in this field came from outside the discipline.

One of the few geographers to pay explicit attention to the question of space was Sack (1980). In his book *Conceptions of space in social thought: a geographic perspective* he emphasized that there are many different conceptions of space, which arise because the conceptual separation between space and its substance 'can occur at different levels of abstraction and from different viewpoints and modes of thought' (Sack, 1980: 4). In an effort to provide a framework for interpreting such views he suggested that they vary depending essentially on the degrees of objectivity and subjectivity that are involved in the conceptualization of the relationships between space and substance. The domain of physical science concerns mainly the realm of objectivity, and the artistic domain that of subjectivity. For Sack (1980: 27), 'Whereas the significance of space in science is determined by the conceptual recombination of space with substance in laws, the significance of space in art lies in its connection to feeling, that is, in the import of the illusion.' Such modes of thought, which recognize that they have to some extent separated subjective from objective and space from substance, and which have attempted to develop a synthesis from them, he terms 'sophisticated–fragmented' thought. In contrast, Sack (1980) suggests that it is possible to identify a second very different pattern of thought, which he terms the 'unsophisticated–fused' pattern, characteristic of children, the practical view and the mythical–magical view. These modes have low levels of abstraction, and space and substance are little differentiated; symbols frequently embody both facts and feelings. Sack's (1980: 201) plea is for geographers to adopt a realist theory of science, combining the insights obtainable from all of these different modes of thought, in order to 'bring us in touch with the variety of human experiences, feelings, emotions and their symbolic form'.

The 1980s have seen some movement towards the development of social theory in which concepts of space play a central role (Giddens, 1981, 1985; Harvey, 1989a). Much of this work has been strongly influenced by the ideas of the social philosopher Henri Lefebvre (1991: 404), who suggests that 'Social relations, which are concrete abstractions, have no real existence save in and through space. *Their underpinning is spatial.*' In particular, Lefebvre (1991) draws four main implications from his hypothesis that space is produced: first, that '(physical) natural space is disappearing' (Lefebvre, 1991: 30); second, that 'every society – and hence every mode of production . . . produces a space, its own space' (Lefebvre, 1991: 31); third, that 'If space is a product, our knowledge of it must be expected to reproduce and expound the process of production' (Lefebvre, 1991: 36); and fourth, that 'we are therefore dealing with history' (Lefebvre, 1991: 46).

Lefebvre (1991) is centrally concerned to produce a reconciliation between what he sees as two kinds of space: on the one hand a *mental* space developed by philosophers and epistemologists and referred to by such terms as literary space or ideological space, and on the other *natural* space, the physical space in which we actually live. This is achieved through his development of a concept of *social* space, which 'is revealed in its particularity to the extent that it ceases to be indistinguishable from mental space (as defined by the philosophers and mathematicians) on the one hand, and physical space (as defined by practico-sensory activity and the perception of "nature") on the other' (Lefebvre, 1991: 27). He suggests that 'such a social space is constituted neither by a collection of things or an aggregate of (sensory) data, nor by a void packed like a parcel with various contents, and that it is irreducible to a "form" imposed upon phenomena, upon things, upon physical materiality' (Lefebvre, 1991: 27).

According to Lefebvre (1991) the basic flaw with most theories of space is that they picture it as a frame or container into which substance is put. He thus argues that

> a space that is apparently 'neutral', 'objective' fixed, transparent, innocent or indifferent implies more than the convenient establishment of an inoperative system of knowledge, more than an error that can be avoided by evoking the 'environment', ecology, nature and anti-nature, culture, and so forth. Rather, it is a whole set of errors, a complex of illusions, which can even cause us to forget completely that there is a total subject which acts continually to maintain and reproduce its own conditions of existence, namely the state (along with its foundation in specific social classes and fractions of classes) (Lefebvre, 1991: 94).

This emphasis on the state, and his assertion that class struggle is inscribed in space, indicate the underlying nature of his project, which is to use his theory of space to contribute 'to the dismantling of existing society by exposing what gnaws at it from within' (Lefebvre, 1991: 420).

### *8.2.2 Relativity and quantum theory*

Considerations of space by physical geographers, in contrast, have been grounded firmly in the traditions of the natural sciences. One of the most significant contributions in this field was Schumm and Lichty's (1965) paper entitled 'Time, space and causality in geomorphology', which had a major influence on the subsequent way in which geomorphologists and physical geographers considered the influence of space and time on environmental processes. Their key argument was that causal relationships in the development of landforms were a function of time and space. For them, 'the distinctions between cause and effect in the molding of landforms depend on the span of time involved and on the size of the geomorphic system under consideration' (Schumm and Lichty, 1965: 110). In illustrating this they suggested that some variables relating to rivers could be dependent at one temporal scale, and yet independent at another. Thus valley dimensions were seen as being dependent during the geologic time span, but independent during the modern and present time spans. In effect, what they did was merely to illustrate that different theories tended to explain landform development at different scales. Nevertheless, as Haines-Young and Petch (1986: 139) comment,

> Despite the mythical nature of their ideas their paper constituted something of a landmark in geomorphology. In terms of the history of the subject its significance lies in the fact that geomorphologists had failed to see that theories can be about different things and answer different questions and yet not be incompatible.

Two particular observations can, though, be made about the way in which Schumm and Lichty (1965) considered space and time. First, rather than developing a comprehensive account of the nature of space and time they simply equated space with area, and second, they saw no problem in treating time as a variable.

Following Schumm and Lichty's (1965) paper, most geomorphologists have tended to ignore philosophical questions concerning the nature of space. However, in one of the few major publications to treat the subject, Thornes and Brunsden (1977) have specifically addressed the question of temporal explanation in geomorphology. Here, they emphasize that 'The fundamental temporal attribute of duration may be compared to the spatial qualities of area or distance in that they are of finite and of measurable magnitude,' although 'time is distinguished by possessing the property of intrinsic direction and in the macroscopic sense being irreversible' (Thornes and Brunsden, 1977: 2). Thus, while it is possible to move in any direction in three-dimensional space, events flow in a sequence from past through present to future; it does not appear to be possible to go back in time. Having pointed out that many traditional explanations of geomorphological phenomena relied on analogies to human experience, they note that 'Equally common in geomorphology are attempts to use time as one of several explanatory variables' (Thornes and Brunsden, 1977: 5). Although in their own analysis they concentrate primarily on a discussion of temporal data and models, their concluding chapter returns to the question of the relationships between time and space, and they outline an approach to data collection, analysis and model building that includes both a temporal and a spatial dimension. As they point out, 'A basic problem in geomorphology, as in other science, is to represent space, time and several other variables together' (Thornes and Brunsden, 1977: 175). The approach they adopt is to consider five different orders of structure: in zero-order structures time and space are considered independently; in first-order structures, they are considered as discrete, with only one of them being allowed to vary; in second-order structures, either space or time is regarded as continuous; in third-order structures, they envisage both space and time as being continuous; and in fourth-order structures, they include several variables in real, rather than presence–absence, form. As they point out with third-order structures, though, 'no way has yet been found of monitoring the data in a continuous fashion. The nearest approach is to replicate a whole landscape block in the laboratory but even then we are not able to capture the results continuously' (Thornes and Brunsden, 1977: 181). What they seek to do is, nevertheless, to involve all spatial and temporal dimensions, as well as numerous variables, in order to re-create reality, and thus provide a full explanation of the real world. While there is some confusion between the terms 'variable' and 'dimensions' in their analysis, their work is of particular significance because it begins to address the representation of phenomena as manifolds in space-time cubes. Although there has been little attempt among geomorphologists to take this further over the last decade, it does represent one of the few occasions where geographers have started to consider issues which have for much of this century provided the source of some of the most exciting philosophical discussions within astronomy and nuclear physics.

Einstein's theories of relativity, and the subsequent emergence of quantum theory, have proved to be of widespread significance, not only for the development of physics, but also for 20th-century philosophy in general. As Russell (1961: 786) has commented, 'What is important to the philosopher in the theory of relativity is the substitution of space-time for space and time.' Given geography's long-standing concern with space, it is remarkable that so few geographers have attempted to grapple with these ideas. Even those geographers who have attempted to address such issues seem to have failed to grasp their essential significance. Bird (1981: 130–1), for example, has commented that 'As post-Einsteinians we all know of the combination of space-time, and Ullman once argued before this Institute the interdependence of space and time, and the substitution of one for the other as "pervasive and fundamental"' (see Ullman, 1974: 136). However, he goes on to state that 'we must acknowledge that space and time possess very different properties' (Bird, 1981: 131). Likewise, much work in time geography, which consciously seeks to include both a temporal and a spatial dimension to its analysis (Carlstein, Parkes and Thrift, 1978) rejects the central conclusion of Einstein's work, which represents a departure from the simple four dimensions of classical physics.

If Einstein was correct, space and time can in no way continue to be thought of as separate concepts. As he commented,

> It is a widespread error that the special theory of relativity is supposed to have, to a certain extent, first discovered, or at any rate, newly introduced, the four-dimensionality of the physical continuum. This, of course, is not the case. Classical mechanics, too, is based on the four-dimensional continuum of space and time. But in the four-dimensional continuum of classical physics the subspaces with constant time value have an absolute reality, independent of the choice of the reference system. Because of this [fact], the four-dimensional continuum falls naturally into a three-dimensional and a one-dimensional (time), so that the four-dimensional point of view does not force itself upon one as *necessary*. The special theory of relativity, on

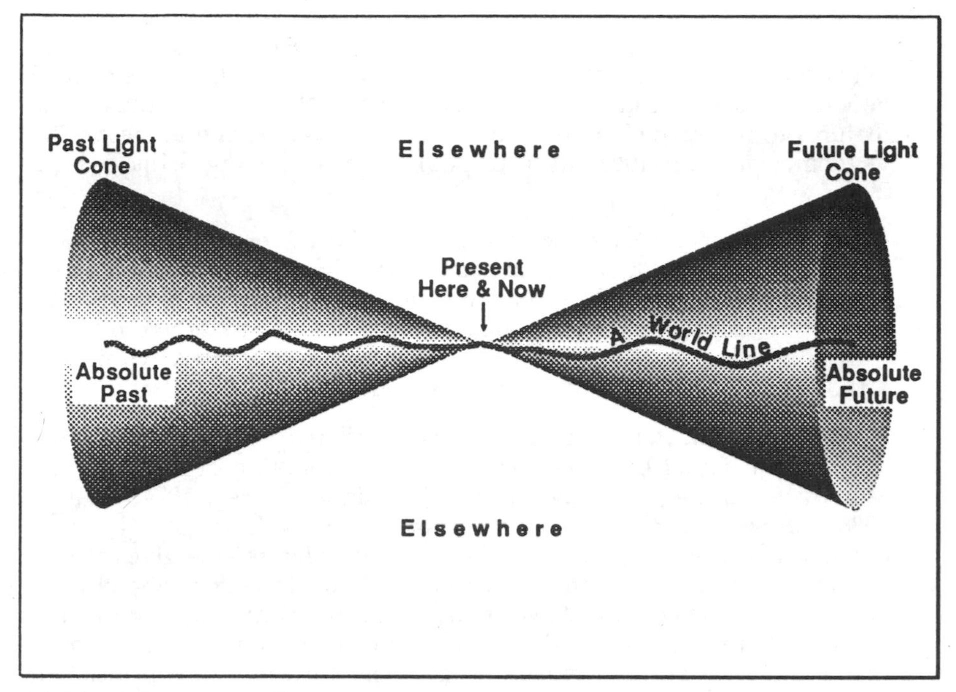


**Fig. 8.1** Space-time diagram.
*Source*: based in part on Shallis (1986: 87)

> the other hand, creates a formal dependence between the way in which the spatial co-ordinates, on the one hand, and the temporal co-ordinates, on the other, have to enter into the natural laws (Einstein, 1964: 281–2).

It thus no longer makes any sense to talk of space and time; rather we should begin to examine the meaning of space-time. It was Minkowski, working at the beginning of this century, who with his development of a four-dimensional tensor calculus, first really provided the basis for a new conceptualization of space-time. As he argued in his address to the 80th Assembly of German Natural Scientists and Physicians at Cologne in 1908,

> The views of space and time which I wish to lay before you have sprung from the soil of experimental physics, and therein lies their strength. They are radical. Henceforth space by itself, and time by itself, are doomed to fade away into mere shadows, and only a kind of union of the two will preserve an independent reality (Minkowski, 1964: 297).

According to Minkowski's arguments, a single point of space-time can be depicted as the intersection of two cones of light rays (Fig. 8.1) (Shallis, 1986). Everything outside the cones is inaccessible, and events lie on lines from the cone of the past, through the present to the cone of the future. Each event can be represented by a point, labelled by four coordinates. This conceptualization, it should be noted, is fundamentally different from the standard cube like space-time diagrams of time geography (see for example Parkes and Thrift, 1980). One of the central features to be grasped about Minkowski's arguments is that, according to them, space cannot endure through time. Following Schlick (1964: 293)

> One may not, for example, say that a point traverses its world-line; or that the three-dimensional section which represents the momentary state of the actual present, wanders along the time-axis through the four-dimensional world. For a wandering of this kind would have to take place in time; and time is already represented within the model and cannot be introduced from outside.

Underlying Einstein's special theory of relativity is the assumption that the velocity of light is independent of the motion of the observer. Moreover, nothing can travel faster than light (Davies, 1974; Sciama, 1986; Hawking, 1988). Thus, superimposing the velocity of light upon any other velocity merely gives the velocity of light. This gives rise to two phenomena: time dilation, the universal physical effect whereby the faster a body moves, the slower it ages; and the contraction of moving bodies, whereby moving bodies become foreshortened in their direction of motion. However, Einstein's special theory of relativity is inconsistent with the Newtonian theory of gravity, according to which, 'gravitational effects should travel with infinite velocity, instead of at or below the speed of light' (Hawking, 1988: 29). In order to resolve this inconsistency, Einstein (1964: 286) developed his general theory of relativity, based on 'the following principle: Natural laws are to be expressed by equations which are covariant under the group of continuous co-ordinate transformations'. The general theory led to the conclusion that space-time is curved because of the distribution of mass and energy in it, and that gravity, rather than being a separate force, is instead a result of this curvature. Thus, according to Hawking (1988: 33) 'Space and time are now dynamic quantities: when a body moves, or a force acts, it affects the curvature of space and time – and in turn the structure of space-time affects the way in which bodies move and forces act.'

Einstein's theories have profound implications for the debate concerning the absolute, or substantialist, and relative, or reductionist, interpretations of space and time. Newton-Smith (1986) thus suggests that both the special and general theories of relativity can be interpreted as supporting either viewpoint. The observation that the field equations of the general theory of relativity can be solved 'even when it is assumed that there is no matter or radiation whatsoever' (Newton-Smith, 1986: 34), can be interpreted as supporting the substantialist view. This would suggest that, although there is no such thing as absolute space and absolute time, it is possible that entirely empty space-times are able to exist. Against this, however, Newton-Smith (1986: 34) points out that

> Drawing attention to the fact that, in general, mathematical equations may have some solutions which are physically sensible and some that

> are not, the reductionist argues that they are extraneous roots which should be cast out. Rather than establishing the possibility of empty space-time they show a deficiency in the General Theory.

Whether or not the general theory of relativity favours reductionism or substantialism therefore remains an open question. What is important, though, is that this debate illustrates that its resolution will have to combine both philosophical and physical arguments. In Newton-Smith's (1986: 35) words, 'no satisfactory philosophical understanding of the nature of space, time or space-time can be achieved by remaining at the purely semantical level. Our question about meaning took us to physics and certain results in physics take us back to meaning'.

Einstein's theories also have importance for our understanding of the simultaneity of events and the direction of time. According to the special theory of relativity, simultaneity is relative, dependent on the choice of a frame of reference in motion. Thus,

> Observers moving relative to each other, but respectively thinking of themselves at rest (which they are each equally entitled to do, according to Einstein), will, if they utter the words 'future' or 'past' as they pass each other, mean different things. For they will be slicing the space-time continuum at different angles (Flood and Lockwood, 1986b: 4).

According to traditional views, this does not make sense, because it does away with the idea that there is a preferred frame of reference, which would enable us to decide on a definite past and future. However, Einstein's special theory implies that there is no real, or ontological difference between the future and the past. This has a further interesting implication, because it suggests that just as we cannot influence the past, it is also not possible to influence the future.

As Davies (1974: 22) has pointed out, though, 'It is a conspicuous fact of nature that the real world exhibits a structural difference between the two directions of time. That is, certain physical processes occur which are apparently asymmetric between these directions.' This is despite the observation that 'all known laws of physics are invariant under time reversal' (Davies, 1974: 26). Two main types of solution to this problem have been suggested, one deriving from the second law of thermodynamics and the other from quantum theory. According to the second law of thermodynamics, 'irreversible change arises from purposeless drifting into the available states' (Atkins, 1986: 86). In other words, there is a natural tendency for matter, energy and coherence to disperse over time in a closed system. One way of defining change is thus through the entropy principle, by which all natural change is seen as occurring in the direction of increasing entropy. According to the second law of thermodynamics, entropy can never decrease in an isolated system. Consequently, all change is in the direction of increasing dispersal of matter, energy and coherence. Hawking (1988), additionally, sees the psychological arrow of time, by which people feel that time passes, as being a direct result of the second law of thermodynamics. While this physical law appears to hold true, it can also be seen as apparently running counter to human endeavour. Hawking (1988: 152), for example, notes that 'The progress of the human race in understanding the universe has established a small corner of order in an increasingly disordered universe.' Moreover, individual human experience is often directed primarily to the creation of order in what appears to be a disordered world. This apparent paradox is usually explained (Davies, 1974; Hawking, 1988) through reference to the closed nature of the system in which the second law of thermodynamics and the principle of entropy are seen to apply. Thus, an increase in order, implying an associated decrease in entropy, in one part of the system/universe, is seen as only being possible at the expense of an entropy increase in the overall system. Once again this is illustrative of a tension between what might be termed the psychological view of time and the physical view, suggesting that the second law of thermodynamics might require some considerable revision.

The second main approach to questions of time asymmetry is that derived from quantum theory, and in particular from Heisenberg's principle of uncertainty, and from Bohr's arguments concerning the relationship between observer and observed. Central to an understanding of quantum theory is the observation that at the subatomic level electrons and photons have both wave-like and particle-like aspects. Moreover, it is possible to construct experiments to illustrate either their wave-like character or their particle-like character, but not both at the same time. Such a situation is usually interpreted as reflecting the existence of two superimposed worlds, which somehow interfere with each other. Heisenberg showed that the more accurately measurements are taken of the position of a particle, the less accurately is it possible to measure its speed, and vice versa (Davies, 1986). More formally, 'the uncertainty in the position of the particle times the uncertainty in its velocity times the mass of the particle can never be smaller than' Planck's constant (Hawking, 1988: 55). According to quantum theory, there is therefore an inherent uncertainty in science. Instead of having velocities and positions, particles have a quantum state, which is a combination of both position and velocity. Furthermore, Bohr suggested that 'no elementary quantum phenomenon *is* a phenomenon until it has been brought to a close by an irreversible act of amplification' (Davies, 1986: 99). In other words, the move from a microscopic world of uncertainty to a concrete macroscopic world is an irreversible change. Such irreversibility suggests that our concept of change is fundamentally associated with what we consider to be real, and it could provide one explanation for the observation of time asymmetry. According to this, time asymmetry might be identified 'with the entry into the consciousness of the observer of the details of where the electron is or, perhaps, how quickly it is moving' (Davies, 1986: 112).

Moreover, because this amplification produces either wave-like or particle-like characteristics, but not both, it implies that what an observer chooses to observe is inextricably linked to what is considered as existing. Once again, as with Einstein's theories of relativity, in order to make sense of quantum theory it is therefore necessary to combine questions of physics with those of philosophy. What is particularly interesting about quantum theory, is that it suggests that it is no longer possible to conclude that there is an objective world of facts separate from those making scientific observations. Typically, the response of physical geographers to such a conclusion is that their world is a macroscopic one, in which the implications of quantum physics are inappropriate. However, given the lack of agreement on the connections between the quantum world and the macroscopic world of everyday experience, this conclusion may not necessarily be appropriate.

### *8.2.3 Interpreting space*

In classical antiquity, a central part of geographical enquiry was concerned with the philosophical understanding of space. This brought the discipline into close allegiance with astronomy, and its most notable expression was Ptolemy's *Geography*. It also involved geographers in questions of cosmology and the place of the human world in the universe, issues which, in the 20th century, have largely become the preserve of physical scientists. The recurrent interest of some geographers, such as Sack (1980, 1986), Harvey (1989a, b) and Thornes and Brunsden (1977), with the nature of space and time nevertheless suggests that this is a theme which might well be reincorporated as a core integrating focus for the discipline, alongside the issues of environmental concern discussed earlier in this chapter. There may well be potential in redeveloping the dialogue between geographers and physicists that was largely broken in the 17th century by the concentration of geographers such as Varenius on the surface features of the earth. Geography's classical concern with profound philosophical questions concerning space and time, and thus with the very essence of human existence in the universe, is one that is as relevant today as it has ever been.

Given the gulf separating the views of physical scientists and geographers concerned with social philosophy on the question of space, it is not easy to see how they can be reconciled. However, three pointers can be identified which might offer a way for geographers to begin to make a significant contribution in this field. First, as the above account has illustrated, the contrasts in the languages of social theory and nuclear physics make it difficult for practitioners from one side to understand those from the other. There is therefore a need for human and physical geographers to begin to learn each other's languages, and instead of decrying the other through ignorance, to seek to understand the meanings of their interpretations.

Second, there is increasing acceptance, not only among social theorists, but also among physicists, that our statements about space(-/and)time are merely intellectual constructs, with which we try to understand our human existence. This has been well expressed by Shallis (1986: 78–9), an astrophysicist, in the following conclusion concerning the relationship between cosmology and time:

> Our cosmology also tells us about the way we perceive our relationship with the cosmos and with nature. That relationship, our cosmology tells us, has ceased to be grounded in the organic and cyclic world of nature in which we live our daily lives and in which time is more explicitly, more immediately present. Cosmologies have always incorporated a society's myths into their presentation and explanation of the world. There is no reason to suppose our cosmology is any different. It too is our myth, but it lacks the symbolic richness, the human ideals of past myths. To that extent it tells us a lot about ourselves and the culture we have built up over the past 300 years or so. . . . (I)f I had to encapsulate what we have learned about time from our cosmology it would be to say that we appear to have abstracted time, to have lost it or simply that we have passed time by.

The conceptualization of space-time is thus a product of our modern society, just as much as a film, a novel or a computer. Shallis (1986), in the above quotation, even hints at the suggestion that it reflects the dehumanization of the social world that lies at the heart of many critiques of contemporary society.

However, to say that the *concept* of space-time is a product of modern society, is not the same as to say that space-time is a product of that society. Herein lies one of the central problems of Lefebvre's (1991), and likewise Harvey's (1989a) and Giddens's (1981, 1985), conceptualization of the production of space. In such terminology, 'space' is used as a shorthand term, an all encompassing concept, which in the end lacks any real meaning. This is in effect admitted by Lefebvre (1991: 73) when he comments that social space 'subsumes things produced, and encompasses their interrelationships in their coexistence and simultaneity'. To move beyond this, geographers need not just to refer to the production of space, but rather to the processes by which specific experienced phenomena are produced in particular spatio-temporal contexts. Indeed, the return by Lefebvre (1991), Harvey (1989b) and Giddens (1981, 1985) to a concentration on space, while going some way to correct an overemphasis in much previous social theory on temporal explanations and interpretations, is in danger of calling upon itself once again the accusation of spatial fetishism. If there is one thing that both nuclear physics and the everyday lived world experience of human existence tell us it is that we cannot understand space and time separa from each other.

Third, our understanding of space(-/and)time is derived from a combination of previous theoretical constructs with our empirical experiences. This is but one illustration of the multifaceted interaction between theory and practice to which Habermas (1974) draws attention. However, it is strikingly evident that much geographical writing has failed sufficiently to address the connections between theory and practice at anything other than a superficial level. Within physical geography, for example, although there is widespread espousal of the need to relate empirical experiments to hypothesis testing and thus theory building, there has actually been remarkably little fundamental change in the theoretical underpinnings of the discipline. It is to this situation that Haines-Young and Petch (1986: 201) have alluded in their assertion that the theoretical framework of physical geography 'seems not to have developed as in other disciplines'. This is in marked contrast to the very different emphasis in some areas of human geography, where there has been a tendency to concentrate on the theoretical side at the expense of the empirical. In part, this avoidance of the empirical world reflects a fear of empiricism born out of the critique of logical positivism, and it is also undoubtedly a result of the increasing financial constraints on the practice of extensive empirical research. However, in seeking to provide a pointer to future geographical research on interpreting the lived world experience of space(-/and)time, it would seem important that this should seek to combine theoretical advances within an empirical context.

## 8.3 The theory and practice of geography

This conceptualization of the relationship between geographical theory and practice has implications not only for research but also for the teaching of the discipline and for the social responsibility of geographers. This penultimate section therefore returns to the focus of the first chapter of the book, and examines some of the implications of a critical stance in geographical practice.

### 8.3.1 Geographical education

Most people who claim to be professional geographers are actively involved in teaching and lecturing at a variety of different levels. It is therefore remarkable that so little attention is paid to the teaching of geography in higher education (but see Gold *et al.*, 1991). It is through teaching that geographers have their greatest influence on society; relatively few people other than professional geographers ever read publications containing the results of geographical research, and there is, for example, very little coverage of the work of geographers in the media. What is taught as geography at the primary, secondary and tertiary levels, and the way in which it is taught, are thus absolutely central to an understanding of the social practice of the discipline.

The Frankfurt School originated in part out of disquiet at the increasing political involvement of the German government in higher education during the 1920s. For Grünberg and his colleagues, the role of higher education was to challenge the existing balance of power and resource allocation. Such sentiments present geographers in the last decade of the 20th century with considerable problems. In the late 1980s there has been growing political involvement at all levels of the education system in many countries of the world, designed to increase the amount of 'useful' knowledge that students are provided with. Universities, schools, departments and individuals are thus all becoming increasingly subject to 'academic auditing', at a time when 'utility' is largely being defined in terms designed to sustain present social, political and economic relations. Teachers openly advocating the adoption of a critical stance to learning, designed to transform these very relations, are therefore likely to find themselves increasingly in conflict with the requirements of institutional survival.

Three main implications for the practice of teaching can be identified in Habermas's critical theory. First, it suggests that there is no such thing as an objective world of facts structured in a lawlike manner. Since the vast bulk of teaching is currently concerned with the passing on of such accepted facts, a critical geography would involve a substantial reorientation of teaching practice. Gold *et al.* (1991: 228) have, for example, commented that 'Our suspicion is that many teachers of geography in higher education perceive the most important educational problem to be that of specifying the geographical content of courses.' Instead of this, they recommend that teaching should become much more student centred, 'concerned with the development of students as geographers and individuals' (Gold *et al.* 1991: 228). To this end they advocate increased staff–student contact, student co-operation, active learning, prompt feedback, attention to time, the communication of high expectations, a respect for diversity of learning experiences, self-evaluation, clear identification of aims, and an educationally informed practice. These ten guiding principles for good teaching practice go some way to encourage the adoption of a critical stance in geographical teaching. However, they need to be underlain by the fundamental recognition of the social construction of what are accepted as facts, and the need for students therefore to question what they are presented with as truth. Higher education is not about the inculcation of accepted facts, but rather about enabling students to develop their own critical approach to the world in which they live.

Second, critical education is about emancipation rather than conformity. It is about providing students with an opportunity to discover their own truths, and their own ways of changing their social and economic conditions. It is about making education an exciting and enabling experience, rather than a chore that has to be undertaken according to externally formulated principles. Critical geographical teaching needs to overcome the tedium that many students feel with the imposed requirements of lectures, reading lists, essays and practicals. Somehow, learning must become instead an all-embracing thirst for emancipatory knowledge. By dehumanizing the learning experience, and institutionalizing it largely as the acquisition of accepted truth, most societies have sought to suppress such a thirst.

Third, having suggested that learning should be critical and emancipatory, it is possible to return to the issue of content. The existence of formal curricula, such as the National Curriculum in England and Wales, provides apparently rigid specifications for the content of geography courses. However, such curricula can be interpreted in a variety of ways. Thus, in referring to the introduction of the National Curriculum in geography, Edwards (1991: 1) has suggested that it 'is a baseline, not a boundary', Morgan (1991: 2) has commented that 'It provides a basis for good practice, but does not breathe life into it', and Rawling (1991: 2) argues that 'it is for individual teachers and schools to mould and adapt the Geography Order from the basis of good practice in geographical education'. Curricula can therefore all be interpreted in a variety of ways. Nevertheless, as the previous chapters have illustrated, geography has the ability to provide people of all ages with a chance to reflect on some of the most important questions of contemporary society, concerning such issues as environmental degradation, climatic change, differential access to resources, famine and poverty. Although geography has always been a wide-ranging discipline, it has a unique role at the primary and secondary levels of education in providing a critical interpretation of the human occupance of the earth, and the differences between places.

### *8.3.2 The choice of research*

Critical theory also has much to say about the practice of research. Above all, it suggests that there is no such thing as value-free, objective science. All research reflects an interaction between observer and observed, and between theory and practice. Moreover, it also reflects the social, economic and political context within which it is practised. Research can never be politically neutral. The responsibility for the research that is undertaken lies ultimately with those who do it. Particularly at times of increasingly stringent financial controls on scientific practice, there is great pressure on individuals and institutions to apply for research grants or contracts in fields that are supported by government agencies or industrial companies. However, the aims of such research are not always consonant with the ethical and moral positions of particular scientists. The classic examples of such conflicts can be seen with the debates over military research related to the US 'Star Wars' programme, and with those pertaining to animal experimentation. Nevertheless, this tension applies to all research, and geographers are faced with equally sensitive decisions about the types of research that they practise. Habermas (1978) thus cautions that empirical–analytic science has a technical interest. The implication of this for geography is that such science will tend to support the social, economic and political system of which it is a part. One good example of such research is the type of applied physical geography discussed earlier in this chapter. Likewise, the commercial explosion of interest in geographical information systems is being fuelled not so much by academic research interests, but rather by the needs of commerce and government. The 'knowledge' contained within such systems has great utility, providing those who possess it with considerable 'power'. In essence, geographers, like all scientists, have a choice over the sorts of research that they can do: it can either be used by those in power to retain their control, or it can be used to change the power relations existing in a particular society.

Traditionally, geographers have conducted a very wide range of research, from reconstructions of the Quaternary environment, through models of erosion processes and settlement systems, to interpretations of 19th-century novels. Indeed, one of the appeals of geography is that it provides an institutional home in which a great diversity of interests and approaches to research can be sustained. One contribution of a critical geography, though, would be to focus attention particularly on the contradictions and inequalities within societies, in order to provide knowledge that would enable them to be resolved. More specifically, their heritage of concern with the relationships between people and environments, suggests that this is a field in which geographers might have much to contribute. One task for a critical geography could therefore be to reveal the contradictions associated with the human exploitation of environments, and in so doing to suggest ways in which these might be resolved.

Resolutions of contradictions, however, can take a variety of forms, and usually reflect the relative balances of power between different social groups. Once again, this therefore involves a question of the people about and for whom research is undertaken. While the source of research funding is undoubtedly an important factor in this decision-making process, it is by no means the only one. Academics are still in many countries remarkably free to undertake whatever research they like. It is a matter for individual choice whether geographers study the most cost-efficient locations for retail location in advanced capitalist societies, or whether they undertake research on the availability of health-care facilities for rural people in central Africa. Critical geography calls for a commitment to research which seeks to reveal to people the conditions of their existence, and thereby to enable them to change them should they so wish.

How this is done in practice is by no means easy. Progression along the institutional career ladder requires that geographers write books and publish papers in refereed professional journals, that the vast majority of people will never read. In contrast, research designed to change the conditions giving rise to exploitation and deprivation requires communication with the agencies responsible for such conditions. It requires dialogue with local community groups, with government officials, with relief agencies and with charities. Moreover, it involves writing in publications that most geographers would never look at, but which might be read by those in positions to implement change. Above all it is suffused with a commitment to practical change.

### *8.3.3 Geographers in society*

Any commitment to social change is underlain by the belief that existing conditions are, for whatever reason, unsatisfactory. For those geographers content with the contemporary world order, the message of critical geography will be irrelevant. However, for those unhappy with the conditions that permit people to die of famine in eastern Africa, to live in cardboard boxes on the streets of London, to be subjected to racial violence in the United States, to have their livelihoods destroyed by floods in Bangladesh, or to die in makeshift air-raid shelters in Iraq and Yugoslavia, then the message of committed practice is one that cannot go unheeded. It is a message of social action.

Geographical teaching and research exist within particular social contexts. Geographical practice therefore begins with the societies in which individual geographers find themselves. Once again, though, there is a central tension between the institutional constraints seeking to limit social change, and the aspirations of geographers concerned with the implementation of change. Most frequently this tension is reflected in conflicts of financial interest. Governments and universities are thus unlikely to be prepared to continue funding teaching and research in geography departments which are not seen to be providing 'useful' graduates and research findings. Indeed, the closure of geography departments in universities such as Michigan and Chicago is ample testimony to this observation.

The public image of geography is therefore of fundamental importance. It is hardly surprising that a discipline that is widely regarded as being merely about states and capitals, capes and bays, and that does not seem to be particularly good at providing students even with this knowledge, is not generally held to be useful. One solution to this is for geographers to become more actively involved in the media, and in non-academic types of writing. Furthermore, it is also necessary for them to begin to take a much more active role in political action, at a range of scales from those of local government to the international forum of political debate. To do this, however, requires that academics working in geography departments accept the need to change the public image of their discipline, and consider this a worthwhile exercise. Moreover, it requires them to believe that the practice of geography is important, and that geography does indeed have a future as an academic discipline.

## *8.4 The future of geography: peoples, environments and places*

One possible conclusion to the current trend toward the division of geography into distinct physical and human parts is that the discipline will become subsumed within, on the one hand, a broadly defined earth science, and, on the other, an equally broad social science. At the level of particular institutions this has already, for example, involved the creation of specific social science and earth science departments, and at the broader level of scientific communities it can also be seen in the increasing interdisciplinary context of some research. Such a trend raises important questions concerning the future role of geography and geographers.

It would seem that given that such a trend exists, many geographers are happy with it. However, if it is to continue, some of the most important problems facing late-20th-century societies may fail to be addressed sufficiently rigorously by the academic community. In particular, the increasing separation between the social and earth sciences seems likely to leave a void of research on critical issues concerning the human utilization of the environment. It is just this void that Stoddart (1987a) has drawn attention to with his plea for geography to focus on questions concerning the human use of land and resources. There is a long tradition of such geographical research, and although many geographers continue to advocate separate human and physical dimensions to the discipline, it would seem that if geography did not already exist, then there would be a need to institutionalize a new discipline concerned with such issues. The recent expansion in the number of environmental departments and courses would seem to be just such an attempt.

This, it should be emphasized, is much more than just a question of institutional politics, and the maintenance of existing disciplinary boundaries. Subjects such as environmental degradation, pollution control, conservation, climatic change and resource management, for example, all require a substantial knowledge of both physical processes and social practice. An understanding of them can only be partial if approached purely from either the physical or the social sciences. As this book has illustrated, geography has a long tradition of research and teaching about the human occupance of the earth. This is a tradition which has ranged from research exclusively on physical processes operating at the surface of the earth, to social interpretations of landscape, and even cosmological considerations of the place of humanity in the universe. It is a tradition which we need to reclaim. When asked what we do, we need to be able to say with confidence that we are geographers; that we tackle some of the most critical issues facing contemporary society.

If human geography is to be subsumed within a broadly defined social science, and physical geography within the earth sciences, it is difficult to sustain an argument for the survival of anything resembling geography as it has recently been practised. In particular, although the comparatively recent discovery of 'space' by social scientists might appear to offer geographers a specific niche, this is but an illusion. No discipline can claim space as it own, not only because all human existence experiences space, but also because that experience is mediated through an experience of time. Space by itself is meaningless. For the same reason, it is difficult to accept arguments that suggest that the contribution of physical geography to earth sciences is that it considers processes as they operate in space. All physical processes have a spatial context.

Nevertheless, people create their own environments, and we can have no knowledge of environments separate from their human construction. It is this construction that makes places. Understanding such construction, however, presents many challenges, and can be approached in a variety of ways. It is these challenges that geographers such as Entrikin (1991) and Johnston (1991c) are now beginning to turn to, in interpreting the coming together of the objective and subjective worlds of reality. Place has become a focus for understanding the interaction of the human world of experience and the physical world of existence. The task of a critical geography is to enable people to reflect upon this interaction, and in so doing to create a new and better world.

# Glossary

This glossary is designed to provide short and succinct summaries of some of the terms used in the book with which readers may not be familiar. This is by no means a simple task, since many of them have a diversity of meanings reflecting their variant usage by different philosophers. Each term is elucidated in greater detail in the text, and the summaries are provided here primarily as an aid to readers to remind them of the basic meaning of a term. The pagination of the main exegesis of terms is given in parenthesis. Italics indicate cross-references. For more detailed expositions see Edwards (1967) and Johnston, Gregory and Smith (1986).

**analytic proposition**: One in which the predicate is part of the subject, such as 'a small rabbit is a rabbit'. Analytic statements merely serve to elucidate and are not informative. According to Wittgenstein, analytic statements are those whose truth is necessary to the linguistic use of the system of concepts on which they are dependent. Any statement that is not analytic is *synthetic*. (p. 34)

**anarchism**: a political movement concerned with the replacement of centralized states by self-sufficient co-operative groups; the rejection of order and authority. (pp. 87–90)

***a posteriori***: A term applying to truths, propositions and concepts, literally meaning 'from what is posterior'. In general usage, *a posteriori* reasoning is considered to be reasoning from effects to causes. According to Leibniz, *a posteriori* truths are those which can be seen to be true only from experience. Kant assumed it to be impossible for *analytic* judgements to be *a posteriori*. (p. 72)

***a priori***: A term applying to truths, propositions and concepts, literally meaning 'from what is prior'. In general usage, *a priori* reasoning is considered to be reasoning from causes to effects. According to Leibniz, *a priori* truths can be demonstrated as being based on identical propositions. The distinction between *a posteriori* and *a priori* is thus one based on what is, and what is not, derived from experience. (pp. 72–3)

**chorography** (or chorology): The investigation of the areal differentiation of the earth. Ptolemy distinguished between geography, the study of the earth as a whole, and *chorography*, the description of selected regions of the earth. This difference was paralleled by Varenius's distinction in the 17th century between general and special geography, and more recently between systematic and regional geography. (pp. 53–4)

**cosmology**: The study of the universe, either through philosophy or science. Philosophically, cosmology is widely seen as a branch of *metaphysics* concerned with universal concepts. The scientific use of the term refers to studies of the astronomical or physical universe as a whole. (p. 49)

**deduction**: The derivation of a particular truth from a general statement. It is a form of inference through which some premises and propositions are taken as basic, and all other propositions are seen as logically following from them. The opposite of *induction*. (pp. 22–3)

**determinism**: The view that all human behaviour is controlled by some external force. It can also be interpreted as the doctrine according to which everything that happens is the inevitable result of a previous state of affairs. In geography, it is most widely associated with the claims of environmental determinism, which assert that human action is determined by the environment. (pp. 92–5)

**dialectic**: A method of seeking truth through the resolution of paradoxes, opposites or contradictions. It is generally considered to have originated with the 5th century BC paradoxes of Zeno of Elea, and was widely used in Plato's Academy and by Aristotle. Hegel introduced the idea of a necessary movement or logic into his conceptualization of the *dialectic*, which sought the achievement of a higher unity through the passing over of thoughts into their opposites. Marx replaced Hegel's idealist conception of the dialectic, with its emphasis on the Spirit, by a materialist dialectic reflecting the material conditions of human existence. (pp. 164–5)

**empiricism**. The claim that experience rather than reason is the source of all knowledge, and through which observations are given primacy over theoretical statements. It provides a central tenet of *positivism*, but should be contrasted with the term empirical which is generally used to refer to substantive studies incorporating the results of observation, but not giving them the privileged status afforded by *empiricism*. (pp. 20–1)

**entropy**: A measure of disorder. More formally, it is a measure of the amount of usable energy within a system, and is approximately equivalent to the degree of organization of the system. According to the second law of thermodynamics, the flow of energy within a physical system will always result in a rise in its total entropy. (pp. 201–2)

**episteme**: according to Foucault, the world views, or structures of thought, held by a society at a particular time, which impose the same norms and postulates on all branches of knowledge. (p. 27)

**epistemology**: The branch of philosophy concerned with the scope and grounds of knowledge, and the reliability of different claims to knowledge. Plato is generally considered to have been the founder of epistemology, through his questioning of what knowledge can be supplied by reason. It focuses on how we know what we know, and is usually contrasted with *ontology*. (p. 21)

**existentialism**: A wide-ranging 19th- and 20th-century philosophical movement, best known through the writings of Kierkegaard, Heidegger and Sartre. Its central concern is with human existence in the world. Themes explored by existentialist philosophers include the relationships between individuals and systems, intentionality, being and absurdity, freedom and choice, and anxiety and dread. Much existentialist writing focuses on the way in which individuals are estranged from the world of externalized things, with the intention of reuniting them with their inherent creativity. (pp. 148–9)

**hermeneutics**: During the medieval period hermeneutics was concerned with the elucidation and interpretation of texts, and particularly with the identification of authentic Christian scripture. Subsequently, it has become more generally used to refer to the wider understanding of meanings. It emphasizes the artistic, rather than scientific nature of understanding, and focuses on questions of consensus, truth, and the hidden meaning in texts. During the 19th century hermeneutics reached its culmination in the work of Dilthey and Max Weber, and more recently it has formed a central element of Schutz's ethnomethodology. (p. 35)

**humanism**: A general term used to refer to philosophies based on the recognition of the value of people and which take human nature, in its broadest conception, as their theme. The main humanist philosophies recently considered by geographers include *existentialism*, *idealism*, *phenomenology*, *pragmatism* and *realism*. (pp. 145–6)

**hypothesis**: A statement established for the purpose of empirical testing. (p. 21)

**idealism**: A philosophy which takes the view that mental and spiritual conditions provide the basis for our understanding of the world. It assumes some spiritual reality beyond what appears evident through common sense. It is directly opposed to *materialism* and *realism*, in that it denies the possibility that material things exist outside our perception of them. (pp. 149–50)

**ideology**: In general usage the term applies to a body of ideas or a way of thinking. It can also be used to refer to the realm of *metaphysics*. In

Marxist terminology it has frequently been used to refer to the concept of false consciousness and to illusions shared by members of a class, which are associated with a specific economic structure. Althusser argues that *ideology* represents a different form of knowledge from *science*. Since its problematic remains within the social realm he suggests that it cannot produce new knowledge, but only knowledge that is a variation on some original knowledge. Ideology therefore performs the social function of enabling societies to regulate their behaviour. (pp. 26–9)

**idiographic**: A concern with the unique or particular; the opposite of *nomothetic*. (pp. 110–11)

**induction**: Argument through which general statements are derived from particular ones. More formally, it is used to refer to arguments in which the truth of premises provides a reasonable basis for belief in a conclusion, although they do not actually entail the truth of such a conclusion. The opposite of *deduction*. (pp. 22–3)

**law**: A general statement accepted as universally true. In *positivism*, laws are produced through the empirical testing of *hypotheses*. (pp. 21–3)

**logical positivism**: A branch of *positivism* developed by members of the Vienna Circle, such as Mach and Schlick, during the early 20th century, with the intention of using science to disprove the claims of *idealism* and transcendental *metaphysics*. It relied heavily on *empiricism* and the principle of verifiability. Through a misinterpretation of Wittgenstein's distinction between *analytic* and *synthetic* propositions, logical positivists argued that the former constitute the domain of formal sciences, whereas the truth of all *synthetic* statements can only be determined by empirical verification in the domain of factual sciences. Mach's logical positivism also rejected Comte's classification of sciences, and instead sought to unify all sciences through the application to them of the methodology of physics. (pp. 33–4)

**materialism**: Philosophy which gives primacy to the material world, and which, in contrast to *idealism*, relegates mental constructs to a secondary or dependent position. Marx's historical, or dialectical materialism sought to explain the dynamics of social change by focusing on the material foundations of social life, termed the economic base or mode of production. Changes in mode of production are seen to arise through tensions or conflicts in the relations of production of an earlier mode. (pp. 164–5)

**metaphysics**: The branch of philosophy that asks questions about the nature of reality, about being and knowing. Metaphysics claims to deal with questions which are beyond the ability of *science* to solve. (pp. 70–3)

**model**: A structure used to interpret the operation of a formal *system*; a structured representation of the real. Models are usually at different scales or levels from the reality they seek to represent. (pp. 155–6)

**nomothetic**: A concern with the general and universal; the opposite of *idiographic*. (pp. 110–11)

**ontology**: the branch of *metaphysics* which focuses on the nature of being and its relationship to human consciousness. Ontology is concerned with the ultimate objects of being; with what exists. It is usually contrasted with *epistemology*. (pp. 175–6)

**paradigm**: According to Kuhn, paradigms are the universally recognized achievements that provide model problems and solutions for members of a scientific community. A paradigm is taken to define both the problems and the methods used to solve them. (pp. 24–6)

**phenomenalism**: The view that there is no distinction between essence and phenomenon, and therefore that we can only record that which is directly experienced. (p. 32)

**phenomenology**: A term used to refer to philosophies concerned with phenomena. However, different interpretations of the meaning of the word phenomena have led to different usages of the term phenomenology. For Hegel, phenomenology sought to establish how mind is known through an examination of its appearance. Modern phenomenology, derived largely from Husserl, aims to reveal phenomena as intuited essences through direct awareness. Phenomenological statements are thus non-empirical descriptions of phenomena. (pp. 146–8)

**positivism**: A 19th-century philosophical movement originating in the work of Saint-Simon, and developed by Comte. Its central claims are that *science* is the only valid form of knowledge, and that observable facts are the only possible objects of knowledge. Built on the assumptions of *phenomenalism* and *empiricism* it is directly opposed to *metaphysics*. (pp. 31–5)

**postmodernism**: A wide-ranging interpretation of contemporary human existence, focusing on the search for difference. It is suspicious of all grand theories that seek to provide neat and overarching conclusions. (pp. 177–80)

**pragmatism**: A rapidly changing philosophical movement in the late 19th and early 20th century, particularly associated with Peirce, James and Dilthey. It emphasized the rejection of traditional 19th-century philosophy, and maintained a central concern with the clarification of meaning. It was characterized by a conception of knowledge as both fallible and fluid. (p. 152)

**rationalism**: A philosophy reflecting a wide ranging collection of beliefs but which places emphasis on *a priori* reasoning in the acquisition of knowledge. It is most widely associated with Descartes, Spinoza and Leibniz, and is usually contrasted with *empiricism*. (p. 21)

**realism**: A philosophy which holds that material objects exist independently of our sense experiences. It is concerned with the identification of causal mechanisms and empirical regularities, and is opposed to both *phenomenalism* and *idealism*. Realism seeks to reveal the causal mechanisms through which particular events are situated within underlying structures. This is achieved within transcendental realism through the process of abstraction. (pp. 175–7)

**science**: In general usage the term refers to knowledge and particularly to systematized knowledge. The scientific method is widely seen as involving the construction of *theories*, which seek to explain observed events by reference to unobserved forces. For Althusser, *science* is a different form of knowledge to *ideology*, in that it is able to produce new knowledge by using concepts as means of production in order to generate its own objects and orders of proof. (pp. 21–4).

**structuralism**: A range of philosophies, exemplified in the contrasting approaches of Lévis-Strauss and Althusser, which hold in common the view that the empirical world of observable phenomena is determined by underlying structures. The explanation of surface phenomena is thus achieved through the description of underlying structure. (pp. 169–71)

**structuration theory**: A theory developed by Giddens which seeks to transcend *hermeneutics*, functionalism and *structuralism*, through the integration of knowledgeable human agents and the wider social structures in which they are implicated. (pp. 172–74)

**synthetic proposition**: A proposition which is informative, because it connects together two different concepts; any statement which is not *analytic*. (p. 34)

**system**: In general usage, a system consists of a set of entities and the relationships between them. It should be contrasted with general systems theory which seeks to derive theoretical statements about the properties that are common to all systems. (pp. 128–9)

**teleology**: Argument by design. The view that developments are due to the purpose they serve. Frequently used to refer to the argument that because the world exhibits order it must have been produced by an intelligent designer. (pp. 49–50)

**theory**: Logically constructed statements designed to provide explanation or understanding, but which vary greatly from philosophy to philosophy. (pp. 20–44)

**topography**: the description of a small tract or place, contrasted by Varenius in the 17th century with *chorography*, which he saw as the description of a region of a medium size. (pp. 46–7)

# Bibliography

**Ackerman E A** (1945) Geographic training, wartime research, and immediate professional objectives. *Annals, Association of American Geographers* **35**: 121–43

**Ackerman E A** (1958) *Geography as a fundamental research discipline.* Chicago, University of Chicago Department of Geography Research Paper No. 53

**Ackerman E A** (1963) Where is a research frontier? *Annals, Association of American Geographers* **53**: 429–40

**Adickes E** (1911) *Untersuchungen zu Kants physicher Geographie.* Tübingen, J C B Mohr

**Adickes E** (1925) *Kant als Naturforscher.* Berlin, W de Gruyter

**Agnew J A** (1989) Sameness and difference: Hartshorne's *The nature of geography* and geography as areal variation. In: Entrikin J N, Brunn S D (eds) *Reflections on Richard Hartshorne's* The Nature of Geography. Washington DC, Association of American Geographers: 121–40

**Alexandrovskaya O** (1983) Pyotr Alexeivich Kropotkin 1842–1921. In: Freeman T W (ed.) *Geographers: biobibliographical studies*, volume 7. London, Mansell: 57–63

**Allen J** (1983) Property relations and landlordism: a realist approach. *Environment and Planning D: Society and Space* **6**: 339–65

**Althusser L** (1969) *For Marx.* Harmondsworth, Penguin

**Althusser L, Balibar E** (1970) *Reading Capital.* London, New Left Books

**Andrews H F** (1984) The Durkheimians and human geography: some contextural problems in the sociology of knowledge. *Transactions, Institute of British Geographers* n.s. **9**(3): 315–36

**Andrews H F** (1986) The early life of Paul Vidal de la Blache and the makings of modern geography. *Transactions, Institute of British Geographers* n.s. **11**(2): 174–82

**Antevs E** (1928) *The last glaciation, with special reference to the ice retreat in northeastern North America*. New York, American Geographical Society (Research Series, No.17)

**Armstrong P** (1985) Charles Darwin 1809–1882. In: Freeman T W (ed.) *Geographers: biobibliographical studies*, volume 9. London, Mansell: 37–45

**Atkins P W** (1986) Time and dispersal: the Second Law. In: Flood R, Lockwood M (eds) *The nature of time*. Oxford, Basil Blackwell: 80–98

**Aujac G** (1978) Eratosthenes *c*.275-*c*.195 BC. In: Freeman T W, Pinchemel P (eds) *Geographers: biobibliographical studies*, volume 2. London, Mansell: 39–43

**Ayer A J** (ed.) (1959) *Logical positivism*. London, George Allen and Unwin

**Babicz J, Büttner M, Nobis H M** (1982) Nicholas Copernicus 1473–1543. In: Freeman T W (ed.) *Geographers: biobibliographical studies*, volume 6. London, Mansell: 23–9

**Bacon F** (1965) *The advancement of learning*. London, Dent

**Bader F J W** (1978) Die Gesellschaft für Erdkunde zu Berlin und die koloniale Erschliessung Afrikas in der zweiter Hälfte des 19. Jahrhunderts bis zur Gründung der ersten deutschen Kolonien. *Die Erde* 109: 38–50

**Bagnold R A** (1941) *The physics of blown sand and desert dunes*. London, Methuen

**Bagnold R A** (1954) Some flume experiments on large grains but little denser than the transporting fluid, and their implications. *Institute of Civil Engineers, Proceedings*, Paper No. 6041: 174–205

**Bagnold R A** (1966) *An experimental approach to the sediment transport problem from general physics*. Washington DC, United States Geological Survey Professional Paper 422I

**Bailey P** (1989) A place in the sun: the role of the Geographical Association in establishing geography in the National Curriculum of England and Wales, 1975–89. *Journal of Geography in Higher Education* **13**(2): 149–57

**Bailey P, Binns T** (eds) (1987) *A case for geography: a response to the Secretary of State for Education from members of the Geographical Association*. Sheffield: Geographical Association

**Baker A R H** (ed.) (1972) *Progress in historical geography*. Newton Abbott, David and Charles

**Baker A R H, Billinge M** (eds) (1982) *Period and place: research methods in historical geography*. Cambridge, Cambridge University Press

**Baker A R H, Gregory D** (eds) (1984) *Explorations in historical geography: interpretative essays*. Cambridge, Cambridge University Press

**Baker A R H, Hamshere J D, Langton J** (eds) (1970) *Geographical interpretations of historical sources*. Newton Abbot, David and Charles

**Baker J N L** (1937) *A history of geographical discovery and exploration*. London, George G Harrap, rev. edn.

**Baker J N L** (1955) The geography of Bernhard Varenius. *Transactions, Institute of British Geographers* **21**: 51–60

**Barlett R A** (1962) *Great surveys of the American West*. Norman, University of Oklahoma Press

**Barnes B** (1982) *T. S. Kuhn and social science*. London, Macmillan

**Barrows H H** (1923) Geography as human ecology. *Annals, Association of American Geographers* **13**: 1–14

**Bartels D** (1968) *Zur wissenschaftstheoretischen Grundlegung einer Geographie des Menschen*. Wiesbaden, Steiner (Geographische Zeitschrift Beihefte Erdkundliches Wissen, 19)

**Bassin M** (1987a) Friedrich Ratzel 1844–1904. In: Freeman T W (ed.) *Geographers: biobibliographical studies*, volume 11. London, Mansell: 123–32

**Bassin M** (1987b) Race contra space: the conflict between German *Geopolitik* and National Socialism. *Political Geography Quarterly* **6**(2): 115–34

**Bataillon C** (ed.) (1983) French geography in the 1940s. In: Buttimer A (ed.) *The practice of geography*. London, Longman: 119–49

**Bauman Z** (1978) *Hermeneutics and social science: approaches to understanding*. London, Hutchinson

**Bayliss-Smith T P** (1982) *The ecology of agricultural systems*. Cambridge, Cambridge University Press

**Bayliss-Smith T P, Wanmali S** (eds) (1984) *Understanding green revolutions: agrarian change and development planning in south Asia*. Cambridge, Cambridge University Press

**Beaumont J R** (1987) Quantitative methods in the real world: a consultant's view of practice. *Environment and Planning A* **19**: 1441–8

**Beaver S H** (1962) The Le Play Society and field work. *Geography* **47**: 225–40

**Beaver S H** (1982) Geography in the British Association for the Advancement of Science. *Geographical Journal* **148**(2): 173–81

**Beck H** (1982) *Grosse Geographen. Pioniere-Aussenseiter-Gelehrte*. Berlin, Dietrich Reimer Verlag

**Beckinsale R P, Beckinsale R D** (1989) Eustasy to plate tectonics: unifying ideas on the evolution of the major features of the earth's surface. In: Tinkler K J (ed.) *History of geomorphology: from Hutton to Hack*. Boston, Unwin Hyman: 205–21

**Beckinsale R P, Chorley R J** (1991) *The history of the study of landforms or the development of geomorphology. Volume 3: Historical and regional geomorphology 1890–1950*. London, Routledge

**Bednarz S W** (1989) What's good about alliances? *Professional Geographer* **41**(4): 484–6

**Bell C** (1974) *Portugal and the quest for the Indies*. London, Constable

**Bennett R J** (1980) *The geography of public finance: welfare under fiscal federalism and local government finance*. London, Methuen

**Bennett R J** (1985) Quantification and relevance. In: Johnston R J (ed.) *The future of geography*. London, Methuen: 211–24

**Bennett R J, Chorley R J** (1978) *Environmental systems: philosophy, analysis and control*. London, Methuen

**Benton T** (1981) Realism and social science. *Radical Philosophy* **27**: 13–21

**Benton T** (1984) *The rise and fall of structural Marxism: Althusser and his influence*. London, Macmillan

**Berdoulay V** (1978) The Vidal–Durkheim debate. In: Ley D, Samuels M S (eds) *Humanistic geography: prospects and problems*. London, Croom Helm: 77–90

**Berdoulay V** (1981) *La formation de l'école française de geographie (1870–1914)*. Paris, Bibliothéque Nationale [comité des Travaux Historiques et Scientifiques]

**Bergmann G** (1957) *The philosophy of science*. Madison, The University of Wisconsin Press

**Berry B J L** (1972) Notes on relevance and policy analysis. *Area* **4**(2): 77–80

**Berry B J L** (1973) A paradigm for modern geography. In: Chorley R J (ed.) *Directions in geography*. London, Methuen: 3–22

**Bertalanffy L von** (1951) An outline of general systems theory. *British Journal of the Philosophy of Science* **1**: 134–65

**Bertalanffy L von** (1956) General systems theory. *General Systems Yearbook* **1**: 1–10

**Bhaskar R** (1978) *A realist theory of science*. Brighton, Harvester, 2nd edn.

**Bhaskar R** (1980) Scientific explanation and human emancipation. *Radical Philosophy* **26**: 16–28

**Bhaskar R** (1986) *Scientific realism and human emancipation*. London, Verso

**Billinge M** (1977) In search of negativism: phenomenology and historical geography. *Journal of Historical Geography* **3**(1): 55–68

**Billinge M** (1983) The mandarin dialect: an essay on style in contemporary geographical writing. *Transactions, Institute of British Geographers* n.s. **8**(4): 400–20

**Billinge M, Gregory D, Martin R** (eds) (1984a) *Recollections of a revolution: geography as spatial science*. London, Macmillan

**Billinge M, Gregory D, Martin R** (1984b) Reconstructions. In: Billinge M, Gregory D, Martin R (eds) *Recollections of a revolution: geography as spatial science*. London, Macmillan: 1–24

**Bird J** (1981) The target of space and the arrow of time. *Transactions, Institute of British Geographers* n.s. **6**(2): 129–51

**Bird J** (1989) *The changing worlds of geography: a critical guide to concepts and methods*. Oxford, Clarendon Press

**Birkenhauer J A C** (1986) Johann Gottfried Herder 1744–1803. In: Freeman T W (ed.) *Geographers: biobibliographical studies*, volume 10. London, Mansell: 77–84

**Bishop P** (1980) Popper's principle of falsifiability and the irrefutability of the Davisian cycle. *Professional Geographer* **32**: 310–15

**Blackmore J T** (1972) *Ernst Mach: his work, life and influence*. Berkeley, University of California Press

**Blaut J M** (1975) Imperialism: the Marxist theory and its evolution. *Antipode* **7**(1): 1–19

**Bloch M** (1954) *The historian's craft*. Manchester, Manchester University Press

**Bloom B S** (ed.) (1956) *Taxonomy of educational objectives: cognitive domain*. New York, McKay

**Blouet B W** (ed.) (1981) *The origins of academic geography in the United States*. Hamden, Archon Books

**Blowers A T** (1972) Bleeding hearts and open valves. *Area* **4**(4): 290–2

**Boal F W, Livingstone D N** (eds) (1989) *The behavioural environment*. London, Routledge

**Boddy M** (1976) Political economy of housing: mortgage-financed owner occupation in Britain. *Antipode* **8**(1): 15–24

**Boehm R G, Kracht J B** (1986) Enhancing high school geography in Texas. *Professional Geographer* **38**(3): 255–6

**Botting D** (1973) *Humboldt and the Cosmos*. London, Sphere Books

**Bowen M** (1970) Mind and nature: the physical geography of Alexander von Humboldt. *Scottish Geographical Magazine* **86**: 222–33

**Bowen M** (1981) *Empiricism and geographical thought from Francis Bacon to Alexander von Humboldt*. Cambridge, Cambridge University Press

**Bowlby S, Lewis J, McDowell L, Foord J** (1989) The geography of gender. In: Peet R, Thrift N (eds) *New models in geography: the political-economy perspective*, volume II. London, Unwin Hyman: 157–75

**Bowman I** (1934) William Morris Davis. *Geographical Review* **24**: 177–81

**Boyne R** (1991) Power-knowledge and social theory. The systematic misrepresentation of contemporary French social theory in the work of Anthony Giddens. In: Bryant C G A, Jarry D (eds) *Giddens' theory of stucturation: a critical appreciation*. London, Routledge: 52–73

**Bradford M G, Kent W A** (1977) *Human geography: theories and their applications*. Oxford, Oxford University Press

**Braithwaite R B** (1960) *Scientific explanation*. New York, Harper Torchbooks

**Breitbart M M** (1981) Peter Kropotkin, the anarchist geographer. In: Stoddart D R (ed.) *Geography, ideology and social concern*. Oxford, Basil Blackwell: 134–53

**Broc N** (1974) L'établissement de la Géographie en France. *Annales de Géographie* **83**: 71–94

**Brookfield H** (1969) On the environment as perceived. In: Board C *et al*. (eds) *Progress in Geography 1*. London, Edward Arnold: 51–80

**Brunhes J** (1925) Human geography. In Barnes H E (ed.) *The history and prospects of the social sciences*. New York, Alfred A. Knopf: 55–105

**Bryant C G A** (1985) *Positivism in social theory and research*. London, Macmillan

**Buchanan K** (1972) *The geography of empire*. Nottingham, Spokesman

**Bunbury E H** (1879) *A history of ancient geography among the Greeks and Romans from the earliest ages till the fall of the Roman Empire*. London, John Murray

**Bunce V J** (1986) Underrated but invaluable: the image of secondary school geography in the 1980s. *Geography* **71**(4): 325–32

**Bunge W** (1962) *Theoretical geography*. Lund, G W K Gleerup (Lund Studies in Geography, Series C1)

**Bunge W** (1966) *Theoretical geography*. Lund, C W K Gleerup, 2nd ed. (Lund Studies in Geography, Series C1)

**Bunge W** (1968) *Fred K Schaeffer and the science of geography*. Harvard Papers in Theoretical Geography, Special Papers Series A

**Bunge W** (1977) A personal report. In: Peet R (ed.) *Radical geography: alternative viewpoints on contemporary social issues*. Chicago, Maaroufa: 31–39

**Bunge W** (1979a) Fred K Schaeffer and the science of geography. *Annals, Association of American Geographers* **69**: 128–32

**Bunge W** (1979b) Perspective on *Theoretical geography*. *Annals, Association of American Geographers* **69**: 169–79

**Bunting T E, Guelke L** (1979) Behavioral and perception geography: a critical appraisal. *Annals, Association of American Geographers* **69**: 448–62

**Burgess E W** (1925) The growth of the city: introduction to a research project. In: Burgess E W, Park R E (eds) *The city*. Chicago, Chicago University Press: 47–62

**Burgess E W** (1964) A short history of urban research at the University of Chicago before 1946. In Burgess E W, Bogue D J (eds) *Contributions to urban sociology*. Chicago, University of Chicago Press: 2–13

**Burton I** (1963) The quantitative revolution and theoretical geography. *The Canadian Geographer* **7**: 151–62

**Burton I, Kates R** (1964) The perception of natural hazards in resource management. *Natural Resources Journal* **3**: 412–41

**Büsching A F** (1762) *A new system of geography*. London, A. Millar

**Butcher H J** (1968) *Human intelligence: its nature and assessment*. London, Methuen

**Buttimer A** (1971) *Society and milieu in the French geographic tradition*. Chicago, Rand McNally for the Association of American Geographers (Monograph Series, No. 6)

**Buttimer A** (1976) The dynamism of lifeworld. *Annals, Association of American Geographers* **66**(2): 277–92

**Buttimer A** (1978) Charism and context: the challenge of *La géographie humaine*. In: Ley D, Samuels M S (eds) *Humanistic geography: prospects and problems*. London, Croom Helm: 58–76

**Buttimer A** (1979) Reason, rationality and human creativity. *Geografiska Annaler* **61B**: 43–9

**Buttimer A** (1981) On people, paradigms and progress in geography. In: Stoddart D R (ed.) *Geography, ideology and social concern*. Oxford, Basil Blackwell: 81–98

**Buttimer A** (ed.) (1983) *The practice of geography*. London, Longman

**Buttmann G** (1977) *Friedrich Ratzel: Leben und Werk eines deutschen Geographen 1844–1904*. Stuttgart, Wissenschaftliche Verlagsgesellschaft m.b.h.

**Büttner M, Jäkel R** (1982) Anton Friedrich Büsching 1724–1793. In: Freeman T W (ed.) *Geographers: biobibliographical studies*, volume 6. London, Mansell: 7–15

**Butzer K W** (1989) Hartshorne, Hettner, and *The nature of geography*. In: Entrikin J N, Brunn S D (eds) *Reflections on Richard Hartshorne's* The nature of geography. Washington DC, Association of American Geographers: 35–52

**Cameron I** (1980) *To the farthest ends of the earth: the history of the Royal Geographical Society 1830–1930*. London, Macdonald Futura

**Campbell J N, Livingstone D N** (1983) Neo-Lamarckism and the development of geography in the United States and Great Britain. *Transactions, Institute of British Geographers* n.s. **8**(3): 267–94

**Capel H** (1981) Institutionalization of geography and strategies of change. In: Stoddart D R (ed.) *Geography, ideology and social concern*. Oxford, Basil Blackwell: 37–69

**Carlstein T, Parkes D, Thrift N** (eds) (1978) *Timing space and spacing time, volume 2: Human activity and time geography*. London, Edward Arnold

**Carnap R** (1935) *Philosophy and logical syntax*. London, Kegan Paul

**Carney J, Hudson R, Lewis J** (eds) (1980) *Regions in crisis*. London, Croom Helm

**Carrothers G A P** (1956) An historical review of gravity and potential concepts of human interaction. *Journal, American Institute of Planners* **22**: 94–102

**Castells M** (1977) *The urban question: a Marxist approach*. London, Edward Arnold

**Chapman G P** (1977) *Human and environmental systems: a geographer's appraisal*. London, Academic Press

**Chappell J E** (1976) Comment in reply. *Annals, Association of American Geographers* **66**(1): 169–73

**Chisholm M** (1967) General systems theory and geography. *Transactions, Institute of British Geographers* **42**: 45–52

**Chisholm M** (1971) Geography and the question of 'relevance'. *Area* **3**(2): 65–8

**Cholley A** (1948) Géographie et sociologie. *Cahiers Internationaux de Sociologie* **5**: 3–20

**Chorley R J** (1962) *Geomorphology and general systems theory*. United States Geological Survey Professional Paper 500-B

**Chorley R J, Beckinsale R P, Dunn A J** (1973) *The history of the study of landforms or the development of geomorphology. Volume Two: The life and work of William Morris Davis*. London, Methuen

**Chorley R J, Dunn A J, Beckinsale R P** (1964) *The history of the study of landforms or the development of geomorphology. Volume One: Geomorphology before Davis*. London, Methuen

**Chorley R J, Haggett P** (eds) (1965) *Frontiers in geographical teaching*. London, Methuen

**Chorley R J, Haggett P** (eds) (1967) *Models in geography*. London, Methuen

**Chorley R J, Kates R W** (1969) Introduction. In: Chorley R J (ed.) *Water, earth and man*. London, Methuen: 1–7

**Chorley R J, Kennedy B** (1971) *Physical geography: a systems approach*. London, Prentice-Hall International

**Chrisman N R, Cowen D J, Fisher P F, Goodchild M F, Mark D M** (1989) Geographical information systems. In: Gaile G L, Wilmott C J (eds) *Geography in America*. Columbus, Merrill: 776–96

**Christaller W** (1933) *Die zentralen Orte in Süddeutschland*. Jena, Fischer

**Christaller W** (1966) *Central places in southern Germany*. Englewood Cliffs, Prentice-Hall

**Cion J** (1908) *Les paysans de la Normandie orientale. Pays de Caux, Bray, Vescin Normand, Vallée de la Seine. Étude géographique*. Paris, A. Colin

**Clark M J, Gregory K J, Gurnell A M** (eds) (1987a) *Horizons in physical geography*. Basingstoke, Macmillan

**Clark M J, Gregory K J, Gurnell A M** (1987b) Introduction: change and continuity in physical geography. In: Clark M J, Gregory K J, Gurnell A M (eds) *Horizons in physical geography*. Basingstoke, Macmillan: 1–5

**Clark M J, Gregory K J, Gurnell A M** (1987c) A concluding perspective. In: Clark M J, Gregory K J, Gurnell A M (eds) *Horizons in physical geography*. Basingstoke, Macmillan: 382–6

**Claval P** (1981) Epistemology and the history of geographical thought. In: Stoddart D R (ed.) *Geography, ideology and social concern*. Oxford, Basil Blackwell: 227–39

**Cloke P, Philo C, Sadler D** (1991) *Approaching human geography*. London, Paul Chapman

**Cochrane A** (1987) What a difference the place makes: the new structuralism of locality. *Antipode* **19**: 354–63

**Collingwood R G** (1956) *The idea of history*. New York, Oxford University Press

**Connerton P** (ed.) (1976) *Critical sociology: selected readings*. Harmondsworth, Penguin

**Cooke P N** (ed.) (1989) *Localities: the changing face of urban Britain*. London, Unwin Hyman

**Cooke R U** (1987) Geomorphology and environmental management. In: Clarke M J, Gregory K J, Gurnell A M (eds) *Horizons in physical geography*. Basingstoke, Macmillan: 270–87

**Cornell T, Matthews J** (1982) *Atlas of the Roman world*. Oxford, Phaidon

**Cosgrove D** (1984) *Social formation and symbolic landscape*. London, Croom Helm

**Cosgrove D** (1990) Environmental thought and action: pre-modern and post-modern. *Transactions, Institute of British Geographers* n.s. **15**(3): 344–58

**Cosgrove D, Daniels S** (eds) (1988) *The iconography of landscape: essays on the symbolic representation, design and use of past environments*. Cambridge, Cambridge University Press

**Costa J E, Graf W L** (1984) The geography of geomorphologists in the United States. *Professional Geographer* **36**(1): 82–9

**Cox K R, Golledge R G** (eds) (1969) *Behavioral problems in geography: a symposium*. Evanston, Northwestern University Press

**Cox K R, Golledge R G** (eds) (1981) *Behavioral problems in geography revisited*. London, Methuen

**Crone G R** (1968) *Maps and their makers: an introduction to the history of cartography*. London, Hutchinson University Library, 4th ed.

**Darby H C** (ed.) (1973) *A new historical geography of England*. Cambridge, Cambridge University Press

**Darwin C** (1888) *The origin of species by means of natural selection, or the preservation of favoured races in the struggle for life*. London, John Murray, 6th edn.

**Darwin C, Wallace A R** (1958) *Evolution by natural selection*. Cambridge, Cambridge University Press

**Daudé R** (1937) La géographie et l'unité de la science. *IX^e Congrès International de Philosophie* Paris, **10**: 56–61

**Daugherty R** (ed.) (1989) *Geography in the National Curriculum: a viewpoint from the Geographical Association*. Sheffield, Geographical Association

**Davies D C W** (1974) *The physics of time asymmetry*. Guildford, Surrey University Press

**Davies P** (1986) Time asymmetry and quantum mechanics In: Flood R, Lockwood M (eds) (1986) *The nature of time*. Oxford, Basil Blackwell: 99–124

**Davis W M** (1884a) Gorges and waterfalls. *American Journal of Science*, 3rd series, **28**: 123–32

**Davis W M** (1884b) Geographic classification, illustrated by a study of plains, plateaus and their derivatives. *Proceedings of the American Association for the Advancement of Science* **33**: 428–32

**Davis W M** (1889a) Topographic development of the Triassic formation of the Connecticut Valley. *American Journal of Science*, 3rd series, **37**: 423–34

**Davis W M** (1889b) Geographic methods in geologic investigations. *National Geographic Magazine* **1**: 11–26

**Davis W M** (1889c) The rivers and valleys of Pennsylvania. *National Geographic Magazine* **1**: 183–253

**Davis W M** (1900) Practical exercises in physical geography. *Proceedings, New York State Science Teachers Association, 5th Annual Conference*: 6

**Davis W M** (1905) Complications of the geographical cycle. *Report of the Eighth International Geographical Congress, Washington 1904*: 150–63

**Davis W M** (1906) An inductive study of the content of geography. *Bulletin of the American Geographical Society* **38**: 67–84

**Davis W M** (1909) *Geographical essays*. Boston, Ginn and Co.

**Davis W M** (1915) The principles of geographical description. *Annals, Association of American Geographers* **5**: 61–105

**Dawkins R** (1976) *The selfish gene*. Oxford, Oxford University Press

**Dawkins R** (1986) *The blind watchmaker*. London, Longman

**Dear M** (1988) The postmodern challenge: reconstructing human geography. *Transactions, Institute of British Geographers* n.s. **13**(3): 262–74

**Dear M, Scott A J** (eds) (1981) *Urbanisation and urban planning in capitalist society*. London, Methuen

**Demangeon A** (1905) *La plaine picarde. Picardie, Artois, Cambrésis, Beauvaisis. Étude de géographie sur les plaines de craie du Nord de la France*. Paris, Armand Colin

**Department of Education and Science** (1980) *A framework for the school curriculum*. London, Department of Education and Science

**Department of Education and Science** (1981) *The school curriculum*. London, HMSO

**Department of Education and Science** (1991) *National Curriculum: Geography draft statutory order*. London, Department of Education and Science

**Department of Education and Science, Welsh Office** (1987) *The National Curriculum 5–16: a consultation document*. London, Department of Education and Science, Welsh Office

**Department of Education and Science, Welsh Office** (1990) *Geography for ages 5 to 16: proposals of the Secretary of State for Education and Science and the Secretary of State for Wales*. London, DES and Welsh Office

**Descartes R** (1968) *Discourse on method, and the meditations*. Harmondsworth, Penguin

**Dickenson J P, Clarke C G** (1972) Relevance and the 'newest geography'. *Area* **4**(1): 25–7

**Dickinson R E** (1969) *The makers of modern geography*. London, Routledge and Kegan Paul

**Dicks D R** (1970) *Early Greek astronomy to Aristotle*. London, Thames and Hudson

**Diffie B W** (1960) *Prelude to empire: Portugal overseas before Henry the Navigator*. Nebraska City, University of Nebraska Press

**Diffie B W, Winius G D** (1977) *Foundations of the Portuguese empire 1415–1580*. Minneapolis, University of Minnesota Press

**Dilke O A W** (1985) *Greek and Roman maps*. London, Thames and Hudson

**Dilthey W** (1913–67) *Gesammelte Schriften*. Göttingen, Vandenhoeck and Ruprecht

**Dilthey W** (1958) *Gesammelte Schriften. Bd. 7. Der Aufbau der geschichtlichen Welt in den Geisteswissenschaften*. Göttingen, Vanderhoeck and Ruprecht

**Dodgshon R A** (1987) *The European past: social evolution and spatial order*. Basingstoke, Macmillan

**Doughty R W** (1981) Environmental theology: trends and prospects in Christian thought. *Progress in Human Geography* **5**(2): 234–48

**Downs R M** (1979) Critical appraisal or determined philosophical skepticism. *Annals, Association of American Geographers* **69**: 468–71

**Downs R M, Stea D** (eds) (1973) *Image and environment: cognitive mapping and spatial behavior*. Chicago, Aldine

**Downs R M, Stea D** (1977) *Maps in minds: reflections on cognitive mapping*. New York, Harper and Row

**Drake E T, Jordan W M** (eds) (1985) *Geologists and ideas: a history of north American geology*. Geological Society of America, Centennial Special Volume 1

**Dreyer J L E** (1953) *A history of astronomy from Thales to Kepler*. New York, Dover

**Driver F** (1988) The historicity of human geography. *Progress in Human Geography* **12**(4): 497–506

**Driver F** (1991) Henry Morton Stanley and his critics: geography, exploration and empire. *Past and Present* **133**: 134–66

**Driver F** (1992) Geography's empire: histories of geographical knowledge. *Environment and Planning D Society and Space* **10**: 23–40

**Dryer C R** (1924) A century of geographic education in the United States. *Annals, Association of American Geographers* 14: 117–49

**Dunbar G S** (1961) Credentialism and careerism in American geography, 1890–1915. In: Blouet B W (ed.) *The origins of academic geography in the United States*. Hamden, Connecticut, Archon Books: 71–88

**Dunbar G S** (1981) Elisée Reclus, an anarchist geographer. In: Stoddart D R (ed.) *Geography, ideology and social concern*. Oxford, Basil Blackwell: 154–64

**Duncan J** (1985) Individual action and political power: a structuration perspective. In: Johnston R J (ed.) *The future of geography*. London, Methuen: 174–89

**Duncan J, Ley D** (1982) Structural Marxism and human geography: a critical assessment. *Annals, Association of American Geographers* **72**: 30–59

**Duncan S S** (1977) The housing crisis and the structure of the housing market. *Journal of Social Policy* **6**(4): 385–412

**Duncan S S** (1989) What is locality? In: Peet R, Thrift N J (eds) *New models in geography*, Volume 2. London, Unwin Hyman: 221–54

**Dunford M, Perrons D** (1983) *The arena of capital*. London, Macmillan

**Dury G H** (1983) Geography and geomorphology: the last fifty years. *Transactions, Institute of British Geographers* n.s. **8**(1): 90–9

**Ebdon D** (1977) *Statistics in geography: a practical approach*. Oxford, Basil Blackwell

**Edwards K** (1991) Implementing the geography National Curriculum in secondary schools. Abstract of paper presented to Launching the National Curriculum in Geography, Royal Geographical Society, 24 September 1991

**Edwards P** (ed.) (1967) *The encyclopedia of philosophy*. New York, Macmillan

**Einstein A** (1964) Autobiographical notes. In: Smart J J C (ed.) *Problems of space and time*. New York, Macmillan: 276–91

**Elkins T H** (1989) Human and regional geography in the German speaking lands in the first forty years of the twentieth century. In: Entrikin J N, Brunn, S D (eds) *Reflections on Richard Hartshorne's* The Nature of Geography. Washington DC, Association of American Geographers: 17–34

**Eliot Hurst M E** (1985) Geography has neither existence nor future. In: Johnston R J (ed.) *The future of geography*. London, Methuen: 59–91

**Emel J, Peet R** (1989) Resource management and natural hazards. In: Peet R, Thrift N (eds) *New models in geography: the political-economy perspective*, volume 1. London, Unwin Hyman: 49–76

**Emery F W** (1984) Geography and imperialism: the role of Sir Bartle Frere (1815–84). *Geographical Journal* **150**(3): 342–50

***Encyclopedia of Islam*** (1971) Volume III *H-Iram*. London, Luzac (New edition, edited by Lewis B, Ménage V L, Pellat Ch, Schacht J)

**Entrikin J N** (1976) Contemporary humanism in geography. *Annals, Association of American Geographers* **66**(4): 615–32

**Entrikin J N** (1980) Robert Park's human ecology and human geography. *Annals, Association of American Geographers* **70**: 43–58

**Entrikin J N** (1981) Philosophical issues in the scientific study of regions. In: Herbert D T, Johnston R J (eds) *Geography and the urban environment*, volume 4. Chichester, John Wiley: 1–27

**Entrikin J N** (1989) Introduction: *The Nature of Geography in perspective*. In: Entrikin J N, Brunn, S D (eds) *Reflections on Richard Hartshorne's* The Nature of Geography. Washington DC, Association of American Geographers: 1–15

**Entrikin J N** (1991) *The betweenness of place: towards a geography of modernity*. Basingstoke, Macmillan

**Eyles J** (1973) Geography and relevance. *Area* **5**(2): 158–60

**Eyles J, Smith D M** (eds) (1988) *Qualitative methods in human geography*. Cambridge, Polity Press

**Farmer B H** (1973) Geography, area studies and the study of area. *Transactions, Institute of British Geographers* **60**: 1–15

**Farmer B H** (1983) British geographers overseas 1933–1983. *Transactions, Institute of British Geographers* n.s. **8**(1): 70–9

**Fawcett C B** (1917) Natural divisions of England. *Geographical Journal* **49**: 129–41

**Febvre L** (1922) *La terre et l'évolution humaine. Introduction géographique à l'histoire*. Paris, L'Évolution de l'Humanité

**Febvre L** (1925) *A geographical introduction to history*. London, Paul, Trench, Tubner

**Fenneman N M** (1919) The circumference of geography. *Annals, Association of American Geographers* **9**: 3–11

**Feyerabend P** (1975) *Against method*. London, Verso

**Feyerabend P** (1978) *Science in a free society*. London, Verso

**Feyerabend P** (1981) How to defend society against science. In: Hacking I (ed.) *Scientific revolutions*. Oxford, Oxford University Press: 156–67

**Fleure H J** (1919) Human regions. *Scottish Geographical Magazine* **35**: 94–105

**Fleure H J** (1947) *Some problems of society and environment*. Institute of British Geographers Special Publication No. 12

**Flood R, Lockwood M** (eds) (1986a) *The nature of time*. Oxford, Basil Blackwell

**Flood R, Lockwood M** (1986b) Introduction. In: Flood R, Lockwood M (eds) *The nature of time*. Oxford, Basil Blackwell: 1–5

**Foucault M** (1966) *Les mots et les choses*. Paris, Gallimard

**Foucault M** (1972) *The archaeology of knowledge*. London, Tavistock

**Foucault M** (1980) *Power/knowledge: selected interviews and other writings 1972–1977*. Brighton, Harvester

**Freeman T W** (1961) *A hundred years of geography*. London, Duckworth

**Freeman T W** (1980a) *A history of modern British geography*. London, Longman

**Freeman T W** (1980b) The Royal Geographical Society and the development of geography. In: Brown E H (ed.) *Geography, yesterday and tomorrow*. Oxford, Oxford University Press: 1–99

**Freud S** (1953–74) *The standard edition of the complete psychological works of Sigmund Freud*. London, The Hogarth Press and the Institute of Psychoanalysis

**Friederich C** (1929) *Alfred Weber's theory of the location of industries*. Chicago, Chicago University Press

**Friedman J** (1992) *Empowerment: the politics of alternative development*. Oxford, Basil Blackwell

**Friis H R** (1981) The role of geographers and geography in the Federal Government: 1774–1905. In: Blouet B W (ed.) *The origins of academic geography in the United States*. Hamden, Archon Books: 37–56

**Fuller G** (1989) Why Geographic Alliances won't work. *Professional Geographer* **41**(4): 480–4

**Gadamer H-G** (1975) *Truth and method*. London, Sheed and Ward

**Galois B** (1976) Ideology and the idea of nature: the case of Peter Kropotkin. *Antipode* **8**(3): 1–16

**Garrison W L** (1959a) Spatial structure of the economy: I. *Annals, Association of American Geographers* **49**: 232–9

**Garrison W L** (1959b) Spatial structure of the economy: II. *Annals, Association of American Geographers* **49**: 471–82

**Garrison W L** (1960) Spatial structure of the economy: III. *Annals, Association of American Geographers* **50**: 357–73

**Garrison W L** (1979) Playing with ideas. *Annals, Association of American Geographers* **69**: 118–20

**Garrison W L, Marble D F** (eds) (1967a) *Quantitative geography. Part I: economic and cultural topics*. Evanston, Northwestern University Studies in Geography, No.13

**Garrison W L, Marble D F** (eds) (1967b) *Quantitative geography. Part II: physical and cartographic topics*. Evanston, Northwestern University Studies in Geography, No. 14

**Gattrell A C** (1985) Any space for spatial analysis. In: Johnston R J (ed.) *The future of geography*. London, Methuen: 190–208

**George P** (1966) *Sociologie et géographie*. Paris, Presses Universitaires de France

**Geuss R** (1981) *The idea of a critical theory: Habermas and the Frankfurt School*. Cambridge, Cambridge University Press

**Giblin B** (1979) Elisée Reclus 1830–1905. In: Freeman T W, Pinchemel P (eds) *Geographers: biobibliographical studies*, volume 3. London, Mansell: 125–32

**Giddens A** (1979) *Central problems in social theory: action, structure and contradiction in social analysis*. London, Macmillan

**Giddens A** (1981) *A contemporary critique of historical materialism*, volume 1: *power, property and the state*. London, Macmillan

**Giddens A** (1985) Time, space and regionalisation. In Gregory D, Urry J (eds) *Social relations and spatial structures*. Basingstoke, Macmillan: 265–95

**Gilbert E W** (1960) The idea of the region. *Geography* **45**: 157–75

**Gilbert E W, Goudie A S** (1971) Sir Roderick Impey Murchison, Bart., K.C.B., 1792–1871. *Geographical Journal* **137**: 505–11

**Gilbert G K** (1902) John Wesley Powell. *Annual Report of the Smithsonian Institute for 1902*: 633–40

**Giraldus Cambrensis** (1951) *The first version of the Topography of Ireland*. Dundalk, Dundalgen Press

**Glacken C J** (1967) *Traces on the Rhodian shore: nature and culture in Western thought from ancient times to the end of the eighteenth century*. Berkeley and Los Angeles, University of California Press

**Glucksmann M** (1974) *Structuralist analysis in contemporary social thought: a comparison of the theories of Claude Lévi-Strauss and Louis Althusser*. London, Routledge and Kegan Paul

**Goetzmann W H** (1966) *Exploration and empire: the explorer and the scientist in the winning of the American West*. New York, A. Knopf

**Gold J R** (1980) *An introduction to behavioural geography*. Oxford, Oxford University Press

**Gold J R, Jenkins A, Lee R, Monk J, Riley J, Shepherd I, Unwin D** (1991) *Teaching geography in higher education: a manual of good practice*. Oxford, Basil Blackwell (Institute of British Geographers Special Publication No. 24)

**Golledge R G, Brown L A, Williamson F** (1972) Behavioral approaches in geography: an overview. *Australian Geographer* **12**: 59–79

**Gorz A** (1979) *Ecology and politics*. Boston, South End Press

**Goudie A S** (1981) *The human impact: man's role in environmental change*. Oxford, Basil Blackwell

**Goudie A** (ed.) (1990) *Geomorphological techniques*. London, Unwin Hyman 2nd ed.

**Gould P** (1979) Geography 1957–1977: the Augean period. *Annals, Association of American Geographers* **69**: 139–51

**Gould P** (1985) *The geographer at work*. London, Routledge and Kegan Paul

**Gould P R, White R** (1974) *Mental maps*. Harmondsworth, Penguin

**Gradmann R** (1931) *Süd-deutschland*. Stuttgart, Bibliothek länderkundlicher Handbücher

**Greer-Wootten B** (1972) *The role of general systems theory in geographic research*. Toronto, Department of Geography, York University (Discussion Paper No. 3)

**Gregory D** (1978) *Ideology, science and human geography*. London, Hutchinson

**Gregory D** (1980) The ideology of control: systems theory and geography. *Tijdschrift voor Economische en Sociale Geographie* **71**(6): 327–42

**Gregory D** (1981) Human agency and human geography. *Transactions, Institute of British Geographers* n.s. **6**(1): 1–18

**Gregory D** (1982a) A realist construction of the social. *Transactions, Institute of British Geographers* **7**(2): 254–6

**Gregory D** (1982b) *Regional transformation and industrial revolution: a geography of the Yorkshire woollen industry*. London, Macmillan

**Gregory D** (1984) Contours of crisis? Sketches for a geography of class struggle in the early industrial revolution in England. In: Baker A R H, Gregory D (eds) *Explorations in historical geography: interpretative essays*. Cambridge, Cambridge University Press: 68–117

**Gregory D** (1985a) Suspended animation: the stasis of diffusion theory. In: Gregory D, Urry J (eds) *Social relations and spatial structures*. Basingstoke, Macmillan: 296–336

**Gregory D** (1985b) People, places and practices: the future of human geography. In: King R (ed.) *Geographical futures*. Sheffield, Geographical Association: 56–76

**Gregory D** (1986) Critical theory. In: Johnston R J, Gregory D, Smith D M (eds) *The dictionary of human geography*. Oxford, Basil Blackwell, 2nd edn: 81–4

**Gregory D** (1989) Areal differentiation and post-modern human geography. In: Gregory D, Walford R (eds) *Horizons in human geography*. Basingstoke, Macmillan: 67–96

**Gregory D, Urry J** (eds) (1985a) *Social relations and spatial structures*. Basingstoke, Macmillan

**Gregory D, Urry J** (1985b) Introduction. In: Gregory D, Urry J (eds) *Social relations and spatial structures*. Basingstoke, Macmillan: 1–8

**Gregory D, Walford R** (eds) (1989) *Horizons in human geography*. Basingstoke, Macmillan

**Gregory K J** (1985) *The nature of physical geography*. London, Edward Arnold

**Gritzner C F** (1986) The South Dakota experience. *Professional Geographer* **38**(3): 252–3

**Guelke L** (1974) An idealist alternative in human geography. *Annals, Association of American Geographers* **64**: 193–202

**Guelke L** (1976) The philosophy of idealism. *Annals, Association of American Geographers* **66**: 168–9

**Guelke L** (1977) The role of laws in human geography. *Progress in Human Geography* **1**: 376–86

**Guelke L** (1981) Idealism. In: Harvey M E, Holly B P (eds) *Themes in geographic thought*. London, Croom Helm: 133–47

**Guirand F** (1968) Greek mythology. In: *New Larousse encyclopedia of mythology*. London, Paul Hamlyn, new edn: 85–198

**Gurvitch G** (1958–60) *Traité de sociologie*. Paris, Presses Universitaires de France, 2 vols

**Habermas J** (1974) *Theory and practice*. London, Heinemann

**Habermas J** (1976) *Legitimation crisis*. London, Heinemann

**Habermas J** (1978) *Knowledge and human interests*. London, Heinemann 2nd edn

**Habermas J** (1982) A reply to my critics. In: Thompson J B, Held D (eds) *Habermas: critical debates*. London, Macmillan: 219–83

**Habermas J** (1984) *The theory of communicative action*. Volume 1: *Reason and the rationalization of society*. Boston, Beacon Press

**Habermas J** (1987a) *The theory of communicative action. Volume 2: Lifeworld and system*. Cambridge, Polity Press

**Habermas J** (1987b) *The philosophical discourse of modernity: twelve lectures*. Cambridge, Polity Press

**Haeckel E** (1869) Über Entwicklungsgang und Aufgabe der Zoologie. *Jenaische Zeitschrift* **5**: 353–70

**Hägerstrand T** (1953) *Innovationsförloppet ur korologisk synpunkt*. Lund, C W K Gleerup

**Hägerstrand T** (1975) Space, time and human conditions. In: Karlquist L, Lundquist L, Snickars F (eds) *Dynamic allocation of urban space*. Farnborough, Saxon House: 3–12

**Hägerstrand T** (1983) In search for the sources of concepts. In: Buttimer A (ed.) *The practice of geography*. London, Longman: 238–56

**Haggett P** (1965) *Locational analysis in human geography*. London, Edward Arnold

**Haggett P** (1990) *The geographer's art*. Oxford, Basil Blackwell

**Haggett P, Chorley R J** (1967) Models paradigms and the new geography. In: Chorley R J, Haggett P (eds) *Models in geography*. London, Methuen: 19–41

**Haggett P, Chorley R** (1989) From Madingley to Oxford: a foreword to *Remodelling geography*. In: Macmillan B (ed.) *Remodelling geography*. Oxford, Basil Blackwell: xv-xx

**Haines-Young R H, Petch J R** (1986) *Physical geography: its nature and methods*. London, Harper and Row

**Hakluyt R** (1903) *The principal navigations, voyages, traffiques and discoveries of the English nation*, volume 1. Glasgow, James MacLehose

**Hall D** (1989) The National Curriculum and the two cultures: towards a humanistic perspective. *Geography* **75**(4): 313–24

**Hall P** (1963) *London 2000*. London, Faber

**Hall P** (ed.) (1966) *Von Thünen's isolated state*. Oxford, Pergamon

**Hall P** (1980) *Great planning disasters*. London, Weidenfeld and Nicolson

**Hall P** (1988) *Cities of tomorrow*. Oxford, Basil Blackwell

**Hall R B** (1935) The geographic region: a resumé. *Annals, Association of American Geographers* **25**: 122–30

**Hammond R, McCullagh P S** (1974) *Quantitative techniques in geography: an introduction*. London, Oxford University Press

**Hamy E-T** (ed.) (1905) *Lettres américaines d'Alexandre de Humboldt 1798–1807*. Paris, Librarie Orientale et Américaine

**Harré R** (1986) *Varieties of realism*. Oxford, Basil Blackwell

**Harris C** (1978) The historical mind and the practice of geography. In: Ley D, Samuels M S (eds) *Humanistic geography: prospects and problems*. London, Croom Helm: 123–37

**Harris C D, Ullman E L** (1945) The nature of cities. *Annals of the American Academy of Political and Social Science* **242**: 7–17

**Harrison R T, Livingstone D N** (1979) There and back again – towards a critique of idealist human geography. *Area* **11**(1): 75–9

**Harriss J, Harriss B** (1979) Development studies. *Progress in Human Geography* **3**: 576–84

**Hart J F** (1979) The 1950s. *Annals, Association of American Geographers* **69**: 109–14

**Hart J F** (1982) The highest form of the geographer's art. *Annals, Association of American Geographers* **72**: 1–26

**Hartshorne R** (1939) *The nature of geography*. Lancaster, Association of American Geographers

**Hartshorne R** (1954) Comment on 'Exceptionalism in geography'. *Annals, Association of American Geographers* **44**: 108–9

**Hartshorne R** (1955) 'Exceptionalism in geography' re-examined. *Annals, Association of American Geographers* **45**(3): 205–44

**Hartshorne R** (1958) The concept of geography as a science of space from Kant and Humboldt to Hettner. *Annals, Association of American Geographers* **48**(2): 97–108

**Hartshorne R** (1959) *Perspective on the nature of geography*. Chicago, Rand McNally

**Harvey D W** (1961) Aspects of agricultural and rural change in Kent, 1800–1900. Unpublished Ph.D. thesis, University of Cambridge

**Harvey D** (1964) *Behavioural postulates and the construction of theory in human geography*. Bristol, University of Bristol Department of Geography Papers Series A: 6

**Harvey D** (1969) *Explanation in geography*. London, Edward Arnold

**Harvey D** (1972) Revolutionary and counter revolutionary theory in geography and the problem of ghetto formation. *Antipode* **6**(2): 1–13

**Harvey D** (1973) *Social justice and the city*. London, Edward Arnold

**Harvey D** (1974) Class-monopoly rent, finance capital and the urban revolution. *Regional Studies* **8**: 239–55

**Harvey D** (1982) *The limits to capital*. Oxford, Basil Blackwell

**Harvey D** (1985a) *The urbanization of capital*. Oxford, Basil Blackwell

**Harvey D** (1985b) *Consciousness and the urban experience*. Oxford, Basil Blackwell

**Harvey D** (1989a) *The urban experience*. Oxford, Basil Blackwell

**Harvey D** (1989b) *The condition of postmodernity: an enquiry into the origins of cultural change*. Oxford, Basil Blackwell

**Harvey D, Chatterjee L** (1974) Absolute rent and the structuring of space by governmental and financial institutions. *Antipode* **6**(1): 22–36

**Harvey P D A** (1980) *The history of topographical maps: symbols, pictures and surveys*. London, Thames and Hudson

**Hawking S W** (1988) *A brief history of time from the big bang to black holes*. London, Bantam Press

**Hay A** (1985) Scientific method in geography. In: Johnston R J (ed.) *The future of geography*. London, Methuen: 129–42

**Hegel G W F** (1977) *Phenomenology of spirit*. Oxford, Clarendon Press

**Heidegger M** (1959) *An introduction to metaphysics*. New Haven, Yale University Press

**Held D** (1980) *Introduction to critical theory: Horkheimer to Habermas*. London, Hutchinson

**Hempel C G** (1965) *Aspects of scientific explanation and other essays on the philosophy of science*. New York, Free Press

**Herbertson A J** (1905) The major natural regions of the world. *Geographical Journal* **25**: 300–10

**Herbst J** (1961) Social Darwinism and the history of American geography. *Proceedings of the American Philosophical Society* **105**(6): 538–44

**Herodotus** (1954) *The histories*. Harmondsworth, Penguin

**Hesse M** (1982) Science and objectivity. In: Thompson J B, Held D (eds) *Habermas: critical debates*. London, Macmillan: 98–115

**Hettner A** (1895) Geographische Forschung und Bildung. *Geographische Zeitschrift* **1**: 1–19

**Hettner A** (1903) Grundbegriffe und Grundsätze der physischen Geographie. *Geographische Zeitschrift* **9**: 21–40, 121–39, 193–213

**Hettner A** (1905) Das System der Wissenschaften. *Preussische Jahrbücher* **122**: 251–77

**Hettner A** (1921) *Die Oberflächenformen des Festlandes*. Leipzig, Teubner

**Hettner A** (1927) *Die Geographie, ihre Geschichte, ihr Wesen und ihre Methoden*. Breslau, Ferdo Hirt

**Hettner A** (1932) Das länderkundliche Schema. *Geographische Anzeiger* **33**: 51–6

**Hettner A** (1972) *The surface features of the earth*. London, Macmillan

**Hill A D, LaPrairie L A** (1989) Geography in American education. In: Gaile G L, Willmott C J (eds) *Geography in America*. Columbus, Merrill: 1–26

**Hoare M E** (1976) *The tactless philosopher: Johann Reinhold Forster (1729–94)*. Melbourne, The Hawthorn Press

**Hoare M E** (ed.) (1982) *The Resolution journal of Johann Reinhold Forster 1772–1775*. London, Hakluyt Society

**Holcomb B, Tiefenbacher J** (1989) National Geography Awareness Week 1987: an assessment. *Journal of Geography in Higher Education* **13**(2): 159–64

**Holmes A** (1944) *Principles of physical geology*. Edinburgh, Nelson

**Holt-Jensen A** (1981) *Geography: history and concepts*. London, Harper and Row

**Holt-Jensen A** (1988) *Geography: history and concepts*. London, Paul Chapman, 2nd edn

**Horkheimer M** (1972) *Critical theory: selected essays*. New York, Herder and Herder

**Horton R E** (1932) Drainage basin characteristics. *Transactions, American Geophysical Union* **13**: 350–61

**Horton R E** (1933) The role of infiltration in the hydrologic cycle. *Transactions, American Geophysical Union* **14**: 446–60

**Horton R E** (1935) *Surface runoff phenomena*, Pt. 1: *Analysis of the hydrographs*. Voorheesville, Horton Hydrologic Laboratory

**Horton R E** (1945) Erosional development of streams and their drainage basins: hydrophysical approach to quantitative morphology. *Bulletin of the Geological Society of America* **56**: 275–370

**Horvath R** (1971) The 'Detroit Geographical Expedition and Institute' experience. *Antipode* **3**(1): 73–85

**Howard M C, King J E** (1985) *The political economy of Marx*. London, Longman, 2nd edn

**Howarth O J R** (1951) The centenary of Section E (Geography). *Advancement of Science* **8**(30): 151–65

**Hoyt H** (1939) *The structure and growth of residential neighborhoods in American cities*. Washington DC, Federal Housing Administration

**Hudson B** (1977) The new geography and the new imperialism, 1870–1918. *Antipode* **9**: 12–19

**Hudson R** (1981) Personal construct theory, the repertory grid and human geography. *Progress in Human Geography* **4**(3): 346–59

**Huntington E** (1925) *The character of races as influenced by physical environment, natural selection and historical development*. New York, Charles Scribner's Sons

**Huntington E** (1945) *Mainsprings of civilization*. New York, John Wiley and Sons

**Huntington E, Cushing S W** (1934) *Principles of human geography*. New York, John Wiley and Sons

**Ibn Khaldūn** (1967) *The Muqaddimah: an introduction to history*. London, Routledge and Kegan Paul, 2nd ed.

**Jackson P** (1981) Phenomenology and social geography. *Area* **13**(4): 299–305

**Jackson P** (ed.) (1987) *Race and racism: essays in social geography*. London, Allen and Unwin

**Jackson P** (1989) *Maps of meaning*. London, Unwin Hyman

**Jackson P, Smith S J** (eds) (1981) *Social interaction and ethnic segregation*. London, Academic Press

**Jackson P, Smith S J** (1984) *Exploring social geography*. London, George Allen and Unwin

**James P** (1967) Continuity and change in American geographic thought. In: Cohen S B (ed.) *Problems and trends in American geography*. New York, Basic Books: 3–14

**James P E** (1934) The terminology of regional description. *Annals, Association of American Geographers* **24**: 78–86

**James P E, Bladen W, Karan P** (1983) Ellen Churchill Semple and the development of a research paradigm. In: Bladan W, Karan P (eds) *The evolution of geographic thought in America: a Kentucky root*. Dubuque, Kendall Hunt: 59–85

**James P E, Jones C F** (eds) (1954) *American geography: inventory and prospect*. Syracuse, Syracuse University Press

**Jaubert P-A** (1975) *La Géographie d'Édrisi*. Amsterdam, Philo Press

**Johnston R J** (1979) *Geography and geographers: Anglo-American human geography since 1945*. London, Edward Arnold

**Johnston R J** (1983) *Geography and geographers: Anglo-American human geography since 1945*. London, Edward Arnold, 2nd edn

**Johnston R J** (1984) The region in twentieth century British geography. *History of Geography Newsletter* **4**: 26–35

**Johnston R J** (1986a) Four fixations and the quest for unity in geography. *Transactions, Institute of British Geogaphers* n.s. **11**(4): 449–53

**Johnston R J** (1986b) Radical geography. In: Johnston R J, Gregory D, Smith D M (eds) *The dictionary of human geography*. Oxford, Basil Blackwell, 2nd edn: 385–6

**Johnston R J** (1987) *Geography and geographers: Anglo-American human geography since 1945*. London, Edward Arnold, 3rd edn

**Johnston R J** (1989) The Institute, study groups, and a discipline without a core? *Area* **21**(4): 407–14

**Johnston R J** (1991a) *Geography and geographers: Anglo-American human geography since 1945*. London, Edward Arnold, 4th edn

**Johnston R J** (1991b) A place for everything and everything in its place. *Transactions, Institute of British Geographers* n.s. **16**(2): 131–47

**Johnston R J** (1991c) *A question of place*. Oxford, Basil Blackwell

**Johnston R J, Gregory D, Smith D M** (eds) (1986) *The dictionary of human geography*. Oxford, Blackwell Reference

**Jumper S R** (1986) The Tennessee experience. *Professional Geographer* **38**(3): 254–5

**Kahn C H** (1960) *Anaximander and the origins of Greek cosmology*. New York, Columbia University Press

**Kaminski W** (1905) *Über Immanuel Kant's Schriften zur physischen Geographie: Ein Beitrag zur Methodik der Erdkunde*. Königsberg, Hugo Jaeger

**Kates R W** (1971) Natural hazard in human ecological perspective: hypotheses and models. *Economic Geography* **47**: 438–51

**Keat R, Urry J** (1981) *Social theory as science*. London, Routledge and Kegan Paul

**Kellner L** (1963) *Alexander von Humboldt*. London, Oxford University Press

**Kelly G A** (1955) *The psychology of personal constructs*. New York, Norton

**Keltie J S** (1886) *Report of the Proceedings of the Society in reference to the improvement of geographical education*. London, John Murray for the Royal Geographical Society

**Keltie J S** (1921) Obituary for Peter Alexeivich Kropotkin. *Geographical Journal 57*: 316–19

**Kennedy B** (1979) A naughty world. *Transactions, Institute of British Geographers* n.s. **4**(4): 550–8

**Kimble G H T** (1938) *Geography in the Middle Ages*. London, Methuen

**Kimble G H T** (1951) The inadequacy of the regional concept. In: Stamp L D, Wooldridge S W (eds) *London essays in geography*. London, Longmans, Green: 151–74

**King C A M** (1959) *Beaches and coasts*. London, Arnold

**King L J** (1979) Areal associations and regressions. *Annals, Association of American Geographers* **69**: 124–8

**Kirby A M, Lambert D M** (1978) *Geography at school and university – is the gap growing?* Reading, University of Reading (Papers on Education in Geography, No. 2)

**Kirk W** (1952) Historical geography and the concept of the behavioural environment. *Indian Geographical Journal*, Silver Jubilee Edition: 152–60

**Kobayashi A, Mackenzie S** (eds) (1989) *Remaking human geography*. London, Unwin Hyman

**Kockelmans J J** (1965) *Martin Heidegger: a first introduction to his philosophy*. Pittsburgh, Duquesne University Press

**Kockelmans J J** (1967a) *Edmund Husserl's phenomenological psychology: a historico-critical study*. Pittsburgh, Duquesne University Press

**Kockelmans J J** (ed.) (1967b) *Phenomenology: the philosophy of Edmund Husserl and its interpretation*. Garden City, Doubleday

**Koffka K** (1929) *Principles of gestalt psychology*. London, Routledge and Kegan Paul

**Köhler W** (1929) *Gestalt psychology*. New York, Horace Liveright

**Kolakowski L** (1972) *Positivist philosophy: from Hume to the Vienna Circle*. Harmondsworth, Penguin

**Krebs N** (1923) Natur- und Kulturlandschaft. *Zeitschrift der Gesellschaft für Erdkunde zu Berlin*: 81–94

**Krumbein W C** (1955) Experimental design in the earth sciences. *Transactions, American Geophysical Union* **36**: 1–11

**Kuhn T S** (1962) *The structure of scientific revolutions*. Chicago, University of Chicago Press

**Kuhn T S** (1970) *The structure of scientific revolutions*. Chicago, University of Chicago Press, 2nd edn

**Kuhn T S** (1977) Second thoughts on paradigms. In: Suppe F (ed.) *The structure of scientific theories*. Urbana, University of Illinois Press: 459–82

**Kuzmiak D T** (1991) A history of the American environmental movement. *Geographical Journal* **157**(3): 265–78

**Lakatos I** (1978) Falsification and the methodology of scientific research programmes. In: Worrall J, Currie G (eds) *The methodology of scientific research programmes, philosophical papers*, volume 1. Cambridge, Cambridge University Press: 8–101

**Lange G** (1961) Varenius über die Grundfrage der Geographie. *Petermanns Geographische Mitteilungen* **105**: 274–83

**Langton J** (1972) Potentialities and problems of adapting a systems approach to the study of change in human geography. In: Board C *et al.* (eds) *Progress in Geography 4*. London, Edward Arnold: 125–79

**Lautensach H** (1952) Otto Schlüters Bedeutung für die methodische Entwicklung der Geographie. *Petermanns Geographische Mitteilungen* **96S**: 219–31

**Leach E** (1974) *Lévis-Strauss*. Glasgow, Fontana/Collins, rev. edn

**Lee R** (1985) Where have all the geographers gone? *Geography* **70**(1): 45–59

**Lefebvre H** (1974) *La production de l'éspace*. Paris, Éditions Anthropos

**Lefebvre H** (1991) *The production of space*. Oxford, Basil Blackwell

**Leighly J** (1955) What has happened to physical geography? *Annals, Association of American Geographers* **45**: 309–18

**Leighly J** (ed.) (1963) *Land and life: a selection from the writings of Carl Ortwin Sauer*. Berkeley and Los Angeles, University of California Press

**Leiss W** (1974) *The domination of nature*. Boston, Beacon Press

**Le Lannou M** (1949) *La géographie humaine*. Paris, Flammarion

**Lenz K** (1978) The Berlin Geographical Society 1828–1978. *Geographical Journal* **144**: 218–23

**Leopold L B** (1953) Downstream change of velocity in rivers. *American Journal of Science* **251**: 606–24

**Leopold L B, Maddock T** (1953) The hydraulic geometry of stream channels and some physiographic implications. *United States Geological Survey Professional Paper* **252**: 1–57

**Leopold L B, Miller J P** (1956) Ephemeral streams – hydraulic factors and their relation to the drainage net. *United States Geological Survey Professional Paper* **282A**: 1–37

**Leopold L B, Wolman M G, Miller J P** (1964) *Fluvial processes in geomorphology*. San Francisco, W H Freeman and Co.

**Lévi-Strauss C** (1953) Social structure. In: Kroeber A L (ed.) *Anthropology today*. Chicago, University of Chicago Press: 524–54

**Lévis-Strauss C** (1963) *Structuralist anthropology*. New York, Basic Books

**Lewis J, Melville B** (1978) The politics of epistemology in regional science. In: Batey P (ed.) *Theory and method in urban and regional science*. London, Pion: 82–100

**Lewthwaite G R** (1966) Environmentalism and determinism: a search for clarification. *Annals, Association of American Geographers* **56**: 1–23

**Ley D** (1977) Social geography and the taken-for-granted world. *Transactions, Institute of British Geographers* n.s. **2**(4): 498–512

**Ley D** (1980) Liberal ideology and the postindustrial city. *Annals, Association of American Geographers* **70**(2): 238–58

**Ley D** (1983) *A social geography of the city*. New York, Harper and Row

**Ley D** (1988) Interpretative social research in the inner city. In: Eyles J (ed.) *Research in human geography: introductions and investigations*. Oxford, Basil Blackwell: 121–38

**Ley D, Samuels M S** (eds) (1978a) *Humanistic geography: prospects and problems*. London, Croom Helm

**Ley D, Samuels M S** (1978b) Introduction: contexts of modern humanism in geography. In: Ley D, Samuels M S (eds) *Humanistic geography: prospects and problems*. London, Croom Helm: 1–17

**Libbey W F, Anderson E C, Arnold J R** (1949) Age determination by radiocarbon content: world-wide assay of natural radiocarbon. *Science* **109**: 227–8

**Lichtenberger E** (1978) Klassische und theoretisch-quantitative Geographie in deutschen Sprachraum. *Berichte zur Raumforschung und Raumplanung* **22**: 9–20

**Linke M** (1981) Carl Ritter 1779–1859. In: Freeman T W (ed.) *Geographers: biobibliographical studies*, volume 5. London, Mansell: 99–18

**Livingstone D N** (1980) Nature and man in America: Nathaniel Southgate Shaler and the conservation of natural resources. *Transactions, Institute of British Geographers* n.s. **5**(3): 369–82

**Livingstone D N** (1984) Natural theology and neo-Lamarckism: the changing context of nineteenth century geography in the United States and Great Britain. *Annals, Association of American Geographers* **74**(1): 9–28

**Livingstone D N** (1985) Evolution, science and society: historical reflections on the geographical experiment. *Geoforum* **16**(2): 119–30

**Livingstone D N** (1987) *Nathaniel Southgate Shaler and the culture of American science*. Tuscaloosa, University of Alabama Press

**Livingstone D N** (1990a) Geography and modernity: past and present. Paper presented to the annual meeting of the Association of American Geographers, Toronto, April 1990

**Livingstone D N** (1990b) Geography, tradition and the scientific revolution: an interpretative essay. *Transactions, Institute of British Geographers* n.s. **15**(3): 359–73

**Livingstone D N, Harrison R T** (1981) Immanuel Kant, subjectivism and human geography: a preliminary investigation. *Transactions, Institute of British Geographers* n.s. **6**(3): 359–74

**Lock G** (1972) Louis Althusser: philosophy and Leninism. *Marxism Today* **16**(6): 180–7

**Lösch A** (1940) *Die raümliche Ordnung der Wirtschaft*. Jena, Gustav Fischer

**Lösch A** (1954) *The economics of location*. New Haven, Yale University Press

**Lovering J** (1985) Regional intervention, defence industries and the structuration of space in Britain. *Environment and Planning D: Society and Space* **3**: 85–107

**Lowenthal D** (1961) Geography, experience and imagination: towards a geographical epistemology. *Annals, Association of American Geographers* **51**: 241–60

**Lurie E** (1960) *Louis Aggasiz: a life in science*. Chicago, University of Chicago Press

**Lynch K** (1960) *The image of the city*. Cambridge, MIT Press

**McCarthy T** (1978) *The critical theory of Jürgen Habermas*. London, Hutchinson

**McCarthy T** (1984) Translator's introduction. In: Habermas J, *Theory of communicative action, volume 1, reason and the rationalization of society*. Boston, Beacon Press: v–xxxvii

**McCarty H H** (1953) An approach to a theory of economic geography. *Annals, Association of American Geographers* **43**: 183–4

**McCarty H H** (1954) An approach to a theory of economic geography. *Economic Geography* **30**: 95–101

**McCarty H H** (1979) Geography at Iowa. *Annals, Association of American Geographers* **69**: 121–4

**McDowell L** (1983) Towards an understanding of the gender division of urban space. *Environment and Planning D: Society and Space* **1**(1): 59–72

**Macintyre S, Tribe K** (1975) *Althusser and Marxist theory*, Cambridge, the authors, 2nd ed.

**McKay D V** (1943) Colonialism in the French geographical movement, 1871–1881. *Geographical Review* **33**: 214–32

**Mackenzie S** (1986) Women's responses to economic restructuring: changing gender, changing space. In: Hamilton R, Barrett M (eds) *The politics of diversity: feminism, Marxism and nationalism*. London, Verso: 81–100

**Mackinder H J** (1887) On the scope and methods of geography. *Proceedings of the Royal Geographical Sociey* n.s. **9**: 141–60

**Mackinder H J** (1895) Modern geography, German and English. *Geographical Journal* **6**(4): 367–79

**Macmillan B** (ed.) (1989) *Remodelling geography*. Oxford, Basil Blackwell

**Magee B** (1973) *Popper*. Glasgow, Fontana/Collins

**Maguire D J, Goodchild M F, Rhind D W** (1991) *Geographical information systems: principles and applications*. Harlow, Longman

**Marcuse H** (1964) *One dimensional man: studies in the ideology of advanced industrial society*. London, Routledge and Kegan Paul

**Marcuse H** (1972) *Counterrevolution and revolt*. Boston, Beacon Press

**Marsh G P** (1864) *Man and nature; or, physical geography as modified by human action*. London, New York, Scribner

**Martin G J** (1989) *The Nature of Geography* and the Schaefer–Hartshorne debate. In: Entrikin N J, Brunn S D (eds) *Reflections on Richard Hartshorne's* The Nature of Geography. Washington DC, Association of American Geographers: 69–90

**Marx K** (1976) *Capital*, volume 1. Harmondsworth, Penguin

**Massey D** (1984) *Spatial divisions of labour: social structures and the geography of production*. London, Macmillan

**Massey D, Meegan R A** (1979) The geography of industrial reorganisation. *Progress in Planning* **10**: 155–237

**Massey D, Meegan R A** (1982) *The anatomy of job loss*. London, Methuen

**Masterman M** (1970) The nature of a paradigm. In: Lakatos I, Musgrove A (eds) *Criticism and the growth of knowledge*. London, Cambridge University Press: 59–90

**Matley I M** (1986) John Dee 1527–1608. In: Freeman T W (ed.) *Geographers: biobibliographical studies*, volume 10. London, Mansell: 49–55

**May J A** (1970) *Kant's concept of geography and its relation to recent geographical thought*. Toronto, University of Toronto Press

**Meinig D W** (1983) Geography as an art. *Transactions, Institute of British Geographers* n.s. **8**(3): 314–28

**Mellor J R** (1977) *Urban sociology in an urbanized society*. London, Routledge and Kegan Paul

**Mercer D C, Powell J M** (1972) *Phenomenology and related non-positivistic viewpoints in the social sciences*. Clayton, Monash Publications in Geography, No. 1

**Meyer-Abich A** (1967) *Alexander von Humboldt in Selbstzeugnissen und Bilddokumenten*. Reinbeck bei Hamburg, Rowohlt Taschenbuch

**Meyer-Abich A, Hentschel C** (1969) *Alexander von Humboldt*. Bonn, Inter-Nationes

**Mieli A** (1938) *La science arabe et son rôle dans l'évolution scientifique mondiale*. Leiden, Brill

**Mill H R** (1930) *The record of the Royal Geographical Society 1830–1930*. London, The Royal Geographical Society

**Mill J S** (1843) *A system of logic, ratiocinative and inductive, being a connected view of the principles of evidence and the methods of scientific investigation*. London, John W. Parker

**Minkowski H** (1964) Space and time. In: Smart J J C (ed.) *Problems of space and time*. New York, Macmillan: 297–312

**Momsen J H, Townsend J G** (eds) (1987) *Geography of gender in the Third World*. London, Hutchinson

**Morgan D O** (1988) Ibn Khaldun. In: Cannon J *et al*. (eds) *Blackwell dictionary of historians*. Oxford, Basil Blackwell: 202–3

**Morgan W** (1991) Implementing the National Curriculum in primary schools. Abstract of paper presented to Launching the National Curriculum in Geography, Royal Geographical Society, 24 September 1991

**Morisawa M** (1985) Development of quantitative geomorphology. In: Drake E T, Jordan W M (eds) *Geologists and ideas: a history of North American geology*. Boulder, Geological Society of America: 79–107 (Centennial Special Volume, 1)

**Morrill R** (1984) Recollections of the 'Quantitative Revolution's' early years: the University of Washington 1955–65. In: Billinge M, Gregory D, Martin R (eds) *Recollections of a revolution*. London, Macmillan: 57–72

**Morrill R** (1985) Some important geographical questions. *Professional Geographer* **37**(3): 263–70

**Mügerauer R** (1981) Concerning regional geography as a hermeneutic discipline. *Geographische Zeitschrift* **69**: 57–67

**Mulkay M J** (1978) Consensus in science. *Social Science Information* **17**: 107–22

**Naish M, Rawling E** (1990) Geography 16–19: some implications for higher education. *Journal of Geography in Higher Education* **14**(1): 55–77

**Needham J, Wang Ling** (1970) *Science and civilization in China, volume 3, mathematics and the sciences of the heavens and the earth*. Cambridge, Cambridge University Press

**Neugebauer O** (1983) *Astronomy and history: selected essays*. New York, Springer-Verlag

**Newton-Smith W H** (1981) *The rationality of science*. London, Routledge and Kegan Paul

**Newton-Smith W H** (1986) Space, time and space-time: a philosopher's view. In: Flood R, Lockwood M (eds) *The nature of time*. Oxford, Basil Blackwell: 22–35

**Norton W** (1984) *Historical analysis in geography*. London, Longman

**Oakes G** (1980) Note: Windleband on history and natural science. *History and Theory* **19**(2): 165–8

**Odum E P** (1963) *Ecology*. New York, Holt, Rinehart and Winston

**Olsson G** (1965) *Distance and human interaction: a review and bibliography*. Philadelphia, Regional Science Research Institute (Bibliography Series No. 2)

**Olsson G** (1975) *Birds in egg*. Ann Arbor, Department of Geography, University of Michigan (Michigan Geographical Publications, No. 15)

**Olsson G** (1978) Of ambiguity or far cries from a memorializing mamafesta. In: Ley D, Samuels M S (eds) *Humanistic geography: prospects and problems*. London, Croom Helm: 109–20

**Olsson G** (1979) Social science and human action or hitting your head against the ceiling of language. In: Gale S, Olsson G (eds) *Philosophy in geography*. Dordrecht, Reidel: 287–308

**Olsson G** (1980) *Birds in egg/eggs in bird*. London, Pion

**Olsson G** (1982) -/-. In: Gould P, Olsson G (eds) *A search for common ground*. London, Pion: 223–31

**Olsson G** (1991) Invisible maps. *Geografiska Annaler* **73B**: 85–92

**Olwig K R** (1980) Historical geography and the society/nature 'problematic': the perspective of J F Schouw, G P Marsh and E Reclus. *Journal of Historical Geography* **6**(1): 29–45

**O'Riordan T** (1976) *Environmentalism*. London, Pion

**O'Riordan T** (1989a) Politics, practice and the new environmentalism. In: Gregory D, Walford R (eds) *Horizons in human geography*. Totowa, Barnes and Noble: 395–414

**O'Riordan T** (1989b) The challenge for environmentalism. In: Peet R, Thrift N (eds) *New models in geography: the political-economy perspective*, volume 1. London, Unwin Hyman: 77–102

**Outhwaite W** (1987) *New philosophies of social science: realism, hermeneutics and critical theory*. Basingstoke, Macmillan

**Owens S** (1986) Environmental politics in Britain: new paradigm or placebo? *Area* **18**(3): 195–201

**Panofsky E** (1939) *Studies in iconology: humanistic themes in the art of the Renaissance*. Oxford, Oxford University Press

**Parkes D N, Thrift N J** (1980) *Times, spaces and places: a chronogeographic perspective*. Chichester, Wiley

**Paterson J H** (1974) Writing regional geography: problems and progress in the Anglo-American realm. *Progress in Geography* **6**: 1–16

**Pattison W D** (1961) Rollin Salisbury and the establishment of geography at the University of Chicago. In: Blouet B W (ed.) *The origins of academic geography in the United States*. Hamden, Archon Books: 151–63

**Peake L** (ed.) (1989) The challenge of feminist geography. *Journal of Geography in Higher Education* **13**(1): 85–121

**Pears D** (1971) *Wittgenstein*. Glasgow, Fontana/Collins

**Pearson L** (1939) *Early Ionian historians*. Oxford, Clarendon Press

**Peet R** (ed.) (1977a) *Radical geography: alternative viewpoints on contemporary social issues*. London, Methuen

**Peet R** (1977b) The development of radical geography in the United States. *Progress in Human Geography* **1**(3): 64–87

**Peet R** (1985) The social origins of environmental determinism. *Annals, Association of American Geographers* **75**: 309–83

**Peet R** (1989) Introduction. In: Peet R, Thrift N (eds) *New models in geography: the political-economy perspective*, volume 1. London, Unwin Hyman: 43–7

**Peet R, Thrift N** (eds) (1989a) *New models in geography: the political-economy perspective*. London, Unwin Hyman

**Peet R, Thrift N** (1989b) Political economy and human geography. In: Peet R, Thrift N (eds) *New models in geography: the political-economy perspective*, volume 1. London, Unwin Hyman: 3–29

**Penck A** (1919) Die Gipfelflur der Alpen. *Sitzungsberichte Preussische Akademie der Wissenschaften* **17**: 256–68

**Penck W** (1924) *Die morphologische Analyse: Ein Kapitel der physikalischen Geologie*. Stuttgart, Geographische Abhandlungen, series 2, volume 2

**Pepper D** (1984) *The roots of modern environmentalism*. London, Croom Helm

**Piaget J** (1971) *Structuralism*. New York, Basic Books

**Pickles J** (1985) *Phenomenology, science and geography: spatiality and the human sciences*. Cambridge, Cambridge University Press

**Pitty A F** (1971) *Introduction to geomorphology*. London, Methuen

**Planhol X de** (1972) Historical geography in France. In: Baker A R H (ed.) *Progress in historical geography*, Newton Abbot, David and Charles: 29–44

**Plato** (1971) *Timaeus and Criteas*. Harmondsworth, Penguin

**Plato** (1974) *The republic*. Harmondsworth, Penguin, 2nd edn

**Pliny** (1855–57) *The natural history of Pliny*. London, Henry G. Bohm

**Pocock D** (ed.) (1981) *Humanistic geography and literature: essays on the experience of place*. London, Croom Helm

**Pooler J A** (1977) The origins of the spatial tradition in geography: an interpretation. *Ontario Geography* **11**: 56–83

**Popper K** (1968) *The logic of scientific discovery*. London, Hutchinson, 2nd ed.

**Popper K** (1970) Normal science and its dangers. In: Lakatos I, Musgrave A (eds) *Criticism and the growth of knowledge*. London, Cambridge University Press: 51–8

**Popper K** (1976) *Unended quest: an intellectual autobiography*. London, Fontana

**Porritt J** (1984) *Seeing green: the politics of ecology explained*. Oxford, Basil Blackwell

**Porteous D J** (1984) Putting Descartes before *dehors*. *Transactions, Institute of British Geographers* n.s. **9**(3): 372–3

**Poster M** (1984) *Foucault, Marxism and history:* mode of production *versus* mode of information. Cambridge, Polity Press

**Potter R B** (1977) Spatial patterns of consumer behaviour in relation to the social class variable. *Area* **9**(2): 153–6

**Potter R B, Unwin T** (eds) (1989) *The geography of urban–rural interaction in developing countries: essays for Alan B. Mountjoy*. London, Routledge

**Potter S R** (1983) Pyotr Alexeivich Kropotkin 1842–1921. In: Freeman T W (ed.) *Geographers: biobibliographical studies*, volume 7. London, Mansell: 63–9

**Powell J M** (1990) Australian geography and the corporate management paradigm. *Journal of Geography in Higher Education* **14**(1): 5–18

**Pred A** (ed.) (1981) *Space and time in geography*. Lund, Gleerup

**Pred A** (1984) Place as historically contingent process: structuration and the time-geography of becoming places. *Annals, Association of American Geographers* **74**(2): 279–97

**Prince H** (1971) Questions of social relevance. *Area* **3**(3): 150–3

**Ptolemy** (1966) *Claudii Ptolemaei geographia edidit C.F.A. Noble cum introductione a Aubrey Diller*. Hildesheim, Georg Olms Verlagsbuchhandlung

**Quaini M** (1982) *Geography and Marxism*. Oxford, Basil Blackwell

**Ratzel F** (1882) *Anthropo-Geographie oder Grundzüge der Anwendung der Erdkunde auf die Geschichte*; volume 1. Stuttgart, Engelhorn

**Ratzel F** (1891) *Anthropogeographie. Die Geographische Verbreitung des Menschen*. Stuggart, Engelhorn

**Ravenstein E G** (ed.) (1898) *A journal of the first voyage of Vasco da Gama 1497–1499*. London, The Hakluyt Society (No. 99)

**Rawling E M** (1991) The National Curriculum in geography: developing the opportunities. Abstract of paper presented to Launching the National Curriculum in Geography, Royal Geographical Society, 24 September 1991

**Reclus E** (1868–69) *La Terre. Description des phénomènes de la vie du globe.* Paris, Hachette

**Redclift M** (1987) The production of nature and the reproduction of the species. *Antipode* **19**: 222–30

**Rees J** (1989) Natural resources, economy and society. In: Gregory D, Walford R (eds) *Horizons in human geography*. Basingstoke, Macmillan: 364–94

**Relph E** (1970) An inquiry into the relations between phenomenology and geography. *Canadian Geographer* **14**: 193–201

**Relph E** (1976) *Place and placelessness*. London, Pion

**Relph E C** (1981) Phenomenology. In: Harvey M E, Holy B P (eds) *Themes in geographic thought*. London, Croom Helm: 99–114

**Renner G T** (1935) The statistical approach to regions. *Annals, Association of American Geographers* **25**: 137–45

**Richthofen F von** (1928) Die Geographie im ersten Halbjahrhundert der Gesellschaft für Erdkunde. *Zeitschrift der Gesellschaft für Erdkunde zu Berlin*: 15–30

**Rickert H** (1962) *Science and history: a critique of positivist epistemology*. Princeton, Van Nostrand

**Robinson A H, Bryson R A** (1957) A method for describing quantitatively the correspondence of geographical distributions. *Annals, Association of American Geographers* **47**: 379–91

**Robinson F** (1982) *Atlas of the Islamic world since 1500*. Oxford, Phaidon

**Robinson J L** (1986) Geography in Canada. *Professional Geographer* **38**(4): 411–17

**Robson B T** (1971) Down to earth. *Area* **3**(3): 137

**Roderick R** (1986) *Habermas and the foundations of critical theory*. Basingstoke, Macmillan

**Rössler M** (1989) Applied geography and area research in Nazi society: central place theory and planning, 1935–1945. *Environment and Planning D: Society and Space* **7**: 419–31

**Rowntree D** (1987) *Assessing students: how shall we know them?* London, Kogan Page

**Russell B** (1961) *History of western philosophy and its connection with political and social circumstances from the earliest times to the present day*. London, George Allen and Unwin.

**Russell R J** (1949) Geographical geomorphology. *Annals, Association of American Geographers* **39**(1): 1–11

**Rylands T G** (1893) *The geography of Ptolemy elucidated*. Dublin, University Press

**Saarinen T** (1966) *Perception of the drought hazard on the Great Plains*. Chicago, University of Chicago Department of Geography Research Paper No. 106

**Saarinen T F** (1969) *Perception of the environment*. Washington, Association of American Geographers Commission on College Geography Resource Paper 5

**Saarinen T F** (1979) Commentary – critique on Bunting–Guelke paper. *Annals, Association of American Geographers* **69**: 464–8

**Sack R D** (1976) Magic and space. *Annals, Association of American Geographers* **66**: 309–23

**Sack R D** (1980) *Conceptions of space in social thought: a geographic perspective*. London, Macmillan

**Sack R D** (1986) *Human territoriality: its theory and history*. Cambridge, Cambridge University Press

**Salter C L** (1986) Geography and California's educational reform: one approach to a common cause. *Annals, Association of American Geographers* **76**(1): 5–17

**Salter C L** (1987) The nature and potential of a geographic alliance. *Journal of Geography* **86**: 211–15

**Samuels M S** (1978) Existentialism and human geography. In: Ley D, Samuels M S (eds) *Humanistic geography: prospects and problems*. London, Croom Helm: 22–40

**Sandbach F** (1980) *Environmental ideology and policy*. Oxford, Basil Blackwell

**Santos M** (1974) Geography, Marxism and underdevelopment. *Antipode* **6**(3): 1–9

**Sauer C O** (1924) The survey method in geography and its objective. *Annals, Association of American Geographers* **14**: 17–33

**Sauer C O** (1925) The morphology of landscape. *University of California Publications in Geography* **2**: 19–53

**Sauer C O** (1941) Foreword to historical geography. *Annals, Association of American Geographers* **31**: 1–24

**Sauer C O** (1956) The education of a geographer. *Annals, Association of American Geographers* **46**: 287–99

**Saussure F de** (1916) *Cours de linguistique générale*. Paris, Payot

**Sayer A** (1979) Epistemology and conceptions of people and nature in geography. *Geoforum* **10**: 19–43

**Sayer A** (1984) *Method in social science: a realist approach*. London, Hutchinson

**Sayer A** (1985a) Realism in geography. In: Johnston R J (ed.) *The future of geography*. London, Methuen: 159–73

**Sayer A** (1985b) The difference that space makes. In: Gregory D, Urry J (eds) *Social relations and spatial structures*. Basingstoke, Macmillan: 49–66

**Scargill D I** (1976) The RGS and the foundation of geography at Oxford. *Geographical Journal* **42**(3) 438–61

**Schaefer F K** (1953) Exceptionalism in geography: a methodological examination. *Annals, Association of American Geographers* **43**: 226–49

**Scheidegger A E** (1961) *Theoretical geomorphology*. Berlin, Springer-Verlag

**Schlick M** (1964) The four-dimensional world. In: Smart J J C (ed.) *Problems of space and time*. New York, Macmillan: 292–96

**Schlüter O** (1906) *Die Ziele der Geographie des Menschen*. Munich, R. Öldenbourg

**Schneider W H** (1990) Geographical reform and municipal imperialism, 1870–80. In: Mackenzie J (ed.) *Imperialism and the natural world*. Manchester, Manchester University Press: 90–117

**Scholten A** (1980) Al-Muqaddas *c*. 945–*c*. 988. In: Freeman T W, Pinchemel P (eds) *Geographers: biobibliographical studies*, volume 4. London, Mansell: 1–6

**Schultz H-D** (1980) *Die deutschsprachige Geographie von 1800 bis 1970: Ein Beitrag zur Geschichte ihrer Methodologie*. Berlin, Geographisches Institute der Freien Universität Berlin (Abhandlungen des Geographischen Instituts Anthropogeographie volume 29)

**Schumm S A** (1956a) Evolution of drainage systems and slopes in badlands at Perth Amboy, New Jersey. *Bulletin of the Geological Society of America* **67**: 597–641

**Schumm S A** (1956b) The role of creep and rainwash on the retreat of badland slopes. *American Journal of Science* **254**: 693–706

**Schumm S A, Lichty R W** (1965) Time, space and causality in geomorphology. *American Journal of Science* **263**: 110–19

**Schutz A** (1962) *Collected papers*. The Hague, Martinus Nijhoff

**Schutz A** (1967) *The phenomenology of the social world*. Evanston, Northwestern University Press

**Sciama D** (1986) Time 'paradoxes' in relativity. In: Flood R, Lockwood M (eds) *The nature of time*. Oxford, Basil Blackwell: 6–21

**Scruton R** (1981) *From Descartes to Wittgenstein: a short history of modern philosophy*. London, Routledge and Kegan Paul

**Semple E C** (1911) *Influences of geographic environment on the basis of Ratzel's system of anthropo-geography*. New York, Henry Holt and Company

**Shallis M** (1986) Time and cosmology. In: Flood R, Lockwood M (eds) *The nature of time*. Oxford, Basil Blackwell: 63–79

**Silk J** (1979) *Statistical concepts in geography*. London, George Allen and Unwin

**Simon H A** (1957) *Models of man: social and rational*. New York, John Wiley

**Simon W M** (1963) *European positivism in the nineteenth century*. New York, Cornell University Press

**Sinnhuber K A** (1959) Carl Ritter, 1779–1859. *Scottish Geographical Magazine* **75**: 153–63

**Slater D** (1973) Geography and underdevelopment – 1. *Antipode* **5**(3): 21–33

**Smailes A E** (1971) Urban systems. *Transactions, Institute of British Geographers* **53**: 1–14

**Smart J J C** (ed.) (1964a) *Problems of space and time*. London, Macmillan

**Smart J J C** (1964b) Introduction. In: Smart J J C (ed.) *Problems of space and time*. London, Macmillan: 1–23

**Smith D M** (1971) Radical geography – the next revolution. *Area* **3**(3): 153–57

**Smith N** (1984) *Uneven development: nature, capital and the production of space*. Oxford, Basil Blackwell

**Smith N** (1987) Danger of the empirical turn: the CURS initiative. *Antipode* **19**: 59–68

**Smith N** (1989) Geography as museum: private history and conservative idealism. In: Entrikin J N, Brunn S D (eds) *Reflections on Richard Hartshorne's* The Nature of Geography. Washington, Association of American Geographers: 89–120

**Smith N K** (1923) *A commentary to Kant's 'Critique of pure reason'*. London, Macmillan

**Smith S J** (1984) Practicing humanistic geography. *Annals, Association of American Geographers* **74**(3): 353–74

**Soja E** (1971) *The political organization of space*. Washington DC, Association of American Geographers Commission on College Geography Resource Paper No. 8

**Soja E W** (1985) The spatiality of social life: towards a transformative retheorisation. In: Gregory D, Urry J (eds) *Social relations and spatial structures*. Basingstoke, Macmillan: 90–127

**Soja E W** (1989) *Postmodern geographies: the reassertion of space in critical social theory*. London, Verso

**Sonnenfeld J** (1972) Geography, perception and the behavioral environment. In: English D W, Mayfield R E (eds) *Man, space and environment: concepts in contemporary human geography*. New York, Oxford University Press: 244–51

**Sorre M** (1961) *L'homme sur la terre: traité de géographie humaine*. Paris, Hachette

**Spencer H** (1864) *First principles*. New York, D. Appleton

**Spencer H** (1882) *The principles of sociology*. New York, D. Appleton

**Spiegelberg H** (1960) *The phenomenological movement: a historical introduction*. The Hague, M. Nijhoff

**Stamp D** (1966) Ten years on. *Transactions, Institute of British Geographers* **40**: 11–20

**Stamp L D** (1947) *The land of Britain: its use and misuse*. London, Longman

**Stamp L D** (1957) Major natural regions: Herbertson after fifty years. *Geography* **42**: 201–16

**Stauffer R C** (1960) Ecology in the long manuscript version of Darwin's *Origin of species* and Linnaeus' *Economy of nature*. *Proceedings of the American Philosophical Society* **104**: 235–41

**Steel R W** (1974) The Third World: geography in practice. *Geography* **59**: 189–207

**Steel R W, Watson J W** (1972) Geography in the United Kingdom, 1962–72. *Geographical Journal* **138**: 139–53

**Stephenson D** (1974) The Toronto geographical expedition. *Antipode* **6**(3): 1–9

**Stewart J Q** (1947) Empirical mathematical rules concerning the distribution and equilibrium of populations. *Geographical Review* **37**: 461–85

**Stoddart D R** (1965) Geography and the ecological approach: the ecosystem as a geographic principle and method. *Geography* **50**: 242–51

**Stoddart D R** (1967) Growth and structure of geography. *Transactions, Institute of British Geographers* **41**: 1–19

**Stoddart D R** (1975a) The RGS and the foundations of geography at Cambridge. *Geographical Journal* **141**(2): 216–39

**Stoddart D R** (1975b) Kropotkin, Reclus and 'relevant' geography. *Area* **7**: 188–90

**Stoddart D R** (1981) The paradigm concept and the history of geography. In: Stoddart D R (ed.) *Geography, ideology and social concern*. Oxford, Basil Blackwell: 70–80

**Stoddart D R** (1986) *On geography and its history*. Oxford, Basil Blackwell

**Stoddart D R** (1987a) To claim the high ground: geography for the end of the century. *Transactions, Institute of British Geographers* n.s. **12**(3): 327–36

**Stoddart D R** (1987b) Geographers and geomorphology in Britain between the wars. In: Steel R W (ed.) *British geography 1919–1948*. Cambridge: Cambridge University Press: 156–76

**Stone K H** (1979) Geography's wartime service. *Annals, Association of American Geographers* **69**: 89–96

**Storm M** (1989) Geography in schools: the state of the art. *Geography* **74**(4): 289–98

**Strabo** (1949) *The geography of Strabo*. London, Heinemann

**Strahler A N** (1950) Davis' concept of slope development viewed in the light of recent quantitative investigations. *Annals, Association of American Geographers* **40**: 209–13

**Stutz F P** (1985) Enhancing high school geography at the local level. *Professional Geographer* **37**(4): 391–5

**Sykes P** (1934) *A history of exploration from the earliest times to the present day*. London, George Routledge and Sons

**Tansley A G** (1935) The use and abuse of vegetational concepts and terms. *Ecology* **16**: 284–307

**Tar Z** (1977) *The Frankfurt School: the critical theories of Max Horkheimer and Theodore W. Adorno*. New York, John Wiley

**Tatham G** (1951) Environmentalism and possibilism. In: Taylor G (ed.) *Geography in the twentieth century: a study of growth, fields, techniques, aims and trends*. London, Methuen: 128–62

**Taylor E G R** (1930) *Tudor geography 1485–1583*. London, Methuen

**Taylor E G R** (1934) *Late Tudor and early Stuart geography 1583–1650*. London, Methuen

**Taylor E G R** (1948) Geography in war and peace. *Geographical Review* **38**: 132–41 (reprinted from *The Advancement of Science* **4**(15), 1947)

**Taylor G** (1937) *Environment, race and migration*. Chicago, Chicago University Press

**Taylor P J** (1976) An interpretation of the quantification debate in British geography. *Transactions, Institute of British Geographers* n.s. **1**(2): 129–42

**Taylor P** (1985) *Political geography: world economy, nation-state and locality*. Harlow, Longman

**Thompson E P** (1978) *The poverty of theory and other essays*. London, Merlin Press

**Thompson J B, Held D** (eds) (1982a) *Habermas: critical debates*. London, Macmillan

**Thompson J B, Held D** (1982b) Editors' introduction. In: Thompson J B, Held D (eds) *Habermas: critical debates*. London, Macmillan: 1–20

**Thompson K** (1976) *Auguste Comte: the foundation of sociology*. London, Nelson

**Thomson J O** (1948) *History of ancient geography*. Cambridge, Cambridge University Press

**Thorne J O** (ed.) (1961) *Chambers's biographical dictionary*. Edinburgh, W. and R. Chambers, new ed.

**Thornes J B, Brunsden D** (1977) *Geomorphology and time*. London, Methuen

**Thünen J H von** (1826) *Der isolierte Staat in Beziehung auf Landwirtschaft und Nationalökonomie*. Hamburg, Perthes

**Tinkler K J** (1985) *A short history of geomorphology*. Totowa, Barnes and Noble

**Townsend J G** (1977) Perceived worlds of the colonists of tropical rainforests, Colombia. *Transactions, Institute of British Geographers* n.s. **2**(4): 430–58

**Trimble S W** (1986) Declining student performance in College geography and the down-writing of texts. *Professional Geographer* **38**(3): 270–3

**Tuan Yi-Fu** (1971) Geography, phenomenology and the study of human nature. *Canadian Geographer* **15**: 181–92

**Tuan Yi-Fu** (1974) *Topophilia: a study of environmental perception, attitudes and values*. Englewood Cliffs, Prentice-Hall

**Tuan Yi-Fu** (1976) Humanistic geography. *Annals, Association of American Geographers* **66**(2): 266–76

**Turner R K** (ed.) (1988) *Sustainable environmental management: principles and practice*. London, Belhaven Press

**Ullman E L** (1941) A theory of location for cities. *American Journal of Sociology* **46**: 853–64

**Ullman E L** (1974) Space and/or time: opportunity for substitution and prediction. *Transactions, Institute of British Geographers* **63**: 125–39

**Unstead J F, Taylor E G R** (1910) *General and regional geography for students*. London, George Philip & Son

**Unstead L D** (1933) A system of regional geography. *Geography* **18**: 175–87

**Unwin P T H** (1987) *Portugal*. Oxford, Clio Press

**Unwin T** (1986) Attitudes towards geographers in the graduate labour market. *Journal of Geography in Higher Education* **10**(2): 149–57

**Unwin T** (1989) From secondary to higher education: overlap or divide? *Area* **21**(2): 173–74

**Verstappen H T** (1983) *Applied geomorphology*. Oxford, Elsevier

**Viaud J** (1968) Egyptian mythology. In: *New Larousse encyclopedia of mythology*. London, Paul Hamlyn, rev. edn: 9–48

**Vidal de la Blache P** (1896) Le principe de la géographie generale. *Annales de Géographie* **5**: 129–42

**Vidal de la Blache P** (1913) Les caractères distinctifs de la géographie. *Annales de Géographie* **22**: 289–99

**Vidal de la Blache P** (1922) *Principes de géographie humaine*. Paris, Colin

**Vitek J D** (1989) A perspective on geomorphology in the twentieth century. In: Tinkler K J (ed.) *History of geomorphology from Hutton to Hack*. Boston, Unwin Hyman: 293–324

**Vorzimmer P** (1965) Darwin's ecology and its influence upon his theory. *Isis* **56**: 148–55

**Wagner M** (1868) *Die Darwinische Theorie und das Migrationgesetz der Organismen*. Leipzig, Duncker and Humblot

**Walford R** (1989) Geography and the National Curriculum: a chronicle and commentary. *Area* **21**(2): 161–6

**Wanklyn H** (1961) *Friedrich Ratzel: a biographical memoir and bibliography*. Cambridge, Cambridge University Press

**Warntz W** (1963) *Geography, geometry and graphics*. Princeton, Princeton University, School of Engineering and Applied Science

**Warntz W** (1984) Trajectories and co-ordinates. In: Billinge M, Gregory D, Martin R (eds) *Recollections of a revolution*. London, Macmillan: 134–52

**Weber A** (1909) *Über den Standort der Industrien*. Tübingen, J C B Mohr

**Wegener A** (1915) *Die Enstehung der Kontinente und Ozeane*. Braunschweig (Die Wissenschaft Bd. 66)

**White G** (ed.) (1961) *Papers on flood problems*. Chicago, University of Chicago Department of Geography, Research Paper No. 70

**Whittlesey D S** (1945) The horizon of geography. *Annals, Association of American Geographers* **35**: 1–36

**Wiener P F** (1949) *Evolution and the founders of pragmatism*. Cambridge Mass., Harvard University Press

**Wilson A G** (1970) *Entropy in urban and regional modelling*. London, Pion

**Wilson A G** (1989) Mathematical models and geographic theory. In: Gregory D, Walford R (eds) *Horizons in human geography*. Basingstoke, Macmillan: 29–47

**Wilson A G, Bennett R J** (1985) *Mathematical methods in human geography and planning*. Chichester, John Wiley

**Wilson L S** (1946) Some observations on wartime geography in England. *Geographical Review* **36**: 597–612

**Windelband W** (1980) History and natural science (Rectoral Address, Strasbourg 1894). *History and Theory* **19**: 169–85

**Wisner B** (1969) Editor's note. *Antipode* **1**(1): iii

**Wittgenstein L** (1961) *Tractatus logico-philosophicus*. London, Routledge and Kegan Paul

**Wittgenstein L** (1967) *Philosophical investigations*. Oxford, Basil Blackwell 3rd ed.

**Woldenburg M J, Berry B J L** (1967) Rivers and central places: analagous systems? *Journal of Regional Science* **7**(2): 129–39

**Wolforth J** (1986) School geography – alive and well in Canada? *Annals, Association of American Geographers* **76**(1): 17–24

**Wolpert J** (1964) The decision process in spatial context. *Annals, Association of American Geographers* **54**: 337–58

**Woodcock G, Avakumović I** (1950) *The anarchist prince: a biographical study of Peter Kropotkin*. London, T. V. Boardman

**Woodward D** (1990) Roger Bacon's terrestrial coordinate system. *Annals, Association of American Geographers* **80**(1): 109–22

**Wooldridge S W** (1955) 'Geographical essays' by W M Davis. *Geographical Journal* **121**: 89–90

**Wooldridge S W** (1958) The trend of geomorphology. *Transactions, Institute of British Geographers* **25**: 29–35

**Wooldridge S W, Morgan R S** (1937) *The physical basis of geography: an outline of geomorphology*. London, Longmans, Green and Co.

**Wright J K** (1925) *The geographical lore of the time of the crusades: a study in the history of medieval science and tradition in western Europe*. New York, American Geographical Society

**Wright J K** (1947) Terrae incognitae: the place of the imagination in geography. *Annals, Association of American Geographers* **37**: 1–15

**Wright J K** (1952) *Geography in the making: the American Geographical Society 1851–1951*. New York, American Geographical Society

**Wright J K** (1966) *Human nature in geography*. Cambridge, Harvard University Press

**Wrigley E A** (1965) Changes in the philosophy of geography. In: Chorley R J, Haggett P (eds) *Frontiers in geographical teaching*. London, Methuen: 3–20

**Zelinsky W** (1974) The demigod's dilemma. *Annals, Association of American Geographers* **65**: 123–43

**Zipf G K** (1949) *Human behavior and the principle of least effort*. New York, Hafner

# Index


*a posteriori* reasoning, 72, 212
*a priori* reasoning, 72–3, 212
Abélard, P, 164
Absolutism, 28
Ackerman, E A, 107–9, 112–3
Action, 30, 39–40, 43, 89, 149–50, 152, 160, 163, 167, 171, 182, 184, 209
Admiralty handbooks, 107
Adorno, T, 29
Aesthetics, 94, 102, 148
African Association, 83
Agassiz, L, 94
Age of Discovery, 60–4
Agency, human, 152, 172–4, 176
al-Idrīsī, 57–8
al-Muqaddasī, 57
Alexander the Great, 47
Alexandria, 48, 53, 55
Alienation, 148, 160, 178
Allen, J, 177
Althusser, L, 28, 160, 169–71
American Geographical Society, 85–6, 122
Analytic propositions, 34, 72, 212
Anarchy, 16, 87–90, 132, 179, 212
Anaximander of Miletus, 47–9, 51
Andrews, H F, 139
*Annales de Géographie*, 81
*Annals of the Association of American Geographers*, 114, 131, 154
Anthropology, 71, 73–4, 169, 170–1, 187
*Antipode*, 162
Appearances, 27
Archaeology, 46–7, 81
Architecture, 177–8
Aristotle, 48–50, 52, 59, 62
Armstrong, P, 90
Art, 115–6, 195
Arts, 21
  Liberal, 160
Association of American Geographers, 10, 96, 102, 115, 133, 168
Astrology, 49, 56, 64
Astronomy, 32, 46, 48–9, 51, 53–4, 56, 58–9, 62–4, 67–8, 70, 76, 194, 198, 203
Athens, 50
Australia, 146, 151

Bacon, F, 22, 67–8, 70, 74
Bagnold, R A, 117
Baker, K, 12
Bakunin, M, 88
Bangladesh, 209
Barlow, R, 63
Barnes, B, 26
Barrow, J, 85
Barrows, H H, 96–7, 129
Bartels, D, 124
Base level, 86
Bassin, M, 92, 108
Bauman, Z, 35, 37, 39
Beckinsale, R P, 97
Bede, 59–60
Bednarz, S W, 10
Behaviour, 142–4
Behavioural environment, 141–2
Behavioural geography, 137–8, 140–4, 146
Benjamin, W, 29
Bennett, R J, 128
Benton, T, 176
Bergmann, G, 113, 123
Berkeley, G, 70
Berkeley, University of California, 87, 101, 125, 141
Berlin, 79–80, 85, 87–8, 108, 113
  Congress of, 84
  University of, 88
Berry, B J L, 119, 124, 128, 133, 148
Bertalanffy, L von, 128
Bhaskar, R, 175
Billinge, M, 18
Biogeography, 100, 129

Biology, 104, 128, 156, 171, 193
influences on geographical thought, 90–2
Bird, J, 198
Blaut, J M, 166
Bodie, 125
Bohr, N, 202
Boston, 131
Boundaries, 109
Bowen, M, 67, 69, 70, 75–6, 89
Bowman, I, 97
Bradford, M G, 120
Braithwaite, R B, 21
Breitbart, M M, 89
Britain, 160–2
education system, 1, 3, 7, 11–15, 21, 206
geography in, 11–15, 83–5, 100–1, 107, 117, 125–7, 139–40, 151
British Association for the Advancement of Science, 84
Brookfield, H, 141
Brown, L, 141
Brunhes, J, 95
Brunsden, D, 20, 197–8
Brussels, 88
Buchanan, J Y, 85
Buchanan, K, 166
Bunbury, E H, 48, 51–2
Bunge, W, 119, 123, 132, 158, 163, 166
Bunting, T E, 143
Burgess, E W, 121
Burnet, T, 70
Burton, I, 125
Büsching, A F, 70
Buttimer, A, 82, 138–9, 147

Cabral, P, 61
Calculus, 199
Calendars, 56
California, 9–10
University of, 87, 101, 125, 141
Cambridge, University of, 84–5, 125
Cameron, I, 83
Canada, 125, 146, 149, 151, 163
geography in, 11
Cannabis, 160
Capitalism, 27, 29, 40, 87, 132, 140, 151, 154, 158–61, 164–8, 171, 173, 178–9, 180–3, 188, 192, 208
Careers, academic, 181
Carnap, R, 33–4
Cartography, 55–6, 58–9, 60–4, 124
Castells, M, 166, 169, 171
Cause, 176–7
Central Intelligence Agency, 108
Chang Hêng, 56
Chapman, G P, 128
Chappell, J E, 150
Chemistry, 32
Chicago, 97, 129, 132, 152
University of, 87, 102, 121, 142, 209
Children, 16
Chinese geography, 55–6, 59
Chisholm, M J, 128
Cholley, A, 139
Chomsky, N, 183
Chorley, R J, 25, 97, 125–9
Chorography, 53–4, 64, 67, 70, 101, 104, 110, 116, 213
Chorology, 98–9, 102–4
Christaller, W, 108, 119–121, 124
Christianity, 59–60, 70, 86
Chronology, 63, 118
Chu Ssu-Pên, 56
Cicero, 50
*Circulation*, 100
Civil Rights movement, 132, 159
Clark University, 102, 162
Class
conflict, 166
domination, 84
Classification, 74, 101, 104, 126
Climate
change, 2, 118, 207, 210
classification, 101
'global warming', 118
influence on human character, 58
Cloke, P, 130, 175
Cognitive interests, 30, 40–1, 43, 69, 135, 156, 180, 182–3
Collingwood, R G, 149, 151
Colombus, C, 61
Colonialism, 64, 74, 80–1, 84, 104, 165–6
Colorado River, 86
Commerce, 95
Commercial geography, 72
Communal action, 89
Communication, 30, 36, 41–2, 69, 151–2, 170, 180, 183, 193, 208
Communications, 160
Communism, 124, 160–1, 167, 182
Comparative method, 74
Computers, 116, 125, 154, 204
Comte, A, 31, 76, 82, 134
Congress of Berlin, 84
Consciousness, 37–9, 71, 173–4
Conservation, 94–5, 188
Constantine, 59
Constraint, 156
Consumer society, 159
Contemplation, 69
Contraction of moving bodies, 200
Conventionalism, 176
Cooke, R U, 190
Copernicus, N, 25, 62
Cosgrove, D, 151, 153, 191
Cosmography, 62–4, 66, 76
Cosmology, 39, 49, 56, 60, 62, 203–4, 210, 213
Costa, J E, 116

Creativity, 21
Critical geography, 182–5
Critical science, 26, 31, 39–44, 153
Critical theory, 28–44, 176, 180, 182, 206–7
Critique, 146, 148, 152, 167, 180
Cultural geography, 96, 141
Cultural landscape, 100–1
Cultural sciences, 35, 37, 43
Culture, 93, 98, 141, 163, 178, 191
Cunningham, W, 64
Curriculum
  development, 15–17
  reform, 9–17, 206–7
Cycle of erosion, 97

Dacey, M F, 119, 124
Daniels, S, 151, 153
Darwin, C, 74, 85, 90–2, 94, 96
Daudé, R, 139
Davies, D C W, 201
Davis, W M, 25, 86–7, 93, 95–7, 129
Dawkins, R, 192
De Planhol, X, 82
Deception, 40
Deconstruction, 178
Deduction, 22–3, 153, 213
Dee, J, 64
Dehumanization, 178, 204
Demangeon, A, 100
Democracy, 28
Democritus, 138
Demolins, E, 82
Descartes, R, 22–3, 32, 64, 70, 138
Description, 71, 98, 104–5, 110, 112, 115, 122, 169
Determinism, 146, 213
  environmental, 58, 90, 92–5, 97, 99, 104, 116, 140, 192
Development studies, 162, 167–8
Dialectic, 164–5, 170, 213
Dialectical materialism, 164–5
Dias, B, 61
Dicaearchus, 48, 55
Difference, 178
Diffusion processes, 121, 144–5
Dilthey, W, 35–7, 39, 41, 136, 152
Disciplines, 2, 18
  as social phenomena, 6–8
  construction of, 5–7
  definitions of, 5–7
Discourse, 26–8
Discovery, 60–6, 83
Disorder, 180
Distanciation, 174
Dodgshon, R A, 193
Dom Henrique, 61
Domination, 156, 173
Doubt, 22
Downs, R M, 138
Draft Order for Geography, England and Wales, 14
Drainage systems, 87
Drapeyron, L, 81
Dreams, 40
Dreyer, J L E, 59
Duncan, J, 172–3
Duncan, S, 174
Dunford, M, 165
Dunn, A, 97
Durkheim, E, 33, 82–3, 139, 169
Dury, G H, 116

Earth Science, 209–10
Earth, shape, 48
Ecology, 78, 92, 96, 101, 188
Economic geography, 107
Economic recession, 2, 154, 158–62
Economic restructuring, 2, 174
Economic theory, 119–20
Economics, 187–8
Ecosystems, 128–9
Education, 7–17, 72, 78, 104, 205–7
  social determinants, 16
  social role, 16–18
Education Reform Bill, 1988, 11–12
Education systems, 7–17
  British, 1, 3, 7, 11–15, 21, 206
  United States of America, 3
Edwards, P, 206
Egypt, 49–50
Einstein, A, 194, 198, 201
Elementary geography, 9–10
Emancipation, 29–30, 40–1, 69, 156, 158, 176, 180, 184–5, 206
Emel, J, 188
Emery, F W, 84
Emotion, 148
Empathy, 36, 39
Empedocles of Acragas, 49
Empirical, 122, 179
  analysis, 78
  events, 21
  facts, 48
  method, 74
  research, 76, 102, 150, 153, 171–2, 174, 181
  science, 37, 67, 69–70, 73, 110
  world, 169
Empirical-analytic science, 31–5, 42–3, 110, 131, 134–5, 146, 150, 153, 161, 175, 181, 183, 187, 190, 207
Empiricism, 20–1, 32, 34, 70–2, 145–6, 150, 169, 174–6, 205, 213
Engels, F, 76
Engineering, 155–6, 190
England, geographical education, 7, 11–15
  Draft Order for Geography, 14
  exploration, 63–4
  National Curriculum, 12–16, 21, 206

Enlightenment, 29, 180
Entrikin, J N, 102, 112, 146, 148, 187, 210
Entropy, 201–2, 213
Environment, 13, 87, 141, 154, 188–94, 210
  change, 118
  degradation, 207, 210
  influences on human behaviour, 92–4
  monitoring of, 2
  perception of, 143
Environmental determinism, 58, 90, 92–5, 97, 99, 104, 116, 140, 192
Environmental science, 188
Environmentalism, 190
Epicurus, 50
Episteme, 27, 214
Epistemology, 21, 31, 34–5, 73, 137, 149–50, 155, 170, 175–6, 179, 196, 214
Eratosthenes, 48–9, 52, 54, 55
*Erdkunde*, 75
Erskine, R, 86
Erosion, 86
Essence, 32, 38, 146–7
Ethics, 207
Ethnography, 80
Ethnology, 84
Euclidian space, 194
Events, 34, 176–7
Evolution, 90–1, 95
Examinations, 16
Exceptionalism, 112–3
Existence, 147
Existentialism, 145–6, 148–9, 214
Expeditions, 80, 83, 86
Experience, 21–3, 35–6, 38, 57, 68, 70, 72–3, 102, 142, 153, 174, 176
Experiments, 22–3, 204
  laboratory, 117, 198
Experimental method, 69
Experimentation, 32–4, 75, 78, 117, 127, 188
Explanation, 21, 31, 34, 110, 112, 115, 119, 122, 126, 130–1, 134, 136, 140–1, 149–51, 154, 161, 169, 172, 176, 194, 197–18
Exploration, 16, 60–4, 69–70, 74–5, 79–80, 83–7, 91, 104, 163
  British, 63–4
  Dutch, 63
  Portuguese, 53, 60–3
  Spanish, 53, 60–3

Facts, 22–3, 31, 34, 72, 83, 98, 153, 182, 188, 195, 203, 206
Fairy tales, 23
Falsifiability, 23
Falsification, 26
Family, 82
Famine, 207–8
Fashion, 132
Fawcett, C B, 101
Febvre, L, 95
Federal Bureau of Investigation, 124
Feelings, 149, 195
Feminism, 168
Fenneman, N M, 6
Fetishism of commodities, 165
Feudalism, 165
Feuerbach, L A, 160
Feyerabend, P, 23
Fichte, J G, 75
Field studies, 101, 121
Fleure, H J, 101, 139
Flower power, 132, 160
Fluvial hydrology, 117
Forster, G, 75
Forster, R, 74–5, 186, 188
Foucault, M, 24
France, 88, 132, 160
  geography in, 80–3, 95, 100, 137–9
Frankfurt School, 19, 28–9, 205
Freedom, 160
Freeman, T W, 100
Frere, B, 83–4
Freud, S, 29, 37, 39–40
Fuller, G, 10

Gaea, 49
Galilei, G, 64, 70
Gama, V da, 61
Gannett, H, 86
Garrison, W L, 119–20, 124
Gazetteers, 56
Geddes, P, 101
Geiger, M, 37
Geikie, A, 118
Gender, 168
General geography, 67–9, 104, 110
General Systems Theory, 126, 128
Generality, 106, 109–15
*Generalplan Ost*, 108
*Genre de vie*, 100
Geographer of Ravenna, 59
Geographers, conflicts with geologists, 84–6, 95–6
Geographical Alliances, 9–10
Geographical Association, 11–13
Geographical information systems, 6, 207
Geographical societies, 79–87, 104
  *Gesellschaft für Erdkunde* (Berlin), 79–80
  Indian Geographical Society, 141
  National Geographic Society, 10
  Paris Geographical Society, 81, 88
  Royal Geographical Society (London), 79, 83–5, 89
*Geographische Zeitschrift*, 99
Geography
  15th century, 60–2
  16th century, 62–4
  17th century, 64–74
  18th century, 74–5

19th century, 75–92
as an integrative discipline, 12, 21, 78, 99, 102
as spatial science, 122–24, 134, 138, 146, 175, 186, 194
behavioural, 137–8, 140–4, 146
Chinese, 55–6, 59
commercial, 72
cultural, 96, 141
economic, 107
elementary, 9–10
employers' attitudes to, 5
feminist, 168
general, 67–9, 104, 110
Greek, 46–50, 54–5, 62, 76
historical, 119, 131, 139, 141, 151–2, 165–6
humanistic, 150–3, 156
in secondary education, 7–17
in Britain, 11–15, 83–5, 100–1, 107, 117, 125–7, 139–40, 151
in France, 80–3, 95, 100, 137–9
in Germany, 74–80, 82, 99–100, 188
in United States of America, 95–8
in universities, 79–87
Islamic, 56–59
low status of, 8–9, 11, 96–8, 106, 114, 119, 158
Marxist, 164–6
mathematical, 72
medieval, 59–64
military role, 52
moral, 72
political, 72, 80, 85, 162
political role, 51–2
primary, 9–17, 80–1, 205, 207
public image of, 2–3, 16, 209
quantitative, 119–29, 132, 154, 162
radical, 133, 157, 159, 162–9, 180–1
regional, 77–8, 85–6, 98–115, 139–40, 187–8
Roman, 50–5, 62, 76
school, 78, 85
social, 151–3
social role, 54
special, 67–9, 110
systematic, 77–8, 99, 103–7, 110–1, 114–5, 119–22
theoretical, 122–9
time, 142, 144–5, 173, 194, 198, 200
urban, 119, 121, 131, 151, 162–3, 166, 179–80
Geography, 16–19,
Project, 15
Geography Awareness Week, United States of America, 10
Geological Society, 85
Geological structure, 101
Geological conflicts with geographers, 84–6, 95–6
Geology, 76, 80, 83–7, 95–6, 100, 104, 116–7, 156
Geometry, 51, 68, 148
Geomorphology, 25, 86–7, 95–7, 100, 108, 116–9, 123, 126, 128, 131, 135, 190, 196–7
Geopolitics, 108
George, P, 139
German geography, 74–80, 82, 99–100, 108
influence on geography in the United States of America, 86, 102, 139
*Gesellschaft für Erdkunde* (Berlin), 79–80
Gestalt psychology, 141
Geuss, R, 29
Ghetto, 133, 159, 162
Giddens, A, 28, 145, 172–4, 177, 179, 194, 204
Gilbert, G K, 86–7, 96, 109, 115
Giraldus Cambrensis, 62
Glacken, C J, 45, 49, 60
Global warming, 118
Glucksmann, M, 169
Goals, 146
God, 60, 78
Goethe, J W von, 75
Gold, J R, 206–7
Golledge, R G, 141
Göttingen, 75, 77, 79
Gould, P, 2, 54, 119
Governments
intervention in higher education, 29, 167, 205
research funding, 161, 181
Gradman, R, 99
Graf, W L, 116
Grand theory, 178
Gravity, 200
Gravity models, 122
Greek geography, 46–50, 54–5, 62, 76
Green politics, 188
Greer-Wootten, B, 128
Gregory, D, 18, 140, 145, 147–8, 153, 165, 169, 171–2, 177–8, 180
Gregory, K J, 123, 190
Gritzner, C F, 7
Gropius, W, 178
Grünberg, C, 28–9, 205
Guelke, L, 115, 122, 143, 149–50
Guillemard, F, 85
Gurvitch, G, 139
Guyot, A, 79, 86, 90, 129

Habermas, J, 24, 28–44, 69, 131, 134, 150–2, 156, 176, 180–2, 190, 192–3, 204, 206–7
Haeckel, E, 92
Hägerstrand, T, 121, 124, 144–5, 173
Haggett, P, 25, 125–8
Haines-Young, R H, 155–6, 197, 205
Hakluyt, R, 63–4

Hall, R B, 101, 109
Happiness, 51
Harré, R, 176
Harris, C, 149
Harris, C D, 121
Harrison, R T, 150
Harriss, B, 167
Harriss, J, 167
Hart, J F, 115
Hartshorne, R, 76, 99, 101–3, 111–2, 114–6, 119, 124, 130, 151
Harvard University, 87, 96, 97
Harvey, D, 129–31, 134, 162–3, 166, 169, 173–4, 179, 203–4
Hawking, S W, 200–2
Hecataeus of Miletus, 47
Hegel, G W F, 35, 38, 43, 76, 78, 113, 140, 151, 164–5
Heidegger, M, 151
Heisenberg, W K, 202
Hempel, C G, 151
Herbertson, A J, 101, 109
Herbst, J, 96–8
Herder, J G von, 75
Hereford Map, 62
Hermeneutics, 35–7, 39, 136, 149–50, 172, 176, 214
Herodotus, 47–8
Hess, R, 108
Hettner, A, 26, 97, 99, 102, 107, 110–1, 113
High School Geography Project, 8
Higher education, 124, 181, 190–1
  funding, 154, 156
  government involvement in, 29, 167, 205
  restructuring of, 161, 188
Hill, A D, 7–8
Hillslope processes, 117
Historical geography, 119, 131, 139, 141, 151–2, 165–6
Historical materialism, 136, 160, 170
Historical sciences, 110
Historical understanding, 147
Historical-hermeneutic science, 31, 35–9, 42–3, 146, 150–3, 183, 187, 191
Historicism, 113
History, 8–10, 12, 27, 47, 51, 58, 63–4, 71–4, 81–2, 89, 93–6, 100, 102, 107, 111, 113, 139, 145, 149, 151, 170, 180, 191, 195
  Marx's conception of, 165
Hitler, A, 93
Holcomb, B, 8, 10
Holism, 89, 172
Homer, 47, 51
Horkheimer, M, 21–2, 29
Horton, R E, 117
Hoyt, H, 121
Hudson, B, 84
Hudson, R, 144
Human condition, 136
Human Ecology, 96, 121, 129
Human geography, 19, 35, 43–4, 86, 95–8, 100, 114, 119–24, 140–53, 186–8, 191–4, 205, 209–10
  division from physical geography, 114, 154, 183, 186–8, 193–4
  integration with physical geography, 86, 127, 134
Human sciences, 89, 99
Human welfare, 101
Human world, 31
Humanism, 89, 121, 136–7, 140, 145–53, 159–60, 162, 175, 214
Humanistic geography, 150–3, 156
Humanities, 19, 21, 137, 178, 194
Humanity, 60
Humboldt, F H A von, 74–8, 85, 89, 91, 94, 99, 104, 110, 186, 188
Hume, D, 22, 70
Hunting, 52, 83
Huntington, E, 94, 192
Husserl, E, 35, 37–9, 41, 146–8, 151–2
Hydrology, 117
Hypotheses, 21, 23, 34, 112, 117, 122, 130, 188, 214

Iberia, reconquest of, 60–1
Ibn Baṭṭūṭa, 57–8
Ibn Ḥawḳal, 57
Ibn Khaldūn, 58
Ibn Khurradādhbih, 57
Iconography, 191
Idealism, 34, 38, 71, 76, 78, 137, 145–50, 164, 175–6, 214
Ideology, 20, 23, 26–9, 40, 156, 170, 179, 214
Idiographic thought, 110–1, 215
*Iliad*, 47
Illusion, 196
Images, 143, 191
Imagination, 21, 141–2, 174
Imperialism, 78–9, 84, 87, 89, 104, 163, 165–6
Indian Geographical Society, 141
Individuality, 99, 148
Individuals, 136, 176
Induction, 22–3, 33–4, 68, 74, 78, 215
Industrial development, 82
Inequality, 162–3, 166, 168, 182
Institute of British geographers, 168
Intention, 149
Intentionality, 147
Interest, 39
Interpretation, 115, 154, 191
Intuition, 37, 73
Iowa, University of, 119–20, 123
Iraq, 209
Isard, W, 148

Islam, 60
Islamic geography, 56–9, 61, 65
Italy, 160
Itineraries, 62

Jackson, P, 121, 151–3
James, P, 101, 121
Jeanneret, C E, 178
Jena, 75, 92
Johnston, R J, 6, 26–7, 92, 100, 112, 140, 144, 168, 187, 210
Joseph, K, 11–12
Jumper, S R, 9

Kant, E, 121
Kant, I, 34, 38, 43, 70–6, 78, 102, 113, 151
Kates, R W, 127, 142, 144
Keltie, J S, 89, 95
Kennedy, B, 128–9
Kent, W A, 120
Kiepart, H, 79
Kimble, G H T, 109
King, C A M, 107, 117
King, L J, 113
King, M L, 132
Kinship, 169–70
Kirk, W, 141
Knowledge, 1, 16–18, 26–44, 70, 72–3, 110, 134–5, 142, 145, 147, 158, 161–2, 170, 175–6, 181, 190
  classification of, 24
  distorted, 40–1
  social construction of, 46
Kockelmans, J J, 38
Koffka, K, 141
Köhler, W, 141
Krebs, N, 98
Kropotkin, P, 88–9, 188
Kuhn, T S, 24–6, 126

Laboratory experimentation, 117, 198
Labour, 41, 89, 165, 191, 193
Lactantius, 59
Lakatos, I, 25–6
Lamarck, J B P A de M; Chevalier de, 92–3
Lambert, J, 38
Land Utilization Survey, 107
*Länderkundliche Schema*, 99
Landform classification, 86
Landforms, 116–19
Landscape, 97–8, 100, 102, 119, 141, 149, 154, 179, 184, 191, 210
Language, 27–8, 30, 36, 42–4, 68, 145, 150–2, 154, 170, 178, 180, 183–4, 193, 203
LaPrairie, L A, 7–8
Latitude, 48–9, 53, 62
Lavoisier, A L, 75
Laws, 21, 23, 31–4, 40, 68, 105, 110, 112, 114–5, 117–9, 121–2, 127, 130, 146, 165–6, 215
Le Corbusier, 178
Le Lannou, M, 139
Le Play Society, 101
Le Play, F, 82, 101
*Lebensraum*, 80, 92, 108
Lefebvre, H, 173–4, 194–6, 204
Legitimation, 93
  crisis, 43
Leibniz, G W, 22–3, 70, 173
Leighly, J, 95, 98
Leopold, L B, 117
Leucippus, 138
Lévi-Strauss, C, 169–70
Ley, D, 137, 145, 151–3, 172
Liberal arts, 160
Liberalism, 76–7
Lichty, R W, 196–7
Life, 27–8
Life-paths, 144–5
Life-world, 42–3, 147, 156
Light, 200
Linguistics, 27–8, 170–1, 183
Linke, M, 78
Linnaeus, C, 74
Literature, 151, 191
Livingstone, D (explorer), 85
Livingstone, D N, 64–5, 93, 150
Locale, 173, 179
Locality, 174
Location, 106, 124, 126
Location theory, 113, 119–20, 138, 140, 151
Locke, J, 22, 70
Logic, 23, 50, 68
Logical positivism, 33–4, 36, 106, 112–4, 122–4, 132, 134–7, 140, 142–4, 146, 148–52, 154–6, 162, 168–9, 172, 175, 178, 182, 188, 190, 205, 215
London, 87, 208
Longitude, 48–9, 53–4, 62
Los Angeles, 179–80
Lösch, A, 120
Lovering, J, 177
Lowenthal, D, 141, 143
Lucretius, 50
Lund, 121, 124, 144
Lyell, C, 90
Lynch, K, 138

Mach, E, 33–4
Mackenzie, S, 168
Mackinder, H, 85, 100, 108
Maconochie, A, 84
Madingley, 125
Madison, University of Wisconsin at, 102, 119–20, 123–4
Magalhães, F de, 61
Magic, 55, 65, 195
Malthus, T, 91

Manifolds, 198
Maps, 9, 46–7, 61–3, 78, 83, 86, 107, 124
  mental, 142–3
Marcuse, H, 29, 133, 160
Marsh, G P, 94
Martin, G J, 124
Martin, R, 18
Marx, K, 29, 41, 76, 78, 132, 148, 160–7, 170–1
  conception of history, 165
Marxism, 28, 43, 124, 134, 137, 152, 157, 159–60, 167–9, 179, 191
Marxist geography, 164–6
Mass media, 3, 10, 160, 209
Massey, D, 174
Masterman, M, 25
Materialism, 41, 92, 177, 215
  dialectic, 164–5
Mathematical geography, 72
Mathematical models, 121
Mathematical sciences, 53
Mathematics, 46, 54, 64, 67–8, 70–2, 117, 122–4, 126, 155, 170–1
Matter, 22, 68–9, 76
May, J A, 74
McCarthy, T, 42
McCarthyism, 124, 132
McCarty, H, 113, 120
McDowell, L, 168
McGregor, J, 13
Meaning, 37–9, 147, 151, 153, 156, 172, 183
Measurement, 46, 48, 51, 54, 118
Mecca, 56
Media, 3, 10, 160, 209
Medieval geography, 59–64
Mellor, J R, 121
Mental maps, 142–3
Mercer, D C, 146
Meta-theory, 175
Metaphysics, 21, 32–3, 38, 70–3, 149, 215
Meteorology, 96–7
Methodology, 6, 32, 106, 130, 134–5, 143–4, 150, 155, 162–3, 169, 176–7, 186
Meyer, K, 108
Michigan State University, 163
Milieu, 82, 100, 139
Military, 83, 86
  action, 69, 79
  conquest, 55
  intelligence, 107
Mill, J S, 22, 33
Mind, 22, 35, 38, 68–9, 76, 141, 150, 175, 178
Minkowski, H, 199–200
Minority rights, 162
Missionaries, 85
Missions, 83–4
Mode of production, 165–6, 170–1, 195
Model Curriculum Standards, 9–10
Models, 9, 97, 118–20, 122, 126–7, 130, 146, 154–6, 161, 176, 188, 197, 215
Modernism, 178–9
Modernity, 42
Moltke, H von, 79
Momsen, J, 168
Money, 173–4
Moral geography, 72
Morality, 72, 152
Morals, 207
Morgan, W, 207
Morisawa, M, 118
Morphology, 82, 97, 100
Morrill, R, 119, 121
Münster, S, 62
Murchison, R I, 83–4
Mysticism, 48
Mythology, 49, 55
Myths, 169–70, 195, 204

Nancy, 81
Narcotics, 132, 160
National Council for Geographic Education, 10
National Curriculum, England and Wales, 12–16, 21, 206
National Geographic Society, 10
National Science Foundation, 8
National Socialism, 29, 108, 113
Natural environment, 92–3
Natural hazards, 142
Natural history, 52, 70
Natural laws, 78
Natural sciences, 27, 29, 30, 32, 35, 41, 43, 69, 89, 96, 99, 110, 114, 126, 139, 154–5, 176–7, 196
Natural selection, 91, 94
Natural theology, 78, 90, 93
Natural world, 21, 31, 155
Nature, 25, 30, 33, 43, 49–50, 54, 60, 74–6, 78, 82, 91, 94, 100, 141, 191, 196
Navy, 118
Nazism, 29, 108, 113
Necessity, 165
Needham, J, 55–6
Negation, 43, 165
Neo-Kantianism, 110, 121
Neo-Lamarckism, 90
Neo-Marxism, 136
Neuchatel, 79, 94
Neumann, K, 79
New University of Brussels, 88
New York, 85
Newton, I, 69–70, 138, 200
Newton-Smith, W H, 200–1
Nomothetic knowledge, 111
Nomothetic thought, 110, 122, 216
Normal science, 24
Normative statements, 26, 32

*Noumena*, 38, 73
Novels, 115, 204

O'Riordan, T, 188–9
Objectivity, 21, 29, 36, 38, 146–7, 150, 152–3, 187–8, 195
Objects, 38, 153, 165, 176–7
Observation, 21–2, 27, 32–3, 36, 68, 75, 78, 81, 153, 202–3, 207
Observatories, 59
*Odyssey*, 47
Olsson, G, 138, 178–9, 184
Olwig, K R, 94
Ontology, 38, 137, 175–6, 179, 201, 216
Outhwaite, W, 175–6
Oxford University, 8–5, 101

Painting, 151, 191
Panofsky, E, 191
Paradigms, 24–2, 116, 126, 131, 216
  scientific revolutions, 24–6
Paris, 81–2, 85, 87
Paris Academy of Sciences, 70
Paris Commune (1871), 88
Paris Geographical Society, 81, 88
Paris, M, 62
Park, R, 121, 152
Parmenides, 164
Paterson, J H, 109
Peet, R, 159, 162, 169, 188, 192
Peirce, C S, 152
Penck, A, 97, 118
Penck, W, 26, 97
Pennsylvania, University of, 122
People and environment, interaction between, 10, 12, 30, 43, 49, 60, 76, 82, 90–5, 104, 128, 141, 186, 188, 208–11
People and nature, 78, 191
Perception, 38, 72–3, 76, 138, 141–3, 147, 174
Perrons, D, 165
Personal construct theory, 144
Pestalozzi, J H, 77, 81
Petch, J R, 155–6, 197, 205
Petrus Apianus, 62
Phei Hsui, 56
Phenomena, 31–2, 38, 73, 99, 100–2, 112, 115, 169, 175
Phenomenalism, 32, 216
Phenomenology, 35, 37–8, 137, 145–8, 152, 156, 216
Philo, C, 130, 175
Philosophy, 34–5, 48, 51, 71, 75–6, 110, 130–1, 134, 137, 143, 145–50, 163–4, 167, 170–1, 175, 181, 183, 196–8, 201, 203
Physical environment, 10, 58, 93
Physical geography, 15, 19, 43–4, 71–4, 86–9, 95–8, 100, 105, 107, 114, 116–9, 124, 126, 128–9, 135, 146, 153–6, 161, 186–8, 194, 196–7, 204–5, 209–10
  applied, 188, 207
  division from human geography, 114, 154, 183, 186–8, 193–4
  integration with human geography, 86, 127, 134
Physical processes, 91, 107, 188–9
Physical sciences, 19, 114, 122
Physics, 32–3, 51, 70, 122, 126, 194, 198, 201, 203
Physiogony, 75
Physiography, 75
Physiology, 82
Piaget, J, 169
Pilgrimage, 56, 62
Place, 13, 82, 89, 98, 100, 102, 147–8, 151, 173–4, 186–8, 207, 209–11
Planning, 107–8, 138, 178, 190
Planudes, M, 61
Plate tectonics, 118
Plato, 48, 50
Pliny, 52–3, 59, 70
Plutarch, 48
Pocock, D, 151
Poetry, 151
Political action, 89, 166–7
Political change, 148
Political control, 55
Political decision making, 3
Political economy, 27–8, 76, 157, 159, 162, 169, 188
Political geography, 72, 80, 85, 162
Political involvement in education, 14–15, 154, 205
Political repression, 76
Political theory, 158
Politics, 1–3, 5, 58, 63, 69, 71, 74, 79, 83, 93, 140, 160, 173, 179, 196
Politics, Green, 188
Pollution, 210
Polo, M, 55
Polybius, 51–2
Popper, K, 21–3, 26
Porteous, D J, 151
Portuguese exploration, 53, 60–3
Posidonius, 51, 54
Positivism, 31–7, 39, 76, 82, 92, 110, 113, 121, 134, 148, 150, 153–6, 216
  Logical, 33–4, 36, 106, 112–4, 122–4, 132, 134–7, 140, 142–4, 146, 148–52, 154–6, 162, 168–9, 172, 175, 178, 182, 188, 190, 205, 215
Possibilism, 92, 95, 141
Poster, M, 28
Postmodernism, 104, 157, 159, 175, 177–180, 184, 216
Poverty, 162, 207
Powell, J M, 146
Powell, J W, 86–7, 96

Power, 18, 26–30, 55, 156, 159–62, 172–3, 178, 193, 207–8
Practice, 28, 41–4, 166, 180, 182, 204–11
Pragmatism, 121, 146, 152, 216
Pred, A, 144–5, 172–3
Prediction, 26, 106, 112, 114–15, 118, 123, 134, 138, 151, 168, 188, 190
Primary geography, 9–17, 80–1, 205, 207
Prince, H, 133
Princeton University, 79, 86–7, 122
Process, 13, 97, 115, 122–3, 131, 161, 187
  in physical geography, 116–118
  laws, 113
  oriented science, 106
  physical, 154, 188–9
*Progress in Human Geography*, 193
*Progress in Physical Geography*, 193
Propositions, analytic, 34, 72, 212
Propositions, synthetic, 34, 72
Psychoanalysis, 39–40, 182
Psychology, 138, 141, 143–4, 171
Ptolemy, 49, 53–4, 58–9, 61–2, 64, 67, 110, 203
Public image of geography, 2–3, 16, 209
Purchas, S, 63–4
Purpose, 146
Pythagoras, 48, 65

Quantification, 117, 123, 127, 135, 151
Quantitative geography, 119–29, 132, 154, 162
Quantitative Revolution, 106, 130–1, 143
Quantum theory, 196, 201–3

Race, 132–3
Racial domination, 84
Racial unrest, 162
Racism, 93, 152, 168, 208
Radical geography, 133, 157, 159, 162–9, 180–1
Radical theory, 158
Radicalism, 132
Ralegh, W, 64
Raleigh Travellers' Club, 85
Rank-size rule, 122
Rationalism, 21, 23, 32, 37, 67, 70–2, 216
Rationality, 42, 142
Ratzel, F, 80, 82, 90, 92–3, 95, 108
Rawling, E M, 207
Ray, J, 70
Reagan, R, 161, 167
Realism, 146, 157, 159, 175–7, 180, 187, 195, 216
Reality, 110, 149–50, 169, 175, 183–4, 187, 198, 202
Reason, 21, 28, 43, 165, 180
Recession, economic, 21, 54, 158–62
Reclus, E, 79, 88–9
Reconquest of Iberia, 60–1
Rees, J, 188–9
Reflection, 69, 72, 136–7, 146–7, 149, 152, 156
  self- 35–7, 39–41, 155–6, 180, 184
Reflectivity, 29
Regional description, 76
Regional geography, 77–8, 85–6, 98–116, 139–40, 187–8
  demise of, 107–16
Regional monographs, 98, 100, 107
Regional Science Association, 120
Regions 6, 52, 67–9, 70, 82, 90–1, 99, 100, 102, 104, 115, 126, 141, 179, 187
  boundaries, 109
  problems of definition, 109
Relations of production, 165
Relativity, 196–203
  general theory of, 194, 200–1
  special theory of, 194, 198, 200–1
Religion, 32–3, 58, 64, 93–4
Relph, E, 146–7
Renaissance, 27, 137, 160
Renner, G T, 109
Representation, 178
Repression, 40
Reproduction, 42, 192
Research, 2, 15–17, 29
  applied, 118, 207
  funding, 174, 181, 190–1, 207–8
  practice, 207–8
Resources, 89, 145, 207, 210
Revolution, 26, 76, 88
Revolutionary practice, 166–7
*Revue de Géographie*, 81
Richthofen, F von, 79, 99, 104
Rickert, H, 110
Rights, minority, 162
Ritter, C, 74, 77–80, 85–6, 88, 90–1, 94, 100, 104, 110, 129, 186
Robson, B T, 133
Roger II of Sicily, 57
Roman geography, 50–5, 62, 76
Romanticism, 75
Rousseau, J J, 72
Royal Geographical Society (London), 79, 83–5, 89
Royal Prussian Academy of Sciences, 79
Royal Society, 70
Russell, B, 22, 50, 73, 198

Saarinen, T F, 141, 143
Sack, R D, 138, 193, 195, 203
Sadler, D, 130, 175
Salisbury, R, 87, 97
Salter, C L, 9–10
Samuels, M S, 137, 145, 148, 151
Sandbach, F, 190
Santos, M, 166
Sartre, J P, 148
Sauer, C O, 97, 101–3, 112, 139, 141
Saussure, F de, 170, 183

Sayer, A, 169, 177
Scarperia, J d'A, 61
Scepticism, 165
Schaefer, F K, 112–6, 119, 122, 124
Scheidegger, A E, 123
Scheler, M, 37
Schelling, F W J von, 75, 78
Schiller, J C F von, 75
Schlick, M, 33, 200
Schlüter, O, 99–200, 107
Schumm, S A, 117, 196–7
Schutz, A, 146–7, 149
Science, 18, 20–44, 59, 67–9, 74–6, 89, 97, 100–2, 104, 106, 109, 111–2, 115–6, 119–24, 126–7, 134, 146, 148–50, 155, 162, 170, 176, 182–4, 195, 217
  classification of, 32–3
  definitions of, 21–30
  human, 89, 99
  natural, 27, 29, 30, 32, 35, 41, 43, 69, 89, 96, 99, 110, 114, 126, 139, 154–5, 176–7, 196
  normal, 24
  physical, 19, 114, 122
  social, 19, 28, 33, 43, 82, 121, 127, 137–40, 147, 149, 151, 159–60, 171, 175–8, 186–8, 194, 209–10
  social production of, 26–8
  spatial, 122–4, 134, 138, 146, 175, 186, 194
  value-free, 89, 161
Scientific communities, 24–6
Scientific method, 22–3, 43, 70, 130, 132, 153, 155–6
Scientism, 29, 31, 41, 115
Scruton, R, 72–3
Seattle, University of Washington at, 119, 121, 123
Second law of thermodynamics, 201–2
Second World War, 103, 107–8, 118, 121, 132
Secondary education, 167, 181
Secondary geography, 9–17, 80–1, 114, 205, 207
Self-deceipt, 40
Self-reflection, 35–7, 39–41, 155–6, 180, 184
Semple, E C, 92–5, 192
Senate Bill 813 (California), 9
Senses, 21
Shaler, N S, 94
Shallis, M, 203–4
Shape, 124
  of the earth, 48
Signs, 178
Simon, H A, 142
Simon, J, 81
Simon, W M, 33
Sion, J, 100
Slater, D, 166
Slavery, 83
Smith, D, 133
Smith, N, 192
Smith, S, 121, 152–3
Social change, 158
Social conformity, 16–17
Social Darwinism, 89, 96
Social geography, 151–3
Social groups, 92
Social justice, 134, 163
Social morphology, 139
Social organisation, 121
Social physics, 121–2
Social processes, 163
Social reform, 89
Social relations, 159
Social reproduction, 173
Social science, 19, 28, 33, 43, 82, 121, 127, 137–40, 147, 149, 151, 159–60, 171, 175–8, 186–8, 194, 209–10
Social structure, 144, 169, 172, 174
Social studies, 9–10
Social systems, 162
Social theory, 158, 168, 173, 177, 179, 184, 192, 203–4
Social welfare, 124, 162
Socialism, 88, 124, 164–5, 167
Society, 1–2, 8, 20–1, 29, 178–80, 192–3, 195
  influence on knowledge, 30–1
Society for Human Exploration, 163
Sociology, 10, 32–3, 96, 104, 121, 129, 138–40, 172–4, 187
Socrates, 50
Soja, E, 17–18, 138, 177, 179
Sorre, M, 139
Space, 13, 71, 73–4, 91, 102, 120, 122–9, 138, 144–5, 148, 163, 173–4, 177, 179–80, 186, 192–205
  absolute, 138, 174, 200
  relative, 138, 173–4, 200
  social production of, 194–6
Space-time, 118, 194–200, 204
Spanish exploration, 53, 60–3
Spatial analysis, 153–4
Spatial distribution, 112
Spatial forms, 163
Spatial models, 137
Spatial relationships, 120, 126
Spatial science, 122–4, 134, 138, 146, 175, 186, 194
Spatial structure, 121, 159
Spatiality, 179–80
Special geography, 67–9, 110
Spencer, H, 33, 90, 92–3, 96
Spiegelberg, H, 37
Spinoza, B, 70
Spirit, 35
Stalin, 160
Stamp, L D, 107

Stanley, H M, 84
State University of Iowa, 113
Statements
singular, 22–3
universal, 22–3
Statistics, 15, 70, 85, 125, 127, 132, 155–6
techniques, 120, 135
Stea, D, 138
Stewart, J Q, 121–2
Stoddart, D R, 25–6, 74, 83, 85, 91, 94–5, 116, 129, 186–8, 193, 210
Stoics, 50
Stone, K H, 108
Strabo, 49, 51–2, 54, 56, 62, 65, 68, 79, 108, 166
Strahler, A N, 117
Structuralism, 43, 151–3, 157, 164, 169–71, 175–6, 184, 217
Structuration, 145, 171–4
theory, 172–4, 176–7, 179, 187, 217
Structure, 97, 165, 169–73, 175–7
Student protest, 132, 160
Stutz, F P, 9
Subject, 38, 153
Subjectivity, 21, 32, 36, 38, 42, 102, 147, 152, 156, 187, 195
Survey, 55
Surveying, 86
Svensson, S, 121
Sweden, 120–1, 124, 142
Switzerland, 88
Symbolism, 191
Symbols, 36, 40, 148, 151, 156, 169, 195
Synthesis, 89
Synthetic propositions, 34, 72, 217
Systematic geography, 77–8, 99, 103–7, 110–1, 114–5, 119–22
Systematic science, 98, 102
Systematic studies, 76
Systems, 3, 23, 42, 124, 126–7, 130, 169, 172, 217
Systems analysis, 186
Systems approach, 128–9, 142
Systems theory, 122–9

Tansley, A G, 128
Taxonomy, 101
Taylor, E, 63, 107
Taylor, G, 94
Taylor, P, 131–2
Teaching, 2, 15–17, 29, 205–7
Technical control, 154
Technical interest, 134–5
Techniques, 6, 118, 155, 190
geographical, 6
statistical, 120, 135
Teleology, 49–50, 60, 78, 90, 94, 129, 217
Temperament, 93
Territoriality, 192–3
Thales of Miletus, 48–9
Thatcher, M, 161, 167
*The Times*, 1, 15
Theological geography, 72
Theology, 46, 49, 51, 62, 64, 72
Theoretical geography, 122–9
Theory, 20–44, 51, 67, 86, 117, 119–23, 130–1, 143–4, 149, 153, 155–6, 163, 166–7, 169, 171, 179–82, 191, 204–11, 217
Grand, 178
meta-, 175
of the four elements, 49
Thermodynamics, 201–2
Things, 27
Third World, 133, 159–60, 166
Thompson, E P, 171
Thompson, K, 33
Thomson, J O, 43, 47
Thornes, J B, 197–8, 203
Thought, 149–50, 165, 170, 183–4
Thrift, N, 159, 169
Thünen, J H von, 120–1, 140
Tiefenbacher, J, 8, 10
Time, 71, 73–4, 97–8, 102, 138, 144–5, 173–4, 179, 194–205
Time assymetry, 201–2
Time dilation, 200
Time geography, 142, 144–5, 173, 194, 198, 200
Time-space, 144–5, 172–3
distanciation, 145
Tinkler, K J, 107, 118
Tolstoy, L N, 20, 41
Topography, 46–54, 56, 58–9, 62–4, 67, 70, 86, 101, 217
Topology, 179
Toronto Geographical Expedition, 163
Townsend, J G, 144, 168
*Transactions, Institute of British Geographers*, 131, 153
Transcendental Philosophy, 39, 147, 156
Transcendentalism, 38, 162, 175
Travel, 83
Travel guides, 56
Travelogues, 70
Trimble, S W, 8
Truth, 16, 20–1, 26–9, 36, 42, 72, 183, 194, 206
absolute, 28
relative, 28
Tuan, Yi-Fu, 146–8

Ullman, E L, 119, 121
Unconsciousness, 178
Underdevelopment, 163
Understanding, 16, 18, 35–6, 102, 115, 130, 136, 146–9, 152–4, 156, 160, 175, 191, 194
historical, 147
Union of Socialist Geographers, 164

Union of Soviet Socialist Republics, 167, 182
Uniqueness, 106, 109–15
United States of America, 160–2, 208
  curriculum reforms, 9
  geography in, 7–11, 85–7, 98–9, 101–4, 107–8, 116–7, 119–25, 132–3, 139–4
  Geological Survey, 86, 118, 125
  Military Academy, 86
Universals, 22–3
Universities
  Berlin, 88
  Cambridge, 125
  Chicago, 87, 102, 121, 142, 209
  Clark, 102, 162
  Harvard, 87, 96–7
  Iowa, 119–20, 123
  Michigan, 102, 209
  New University of Brussels, 88
  Oxford, 84–5, 101
  Pennsylvania, 122
  Princeton, 79, 86–7, 122
  State University of Iowa, 113
  University College London, 84
  University of California, Berkeley, 87, 101, 125, 141
  University of Washington at Seattle, 119, 121, 123
  University of Wisconsin at Madison, 102, 119–20, 123–4
  Wayne State, 163
University geography, 79–87
University restructuring, 156
Unstead, L D, 101
Urban architecture, 178
Urban change, 82
Urban geography, 119, 121, 131, 151, 162, 166, 173, 179–80
Urbanism, 163
Urry, J, 140

Value judgements, 32
Value-free science, 89, 161
Values, 146, 149, 162
Varenius, 66–9, 71, 74, 76, 104, 110, 203
Verifiability, 23
Verification, 33, 150
Vidal de la Blache, P, 81–2, 95, 98, 100, 104, 139, 169
Vienna Circle, 33–4, 113
Vietnam War, 132–3, 159, 162
*Volk* movement, 108

Wagner, M, 92
Waldseemüller, M, 62
Wales, geographical education, 7, 11–15
Walford, R, 11
Wang Ling, 55–6
Wappaeus, J, 79
War, 79, 81, 83, 103, 107–8, 118, 121, 132–3, 159, 162
Warntz, W, 122
Washington, D.C., 86
Washington, G, 86
Wayne State University, 163
Weber, A, 120, 140
White, G, 142
Whittlesey, D S, 112
Williamson, F, 141
Windelband, W, 110–1
Wine, 47
Wittgenstein, L, 18, 34, 183
Woldenburg, M J, 128
Wolforth, J, 11
Wolpert, J, 142
Wooldridge, S W, 97, 117
Words, 27
Work, 27–8, 30, 82, 150
World-views, 27
Wright, J K, 141

X-rays, 25
Xenophon, 50

Yugoslavia, 209

Zeno, 50, 164
Zipf, G, 121–2